·A·
Mathematics
Sampler
Topics for
the Liberal Arts

THIRD EDITION

William P. Berlinghoff
COLBY COLLEGE

& Kerry E. Grant
SOUTHERN CONNECTICUT
STATE UNIVERSITY

ARDSLEY HOUSE PUBLISHERS, INC., NEW YORK

Address orders and editorial
correspondence to:
Ardsley House Publishers, Inc.
320 Central Park West
New York, NY 10025

ISBN: 1-880157-02-0

Printed in the United States of America

10 9 8 7 6 5 4 3 2 1

To Phyllis and Sandy

CONTENTS

PREFACE

This book grew out of a liberal-arts course at Southern Connecticut State University taken by nonscience majors, usually in their first year. Our experiments with the form and content of this course for more than a decade seemed to be most successful when several different mathematical areas were introduced to the students and covered in some depth. From these experiments emerged text units, which ultimately became chapters in this book. Thus, *A Mathematics Sampler* is a collection of independent chapters covering a broad spectrum of mathematics, old and new, pure and applied, traditional and unusual. The Second Edition of this book was adopted by more than 40 colleges and universities in its first two years in print. The course on which it is based was cited as an innovative approach to liberal-arts mathematics in Lynne Cheney's 1989 report, *50 HOURS: A Core Curriculum for College Students*, published by the National Endowment for the Humanities.

New Features   As a result of our teaching experiences and those of our colleagues around the country who have been kind enough to share their thoughts with us, we have revised the content of the text as follows:

- A chapter on graph theory (Chapter 9) has been added.

- The treatment of statistics in Chapter 4 has been thoroughly reworked and expanded.

- An optional section has been added to Chapter 8; it discusses the 4-space generalizations of cylinders and cones.

- Appendix B on the history of mathematics has been reworked to acknowledge the contributions of outstanding woman mathematicians and to reflect the emerging trends of the latter part of the 20th century.

Writing Exercises

In our view, the most important new feature of this edition — a feature that sets *A Mathematics Sampler* apart from all other books for this audience — is *the writing exercises that are spread throughout the book*! In recent years there has been widespread national interest in incorporating writing as a tool for learning science and mathematics. Writing assignments have always been an important part of the course as we have taught it. In this edition of the text we have included numerous writing exercises, many of them already class-tested by us. These exercises come in two forms: **(1)** At the end of virtually every section there is a short list of questions under the heading of **Writing Exercises**, most of which call for answers of more than a sentence and less than a page. These are suitable for use as homework or as in-class exercises to help focus on the ideas at hand, or even for incorporation into course journals. **(2)** At the end of each chapter there is a section entitled **Topics for Papers**. These are descriptions of longer assignments for more formal papers of 4–6 pages in length, or perhaps even more at times. They are intended to encourage the student to draw together the main ideas of the chapter and/or to extend those ideas to some independent exploration.

Necessary Preparation

The wide variety of student backgrounds in a first-year course for non-majors makes it unwise to require significant skill or sophistication as a prerequisite. We assume only that the student has had a first course in high-school algebra and is familiar with some elementary geometry and perhaps (though these are reviewed) some set-theoretic ideas. This book will not remedy a student's lack of basic skills. On the other hand, it *will* reinforce those skills by drawing on them in solving interesting problems relevant to the understanding of real mathematics and a real world.

LINK Sections

A primary function of an introductory mathematics course in a student's liberal education is to exhibit the interrelationships between mathematics and the various other fields of human knowledge. Too often we mathematicians tend to think of those interrelationships merely as applications of mathematical methods and results to other subject areas. While this is one kind of link between mathematics and other subjects, there are many others — links of analogy, of common historical influence, of personal achievement. These links provide a sense of the underlying unity of all knowledge and shed some light on both the power and the limitations of mathematics as a tool for understanding the world of our common experience and as an area of individual aesthetic achievement. *A Mathematics Sampler* provides LINK sections at the end of each topic, connecting specific mathematical ideas with many other subject areas. The connections examined are as varied as the fields explored — art, chemistry, coding, demographics, fiction, genetics, management, marketing, music, philosophy, politics, psychology, and social planning. These LINKs are intended to kindle a spark of interest, rather than to teach a technique, so they are brief, informal, and open-ended. They may be used for outside reading, as a basis for class discussions, or as leads for independent student projects.

Synopses of the Chapters   The book begins with an introduction to problem solving in mathematics, followed by an exploration of eight different mathematical topics. Each chapter after the first starts "from scratch" and proceeds to develop a significant mathematical idea, illustrating what mathematicians do in that area. *These eight chapters may be covered in any combination and in any order.* Some chapters have several natural stopping places to allow for either introductory coverage or in-depth exploration of the topic. Although the book contains enough material for a full-year (6-credit) course, it is most frequently used as a one-semester text, thereby affording considerable flexibility in the choice of topics to be covered. Three or four chapters, used in conjunction with Chapter 1 and the history appendix, provide ample material for a 3-credit course. In particular:

- **Chapter 1,** on problem-solving, sets the tone for the entire book. We recommend it as an appropriate first chapter in any choice of topics.

- **Chapter 2** investigates the topic of perfect numbers as a way of exhibiting the power and elegance of number theory. The students see that analytical observation, educated guesswork, and trial-and-error investigation are good sources of mathematical conjecture, but that patterns can be deceiving. To be most effective, this chapter should be covered in its entirety, although some of the proofs along the way are easily omitted.

- **Chapter 3** uses geometry as the vehicle for examining the axiomatic structure of formal mathematics. After a brief overview of the principles of Euclidean geometry, the historical controversy over Euclid's parallel postulate is used as a steppingstone to consistency and independence in formal axiom systems, culminating in a discussion of the modern view of the relationship between pure mathematical systems and real-world models.

- **Chapter 4** begins with a brief treatment of set language and notation, motivated by their utility in discussing probability questions. (Section 4.2 may be used independently as a review of or an introduction to basic set-theoretic ideas.) The next two sections present the fundamental idea of probability and some combinatoric principles. The chapter may be terminated after Section 4.5 if those ideas are sufficient for the needs of the course. Sections 4.6 and 4.7 provide a further exploration of probability, and LINK Section 4.8 is another natural intermediate stopping place. Sections 4.9 through 4.12 examine some fundamental ideas of statistics, including normal distributions and the Central Limit Theorem.

- **Chapter 5** introduces students to microcomputers. Assuming no prior computer knowledge, this chapter develops enough BASIC programming so that students can actually construct and run some simple interactive "video games" by the end of Section 5.6. A cursory treatment of this topic might legitimately stop after Section 5.4 or 5.5, but we recommend including Section 5.7 in any event.

- **Chapter 6** begins with the elementary notions of *set* and *1-1 correspondence*, and progresses to an explanation and proof of Cantor's Theorem. Discussions of paradoxes, the Continuum Hypothesis, and philosophies of mathematics are provided. The proof of Cantor's Theorem in Section 6.7 may be omitted without seriously affecting the integrity of the chapter. A brief treatment might bypass the computational techniques of Section 6.4 and end early in Section 6.7, but unless done with care, it runs the risk of obscuring the entire point of Cantor's work.

- **Chapter 7** takes as its topical goal the explanation and proof of Lagrange's Theorem, exhibiting the power and utility of abstraction. Tables for finite groups are used to examine properties of operations. Section 7.8, the proof of Lagrange's Theorem, may be omitted without disturbing the continuity of the chapter. A brief treatment of this chapter could end with Section 7.5 or 7.6.

- **Chapter 8** introduces four-dimensional geometry by way of analogy with lower dimensions. The earlier sections provide practice and review of the concepts of *coordinate*, *interval*, *path*, and *distance* in 1-, 2-, and 3-space, which are generalized so that visual intuition may be replaced naturally by coordinate-algebraic techniques in making the transition to 4-space. Sections 8.6 and 8.7 extend these ideas to several different types of figures; one or both of these sections may be omitted, if desired.

- **Chapter 9** investigates the twin existence problems of an Euler path and a Hamilton circuit in a graph. The problems are easily understood and appear to be quite similar in form. The radical dissimilarity of results, by contrast, provides an instructive insight into the nature of mathematics. The LINK section illustrates how graph theory is applied to project management.

- **Appendix A** is a self-contained, concise reference unit on the basic principles of mathematical logic. It can be taught as a short unit in its own right, it can be used as an independent study assignment, or it can just be consulted as needed for definitions or examples during the course.

- **Appendix B** is a compact history of mathematics in relation to the major events that shaped the development of Western civilization, from the earliest known evidence of mathematical achievements through the latter part of the 20th century. The authors recommend it as an early outside-reading assignment. This material can then be used throughout the course as a historical context in which to place the ideas covered in the other chapters.

Answers and/or hints to most of the odd-numbered exercises are provided in the back of the book. Examples are numbered by section; figures and tables are numbered consecutively by chapter. We have adopted the typographical convention of signaling the end of each example with the symbol "□."

Gratitude   As preparation of this text progressed through various stages, many people provided help and encouragement. We are grateful to Ross Gingrich, Michael Meck, Dorothy Schrader, Michael Shea, J. Philip Smith, and other colleagues at Southern Connecticut State University for their expert advice, to Martin Zuckerman of City College of the City University of New York for a wealth of helpful suggestions, and to Joseph Moser and his colleagues at West Chester University for their comments on earlier editions of this book. Special thanks to H. T. (Pete) Hayslett, Jr., and Dexter Whittinghill of Colby College for their valuable assistance during the expansion and reworking of the statistics sections of Chapter 4, and to Claudia Henrion of Middlebury College for steering us to relevant material on women in mathematical history.

William P. Berlinghoff
Kerry E. Grant

TO THE STUDENT

As you begin this topical tour of mathematics, we who seek to guide you have a few words of advice and perspective to offer. The mathematics in this book is not highly technical, but the concepts often are challenging. Mastery of them will at times test your patience and perseverance. We hope you will find our writing style comfortable and our explanations clear. But do not expect to read this book — or any mathematics book — like a novel. Expect to read and reread thoughtfully. Build a habit of having paper and pencil at hand to answer questions raised in the text, to work through exercises, and to create examples of your own. Examine your own understanding frequently. Do not go on to new material until you understand the old, or at least until you know exactly what it is that you do not understand. And do not hesitate to ask questions of your instructor!

A word of warning: Just doing the assigned exercises is *not* your main job in this course! If you regard any text material that does not explain "how to" as superfluous, you have fallen into a dangerous trap. The primary purpose of this book is not to refresh old mathematical techniques or to teach you new ones, even though both of those things may well occur in the course. Rather, we seek to show you some mathematical ideas you may not have seen before and may not even regard as part of mathematics. The text discussions you will read and the exercises you will do are important details in these larger pictures, like pieces of a jigsaw puzzle that make little sense until you begin to see the outlines of the picture as a whole. As you work your way along, then, spend some time thinking about how these details fit into the larger picture of the topic you are studying. When you recognize and understand the broad view of the landscape, you will have arrived at your destination. We hope you enjoy the trip!

W. P. B.
K. E G.

CHAPTER

PROBLEMS AND SOLUTIONS

1.1   What Is Mathematics?

This book is written by two mathematicians. Because we are mathematicians and because mathematicians are fond of precise definitions and careful logic, we would like to begin by defining "mathematics" for you. Then we could proceed to unfold that definition to exhibit in logical sequence the various topics covered in the rest of the book. We would like to do that—but we can't. Mathematics, a subject utterly dependent on definitions and logic, cannot itself be defined in a clear, comprehensive sentence or two, or even in a short paragraph. In fact, mathematicians themselves often delight in giving examples to show that any proposed definition of their subject is deficient in one way or another. The many diverse ideas, methods and results of modern mathematics defy simple description. Faced with this ironic situation, how, then, are we to make any sense at all of the label "mathematics"?

Let us look first at what mathematics has been. At different times and different places in the early history of mankind, civilization brought with it the need for counting and measuring. Traders needed to tally their goods and profits; builders needed to cut their stone and wood to exact sizes; landowners needed to mark precisely the edges and corners of their fields; navigators needed to fix their positions by using the stars to compute distances and angles; the list goes on and on. These specialized tasks gave rise to skilled craftsmen of various sorts—counting-house clerks, draftsmen, surveyors, astronomers, etc. Their crafts, though different, had some common features: They all used numbers and/or basic shapes, and they all sought precise quantitative descriptions of some part of their world. This array of

Accounting by the Passamaquoddy Division of the Abnaki Tribe.

specialized counting and measuring techniques was all of mathematics for the first several thousand years of recorded history.

Greek philosophical speculation during the five or six centuries before Christ brought about a major change in mathematics. As the Greeks tried to solve the mysteries of truth, beauty, goodness, and life itself, they turned human reason into formal logic and then applied that logical system to their observations of the world. Their belief that logic was the key to understanding reality led them to apply it to all areas of thought, especially to mathematics. As they systematically recast the techniques of counting and measuring into logical systems based on "self-evident" assumptions about the real world, mathematics changed from craft to science. Complex mathematical statements were deduced from simpler ones, and they all were applied to solve the problems of the physical world. Architecture, surveying, astronomy, mechanics, optics—all these fields grew from mathematics applied to solve real-world problems, and those solutions provided observable confirmation that the physical world behaved as mathematics predicted it would. Thus, for more than two thousand years mathematics was regarded as the "queen of sciences," the only reliable means for finding scientific truth.

Then came the 19th century. The development of the non-Euclidean geometries in the early 1800s provided mankind with three separate geometric systems, each one logically correct by itself, each one providing a perfectly good way of describing physical space, and each one contradicting the other two! If mathematics were truly a science—that is, if mathematical truths were experimentally verifiable facts about the real world—then this paradoxical situation could never occur. Mathematics, therefore, must be a study that is essentially independent of the real world, and hence its pursuit

¶ Ein ander Exempel.

Ein Fraw oder Haußmutter gehet auff
den marckt/ kaufft vberhaupt ein Körblin
mit Rebnerbyrn/darumb gibt sie achtzehen
pfenning / so sie heim kompt / findet sie im
körblin hundert vnd achtzig byrn/Ist die
frag/wie vil byrn sie vmb ein pfenning ha/
be? Thu/als obgelert/ so kompt dir zehen/
Also vil byrn hat sie vmb einen pfen=
ning / Vnd ist wolfeyl
drumb.

The Problem of the Market Woman.
From Köbel's Rechenbüchlein of
1514 (1564 edition). Traders needed
to tally their goods and profits.

need not be bound by human experience. Mathematicians suddenly found themselves free to explore any questions their minds could ask, unhampered by worries about whether their results had anything to do with reality. They found themselves in "a world of pure abstraction;... 'the wildness of logic' where reason is the handmaiden and not the master."[1] Mathematics became a subject essentially different from the (other) sciences: "In other sciences the essential problems are forced upon the subject from external sources, and the scientist has no control over the ultimate end. The mathematician, however, is free to prescribe not only the means of realizing the end, but also the end itself."[2]

Mathematics responded to this newfound freedom with a mushroomlike expansion in all directions. Results began to appear at such a rate that by the 1980s mathematicians were publishing more than 200,000 *new* theorems every year! Some of these theorems solve old problems, but most of them solve new ones, problems suggested by results published only a short time earlier or even suggested for the first time in that same paper. Some of these theorems are profoundly significant, some are utterly useless, most are somewhere in between—and no one really knows (yet) which is which. Modern mathematics has become an art, the art of posing and solving problems

---

[1]Marston Morse, "Mathematics and the Arts." *The Yale Review*, 40 (4): 1951, p. 612.

[2]Raymond G. Ayoub, in a book review of *Mathematics: The Loss of Certainty. MAA Monthly* 89 (9): 1982, p. 716.

Early Navigation. The Roman Navy, from Livy, Venice, 1493. Navigators needed to fix their position by using the stars to compute distances and angles.

by logical reasoning. To the "pure" mathematician, the significance of those problems outside of mathematics is irrelevant; it is enough to seek them out and put them to rest.

Thus, the bewildering variety of topics currently labeled as "mathematics" is unified by a common theme—problem solving. Of course, solving problems is not limited to mathematics; it is part of almost every human activity. Nevertheless, the problem-solving process is most clearly defined in the simple, abstract world of mathematics. It is at the heart of all mathematical activity, from primitive tallying and measuring through its many scientific applications to its most modern abstract speculations. All the mathematical topics you will see in the later chapters of this book are united by that one common bond. There are many differences among them, differences not only in the topics themselves, but also in the ways they are approached. Different mathematicians study different topics in different ways, and if you are to learn something about mathematics as a whole, then you must come to terms with that basic fact. But you should also see that all these different things are appropriately labeled "mathematics" because they all are held together by a common bond. That bond is a way of thinking, a way of using reason to approach and solve problems of any sort. In the rest of this chapter we examine that way of thinking—the problem-solving process itself.

EXERCISES 1.1

1. Write a one-sentence definition of "mathematics" that is broad enough to include all the mathematics you have seen in elementary school and high school, but is specific enough to exclude biology, music, and social studies. Keep this definition until the end of the course; at that time, check to see if it includes all the mathematics you have covered.

2. Read Appendix Sections B.2, B.3, and B.8.

WRITING EXERCISES

1. Look up the definition of "mathematics" in a dictionary, and comment on whether or not it covers all the mathematics you have seen in elementary school and high school.

2. Does the term "quantitative thinking" encompass all of mathematics? Why or why not?

1.2   Problem Solving

Any method of discovering or inventing something, any approach to solving a problem or finding a proof, is called a *heuristic* (a noun disguised as an adjective). Writings on heuristics date back to early Christian times. Pappus, a 3rd-century Greek geometer, devoted quite a bit of attention to problem-solving methods. In the 17th century, Descartes and Leibniz both tried to formulate systematic treatments of heuristic processes. However, in later years intellectual activity became more specialized, and this odd mixture of logic, philosophy, and psychology was largely neglected. Recently there has been renewed interest in problem solving, especially among mathematicians and mathematics educators. Much of this interest is due to the remarkable work of George Pólya (1888–1985). Pólya, a Stanford University mathematician, spent much of his long professional life teaching and writing about heuristic methods. The first of his several books on the subject appeared in 1945, a small, insightful volume entitled *How to Solve It*. Much of the material in the rest of this section is based on that book.

Pólya begins with a four-part overview of how to solve any problem:

Early Methods of Measuring Distance. This was essentially the method used by the Ancient Greeks. From Belli's Libro del Misvar con la vista, *Venice, 1569.*

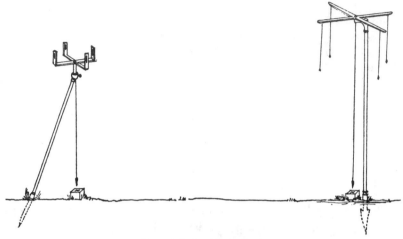

Surveyor's Tools. The 19th Century Surveyor's Cross (on the left) and the Roman 1st Century A.D. Groma. Landowners needed to mark the precise boundaries of their fields.

(1.1)   Understand the problem.
Devise a plan.
Carry out the plan.
Look back at the completed solution.

Taken all by itself, this list is of little use, except perhaps to say, "Slow down! A worthwhile problem is not likely to be solved at first glance." But if we examine each of the four parts, expanding on them a bit, there emerges a truly valuable scheme to help any problem solver in any situation. Let us look at each part in turn.

   Understand the Problem: Exactly what are we trying to find or prove? What facts do we have at our disposal? What parts of the context of the problem might be relevant to what we are trying to find or prove? In order to answer these questions you will have to *check the definitions* of all the words used in the problem to make sure you understand them thoroughly. A useful way to check that is to try to *restate the problem* in as many different forms as you can think of. Sometimes it is helpful to *draw a diagram* to give your visual intuition an image of the problem as a whole. You might also want to *introduce appropriate notation* to symbolize essential pieces of the data or the question for ease in remembering and working with them.

Devise a Plan: This is the hard part, but usually also the part that's the most fun. The main trouble here is having to cope with the uncertainty of not knowing what to do first, or how long the job will take, or even if you will be able to solve the problem at all. In this aspect of problem solving, the main difference between college students and professional mathematicians is not that the mathematicians are brighter or quicker, but rather that they have developed more patience. Pólya provides some comforting advice for all of us:

> The first rule of discovery is to have brains and good luck. The second rule of discovery is to sit tight and wait till you get a bright idea.[3]

Of course, while you "sit tight," there are many things you can do to help you get a bright idea. You can *look for a pattern* in the data you have. If you don't have much data, you might *construct examples* to generate some. You can find a similar problem that has already been solved and try to *argue by analogy*. You might try to *solve a simpler problem* by assuming additional useful information, then try to generalize that solution. If the problem asks you to find something, you might *approximate the answer*, then check to see how far off you are and why; some approaches of this kind are called "trial-and-error" methods. If the problem asks you to prove something, it is often useful to *reason backwards from the desired conclusion*. Finally, you might check to *see if you used all the data*. Remember: none of these techniques is guaranteed to produce a plan for solving the problem, but some combination of them usually works *if* you have the patience and perseverance to work at them until you get the right "bright idea."

Carry Out the Plan: Once you have devised a plan to solve the problem, you must carry it out step by step, checking each step carefully to see that you have not made a reasoning error or introduced some unjustified assumption. This part of the problem-solving process is not nearly as much fun as the excitement that comes when you first see your plan, but it must be done to insure that you actually have a correct solution. (It's like having to split the firewood after cutting down the tree.) Sometimes there are surprises in store for you at this stage; sometimes the plan turns out to be less complete or accurate than you first thought. Good plans often yield dividends when you carry them out; the step-by-step logical development may turn up unexpected connections with other ideas, other problems. And in the event that your plan is wrong in some fundamental way, learning to cope with the disappointment and frustration as you force yourself to search for another plan is itself a valuable lesson.

---

[3]George Pólya, *How to Solve It*, 2nd ed. Princeton: Princeton University Press, 1957, p. 172.

Look Back at the Completed Solution: This is a critical part of the process for any researcher or serious problem solver. Once a problem is solved, it sometimes is possible to find an easier way to comprehend the whole situation. Often you can check your solution that way; you might also check it by finding a simple approximation of the answer. Looking back also gives you a chance to see if you can *generalize the solution* to get a stronger result, and to look for any new questions suggested by your solution.

For convenient reference, we provide here a summary list of the dozen specific problem-solving techniques discussed in this section. You might find it useful to refer back to these suggestions as you wrestle with the questions and exercises in the other chapters.

(1.2)
> Check the definitions.
> Restate the problem.
> Draw a diagram.
> Introduce appropriate notation.
> Look for a pattern.
> Construct examples.
> Argue by analogy.
> Solve a simpler problem.
> Approximate the answer.
> Reason backwards from the desired conclusion.
> See if you used all the data.
> Generalize the solution.

EXERCISES 1.2

Problem A: What is the sum of the measures of the angles of a decagon?

1. List and define all the important terms in Problem A.

2. Draw a diagram for Problem A, labelling important features.

3. Restate Problem A using the diagram and labels of Exercise 2.

4. Find two simpler problems analogous to Problem A.

5. Solve the two problems you found in Exercise 4.

6. Devise two different plans for solving Problem A.

7. Solve Problem A.

8. Generalize Problem A.

9. Solve the generalized problem you stated in Exercise 8.

Problem B: Prove that the product of three consecutive integers is a multiple of 6.

10. List and define all the important terms in Problem B.

11. Restate Problem B in a different form.

12. Construct four different examples that satisfy the conditions of Problem B.

13. Reason backward from the conclusion in Problem B; i.e., what might you prove about the product of the integers that would allow you to conclude that this product is a multiple of 6?

14. Devise a plan for solving Problem B.

15. Solve Problem B.

16. Formulate two problems analogous to Problem B.

Problem C: A piece of music lasting $22\frac{1}{2}$ minutes is recorded in a single band on a $33\frac{1}{3}$ r.p.m. record. When the record is played, the music starts when the needle is 30 centimeters from the center of the record and stops when it is 5 centimeters from the center. If the groove containing the recorded music were in a straight line, how long would the line be?

17. List and define all the important terms in Problem C.

18. Draw a diagram for Problem C. (A diagram need not be a work of art.)

19. Restate Problem C in a different form.

20. Estimate the answer to Problem C.

21. Determine whether your estimate in Problem 20 is likely to be greater than or less than the correct answer. Explain.

Problem D: Assuming that the earth is a sphere with radius 6370 kilometers, what is the average width of a time zone at 60° North Latitude?

22. List and define all the important terms in Problem D.

23. Draw a diagram for Problem D. (You may find that several related diagrams are desirable.)

24. Restate Problem D in a different form.

25. Reason backwards; what intermediate problem or problems could you solve along the way to solving Problem D?

26. Devise a plan for solving Problem D.

WRITING EXERCISES

1. What does *analogy* mean? Give a specific example of an analogy (drawn from anywhere, not necessarily from mathematics) and explain it as if you were trying to explain this concept to a high-school junior.

2. Explain the meaning of the word *pattern* as it is used in reference to problem solving. Go beyond a mere dictionary definition to clarify what it means to talk about a pattern of ideas. Provide an example to illustrate your explanation.

3. Do you think an approximate answer to a mathematical question might ever actually be *better* than an exact one? Why or why not? Illustrate your answer with an example, if appropriate.

1.3   It All Adds Up

We have spoken in general terms about problem solving in the first two sections. Let us now see how some of the suggested techniques can be used to solve a problem.

(1.3)   Find the sum of all positive integers less than 1,000,000.

First let's make sure that we *understand the problem*. Although there

Problem. *How many rectangles are determined by the lines on a standard chessboard? Illustration from Jacobus de Cessolis,* Game of Chess, *Westminster, about 1483.*

is nothing complicated or tricky about this problem, it would be a mistake to ignore this first phase of problem solving. The greatest source of error or failure in problem solving probably comes from failure to understand the problem.

In this case we need to *check the definitions* of the terms: *sum, positive integers,* and *less than.* "Sum," of course, means the result of addition, so we know we are dealing with an addition problem. We recall that the "positive integers" are another name for the natural numbers, 1, 2, 3, 4, ... It is not easy (and not necessary) to give a precise definition of "less than" here; it is important that we know what it means. In particular, we know that the largest number we will include in our sum is 999,999 (not 1,000,000).

We can *restate the problem*, then, as, "What is the sum of all positive integers from 1 to 999,999, inclusive?" Or we might state it using the notation of arithmetic:

$$\text{Find } 1 + 2 + 3 + \ldots + 999,999.$$

The suggestion of *drawing a diagram* might seem inappropriate for this problem because it deals only with numbers and their sum. Frequently, though, numerical problems translate into geometric ones, or vice versa. In this problem, picture a side view of a staircase with 999,999 steps, each 1 unit by 1 unit. (See Figure 1.1.) Because 1 square unit is under the first step, 2 square units are under the second step, and so on, we can restate our problem geometrically by asking

(1.4)   What is the area of the figure indicated in Figure 1.1?

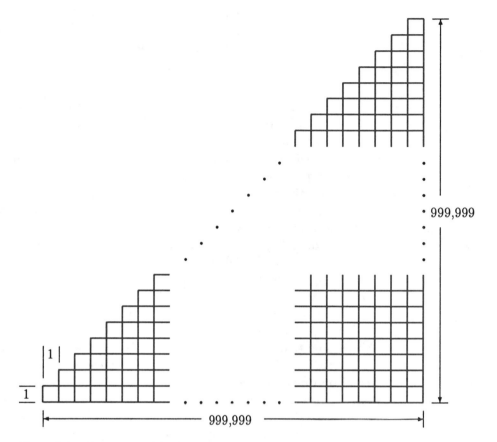

Figure 1.1 *How many 1-by-1 squares are represented by this figure?*

We now turn to the second phase, *devising a plan.* You might think that the plan is predetermined in this problem; there's only one way to add, isn't there? Not at all. We might plan to write down all 999,999 integers on paper and add them by hand, or we might plan to punch the numbers into a hand calculator, or we might plan to program a computer to do the problem. The first two plans would take an enormous amount of time, and the third, though practical, relies on machinery and skills we will not presume to have at our disposal.

Several of the suggestions in the previous section apply here. We could *solve a simpler problem* by changing the upper limit from 1,000,000 to something more manageable. Maybe we can learn from some trivial analogous problems, like the following:

(1.5)
$$
\begin{aligned}
1 + 2 &= 3 \\
1 + 2 + 3 &= 6 \\
1 + 2 + 3 + 4 &= 10 \\
1 + 2 + 3 + 4 + 5 &= 15
\end{aligned}
$$

We can *look for a pattern* in these problems and their answers. Notice in Display (1.5) that 6 is 3 greater than 3, 10 is 4 greater than 6, and 15 is 5 greater than 10. This is a pattern that we can predict to continue; it simply reflects the fact that the next successive (positive) integer is added in each step. However, that doesn't appear to be much help in devising a plan (other than the ones we have already identified).

We might *approximate the answer* by using the fact that we are adding approximately 1,000,000 integers whose average magnitude is about 500,000 (half of 1,000,000). This would suggest an approximate answer of 1,000,000 times 500,000, or 500,000,000,000. This is an interesting piece of information, but it seems unlikely that we can count on it to be an exact answer, and we have no way to determine the size of the error in the approximation.

Let us examine our reasoning in finding this approximation, though, and see if we can develop a precise answer. Our approximation is the product of two quantities—the number of integers being added and the average of these numbers. If we had precise values for these two quantities, our answer would be precise. Here is a new plan, then!

Of course, we know the number of integers in the sum; it is exactly 999,999. But what is the *exact average* of all the numbers being added? The usual way to find an average of a set of numbers is to add them up and divide by the number of numbers in the set. (The average of 2, 5, and 11 is 6 because $(2 + 5 + 11) \div 3 = 6$.) But we don't know the sum of these numbers; if we did, we would already have our answer, and we wouldn't need the average. On the other hand, we are not dealing here with some arbitrary collection of numbers, but with a set that has a *very special pattern*.

Let us turn to *a simpler problem* and look at the problems in Display (1.5), where we do know the sums and can easily figure out the averages. These are recorded in Table 1.1.

Numbers	Average
1, 2	$1\frac{1}{2}$
1, 2, 3	2
1, 2, 3, 4	$2\frac{1}{2}$
1, 2, 3, 4, 5	3

Table 1.1   Some simple averages.

Here we see a striking pattern, but one that is hardly surprising when we think about it. As the number of integers increases by one, the average increases by one half. Another way to state the relationship, in a general form, is to say that the average of the integers 1, 2, 3, ..., n is simply the average of the first and the last numbers: $(1 + n) \div 2$.

Now we have a plan that can be carried out in a reasonable amount of time, without a computer. We *carry out the plan* by computing the average of the integers 1, 2, 3, 4, ..., 999,999, which is $(1 + 999,999) \div 2$, or 500,000. (Our estimate happened to be exact.) Then we multiply by the number of integers, 999,999, and obtain the desired answer, 499,999,500,000.

If we *look back at the completed solution*, we can easily *generalize this solution* to the sum of the set of all positive integers from 1 to n, inclusive, where n can represent any positive integer. In arithmetic notation,

(1.6)
$$1 + 2 + 3 + 4 + \ldots + n = \frac{n \cdot (n + 1)}{2}$$

As you reflect back on this problem and its solution, note that the "obvious" plan of solution is not at all the simplest. And note that one of the suggested techniques (solving a simpler problem) proved to be helpful, *not* the first time we used it, but rather, the second time. In other words, be open to trial and error *and retrial.*

EXERCISES 1.3

1. Devise a plan to solve the problem of this section as restated in its geometric form (1.4). (*Hint*: Work with two copies of Figure 1.1.) Carry out the plan and solve the problem.

2. What is the sum of all positive integers from 1000 to 100,000, inclusive?

3. Generalize the problem of Exercise 2 and solve this problem.

4. How many squares are determined by the lines on a standard chessboard?

5. Generalize the problem of Exercise 4 to a square analogous to a chessboard. Restate this in arithmetic terminology.

6. Generalize the problem of Exercise 5 to a three-dimensional analogue. Restate the generalization in arithmetic terminology.

7. How many rectangles are determined by the lines on a standard chessboard?

8. Give two different generalizations of the problem of Exercise 7.

9. Fifty points are marked on a circle, and all possible chords are drawn with these as endpoints. How many chords are there?

10. Generalize the problem of Exercise 9 and solve this generalized problem.

11. (**The Towers of Hanoi Problem**) In a hidden monastery near Hanoi, embedded in the floor of a room are three silver spikes;

on one of them there were sixty-four gold disks, shaped like large washers, graduated in diameter to form a conical tower. The monks work tirelessly at the task of transferring the disks to re-form the tower on a different spike, according to the following rules: Only one disk at a time may be moved; each disk must be on one of the spikes except when being moved; and a large disk may not be placed on top of a smaller disk. What is the minimum number of moves that the monks must make to complete their task? Assuming that they make no wasted moves and move one disk every second, how long will it take them to finish?

12. Generalize the Towers of Hanoi Problem (Exercise 11) in two different ways.

13. Explore an alternative way of arriving at the formula (1.6) for the sum of the first n positive integers, starting with a simpler problem, as follows:

   (a) Observe that the sum
   $$1 + 2 + 3 + \ldots + 10$$
   can be found by "peeling off" the first and last numbers in the string and adding them, "peeling off" and adding the first and last numbers of the remaining string, and so on, until all the numbers have been used up. That is, we can write the desired sum in the form
   $$(1 + 10) + (2 + 9) + \ldots + (5 + 6)$$
   Now, each summand in parentheses is

11, and there are 5 of them. How is 11 related to 10? How is 5 related to 10?

   (b) What is the analogous form for the sum of the first 20 numbers? What is the value of each two-number summand? How many are there? How are these two answers related to 20?

   (c) What are the analogous answers to Part (b) for the sum of the first 100 numbers?

   (d) What pattern seems to be emerging from Parts (a–c)? Is this pattern disturbed if the largest number in your sum is odd?

   (e) Can you relate the pattern you see to Statement (1.6)? If so, how?

WRITING EXERCISES

1. Write a dialogue in which you, as questioner, lead a hypothetical friend (roommate, companion marooned with you on a desert island, or whomever) to discover the pattern that occurs when you add up successive even positive integers, starting with 2. Get your "friend" to describe the general pattern *in words* as clearly as he/she can.

2. Try to devise a problem for which drawing a diagram would *not* be a helpful problem-solving technique. If you can think of one, explain it and explain why a diagram would not be helpful. If you can't think of one, justify the claim that diagrams are helpful for *all* problems.

1.4   The Mathematical Way of Thinking

Each of the other chapters in this book focuses on a particular mathematical problem. Despite differences among the areas they explore, the methods they develop, and the results they obtain, all these chapters have a fundamental similarity in the way their problems are approached. This general methodology is sometimes called *the mathematical way of thinking.*

It is based on the assumption that the key to mathematical truth is logical reasoning. The correctness of mathematical statements and solutions to problems can only be proven true by checking step by step the logical validity of the arguments used. But constructing these proofs is only half the story, usually the second half. The first (and more interesting) half of the mathematical way of thinking is the problem solving itself, which was discussed in general in Section 1.2. Its main features appear in every chapter, emphasizing the underlying unity of this vast, sprawling field called mathematics.

As you proceed to sample the various topics in the chapters to follow, you will see some of the diversity of mathematics. Be forewarned, too, that the common methods of problem solving are not bound to appear exactly as you saw them in this chapter. For instance, restating the problem will not always be a shift from algebra to geometry. It may be a shift from a formula to a graph, from words to symbols, or from space to time.

In the preceding section we saw a problem for which an appropriate diagram was a geometric figure whose area represented the solution to the problem. At the beginning of Chapter 4 the appropriate diagram to help understand sets is a Venn diagram, for which the important feature is overlapping circles. Later in the same chapter, counting problems are solved by using tree diagrams, in which the most important feature is the number of branches. In Chapters 3 and 8 the diagrams are geometric figures that are literally the topic of discussion, whereas in Chapter 6 an appropriate diagram is an array of fractions, organized so as to communicate visually a property of those fractions not otherwise obvious. In many chapters, "diagrams" may be charts, graphs, tables, or some other visually structured aid.

Patterns and examples are used to solve the problems in virtually every chapter, but, of course, the structure of the patterns and the content of the examples depend on the problem. For example, Chapter 2 looks at patterns in the prime factors of certain numbers. Chapter 7 draws conclusions from patterns of operations such as addition and multiplication. Chapter 8 finds a way of "seeing" figures and distances in four-dimensional space by looking at the patterns for those ideas in two- and three-dimensional space.

The technique of solving a simpler problem appears in every chapter. Sometimes "simpler" means *smaller* numbers (as in Chapter 2) and sometimes it means *fewer* numbers (as in Chapter 4). Sometimes it means a *lower* dimension (as in Chapter 8), sometimes it means a *shorter* task (as in Chapter 5), and sometimes it means *finite* rather than *infinite* (as in Chapter 7). Be prepared to shift your ideas and labels on what is the "simpler" problem and what is the "general" one. Often you will find the "general solution" of one section in a chapter becomes the springboard "simpler problem" of the next section.

Your understanding of mathematics as a liberal art with a methodology all its own will be greatly enhanced if, as you work through the details of each topic, you watch for the many instances of the heuristics used. And remember to use these problem-solving techniques yourself as you attempt the exercises. Mathematics is not a spectator sport!

Topics for Papers – Chapter 1

1. Write a detailed explanation of the Tower of Hanoi Problem (Exercise 11 of Section 1.3) that illustrates how to use as many as possible of the dozen problem-solving techniques listed at the end of Section 1.2. (*Note*: Section 1.3 itself is a somewhat analogous treatment of the problem of summing successive positive integers; it might be a helpful guide for you.)

2. George Pólya is one of the most interesting personalities in 20th-century mathematics. Write a human-interest biography of Pólya that highlights his personal qualities, as if you were writing an article for a popular magazine. Be sure to provide a bibliography that lists any sources you used in writing the paper, and be careful that you do not simply copy or paraphrase the material from your sources. Any material not explicitly quoted should have been mentally digested by you, so that your own words describe what really are your own ideas when you write them down.

For Further Reading

1. Averbach, Bonnie, Patricia Brewer, and Orin Chein. *Mathematics: Problem Solving Through Recreational Mathematics*. San Francisco: W. H. Freeman and Co., 1980.

2. Gardiner, A. *Discovering Mathematics: The Art of Investigation*. New York: Oxford University Press, 1987.

3. Hayes, John R. *The Complete Problem Solver*. Hillsdale, NJ: L. Erlbaum Associates, 1989.

4. Hughes, Barnabas. *Thinking Through Problems: A Manual Of Mathematical Heuristic*. Palo Alto: Creative Publications, 1976.

5. Pólya, George. *How To Solve It*, 2nd ed. Princeton: Princeton University Press, 1957.

6. ——— . *Mathematical Discovery* (2 vols.). New York: John Wiley and Sons, inc., 1962, 1965.

7. Schoenfeld, Alan H. *Mathematical Problem Solving.* Orlando, FL: Academic Press, 1985.

8. Whimbey, Arthur. *Problem Solving and Comprehension.* Philadelphia: Franklin Institute Press, 1982.

9. Wickelgren, Wayne A. *How To Solve Problems.* San Francisco: W. H. Freeman and Co., 1974.

MATHEMATICS OF PATTERNS: NUMBER THEORY

2.1 What Is Number Theory?

One of the most intriguing fields of mathematics is the theory of the **natural numbers**, that is, of the numbers 1, 2, 3, 4, 5, ... Throughout the Greek era (600 B.C. to A.D. 400) this field embraced all of theoretical mathematics, including much of what we would now call geometry or algebra. Number theory as a specialized discipline first developed in the 17th century with the leisure-time activities of Pierre de Fermat; it became firmly established by Carl Friedrich Gauss in 1801. Despite the length of its history and the ingenuity of its many brilliant devotees, number theory remains "the last great uncivilized continent of mathematics."[1] Its jungles of primes, multiples and factorizations draw nourishment from countless streams of results whose common headwaters lie in unexplored territory that always seem to be just beyond the next problem.

The purpose of this chapter is to provide a taste of number theory and with that taste some flavor of "pure" mathematics. The main topic will be *perfect numbers*, together with some necessary preliminaries and several related topics. Number theory uses the familiar operations of arithmetic (addition, subtraction, multiplication, division), but more as the starting point of intriguing investigations than as topics of primary interest. Number theory is more involved in finding relations, patterns and structures of natural numbers. For example, consider this "parlor trick" with numbers:

---

[1] Eric Temple Bell, "The Queen of Mathematics." In *The World of Mathematics* (James R. Newman, ed.) New York: Simon & Schuster, Inc., 1956, p. 499.

Greek Multiplication on a Wax Tablet.

Step 1. Pick any three-digit number (e.g., 239).

Step 2. Form a six-digit number by repeating the three-digit number chosen (e.g., 239,239).

Step 3. Divide this number by 13, obtaining a quotient and a remainder.

Step 4. Divide the quotient obtained in *Step 3* by 7, obtaining a second quotient and remainder.

Step 5. Divide the quotient obtained in *Step 4* by 11.

This last quotient, found in *Step 5*, will equal the original number!

If you repeat the process several times using several different starting numbers, you should be persuaded that it always works. [Of course, unless you try all 900 possible numbers, or discover some general proof, you don't know for sure that it *always* works.] Clearly, this is not an accident. But what is the explanation?

At this point, rather than produce the answer, let us consider some related questions. Is there a different set of divisors (other than 13, 7, and 11) that would produce the same result? Is there a similar procedure for repeated four-digit numbers (e.g. 48,764,876)? for five-digit numbers? etc. Is there an analogous procedure for numbers repeated a third time (e.g. 479,479,479)? a fourth time? etc. The answers to all these questions are well within your grasp, and they will be addressed in the exercises of Section 2.3. For now, note that these are the kind of generalizations that mark number theory. Indeed, the hallmark of mathematicians, in general, is not primarily whether they can answer these or similar questions, but the fact that they pose the questions in the first place.

Turning to another example, consider the following:

$$1 + 3 = ?$$
$$1 + 3 + 5 = ?$$
$$1 + 3 + 5 + 7 = ?$$

If there were such persons as "arithmeticians," they would, presumably, be interested only in the answers to these addition problems: 4, 9, and 16, respectively. However, number theorists would note the similarity in the three problems (sums of odd numbers) and in their answers: 4, 9, and 16, which are, respectively, 2×2, 3×3, and 4×4. Struck by this curiosity, they would go on to ask whether these are isolated instances or part of some general pattern.

You may already have noticed a specific pattern; namely, that the sum of consecutive odd numbers, starting with 1, is a square. If you test this conjecture for several more cases, you will find it verified. You might even make a more precise observation about this pattern, which could be formulated as:

(2.1)   The sum of consecutive odd numbers, starting with 1, is the square of the number of odd numbers added.

Here we begin to have problems with language; Statement (2.1) is wordy and so may tend to confuse, rather than to clarify, the pattern we have been studying. Fortunately, we can use letters to represent numbers. You have encountered this process in algebra, where letters are used to label unknown numbers. (e.g., "Let Mary's age equal x.") The letters are then used like numbers in equations, the equations are manipulated in accordance with the rules of arithmetic, and the number (solution) is determined. Letter representation can also convey information about a large, sometimes infinite, set of numbers. For example, we can abridge the verbal statement,

The sum of any two numbers is unchanged when we reverse the order in which the two are added

to

$$x + y = y + x$$

To be precise, the latter formulation should include a verbal or symbolic phrase indicating the set of numbers whose elements can replace x and y

Arithmetica *Carrying a Weight-driven Clock. From a 16th Century engraving by Martin de Vos.*

in the statement. *In this chapter we shall make an agreement that, unless otherwise stated, any letter used to represent a number may be replaced by any natural number.* This means that any general statement or formula without explicit restriction is understood to be true for all natural numbers, but is not necessarily true, or even meaningful, for other numbers.

Returning to our example, we can replace the verbal statement (2.1) by

(2.2)   If the first n odd numbers are added, the sum is n^2.

We can symbolize even further by representing "the first n odd numbers added" as

$$1 + 3 + 5 + \ldots$$

This would be ambiguous, though, for it doesn't indicate where we intend the sequence of numbers to stop. To correct this, we need an expression to represent the nth odd number. You can figure that out yourself, if you study the following three sequences:

$$1, \quad 2, \quad 3, \quad 4, \quad 5, \quad 6, \quad 7, \quad \ldots$$
$$2, \quad 4, \quad 6, \quad 8, \quad 10, \quad 12, \quad 14, \quad \ldots$$
$$1, \quad 3, \quad 5, \quad 7, \quad 9, \quad 11, \quad 13, \quad \ldots$$

The first sequence is merely the sequence of natural numbers, which we use as a reference sequence. The second, of course, is the sequence of even numbers, and the third is the sequence of odd numbers. The even numbers form a "bridge" to clarify the connection between the first and third sequences. If we continued listing the terms in each sequence, which numbers would line up below 50? By generalizing the pattern begun, you should see that the 50th even number is 100 and the 50th odd number is 1 less, or 99. Similarly, the 123rd even number must be 246 and the 123rd odd number must be 245.

In general, the nth even number is $2n$ and the nth odd number is $2n - 1$; we can use this to rewrite the generalization (2.2) as:

(2.3) $$\boxed{1 + 3 + 5 + 7 + \ldots + (2n - 1) = n^2}$$

The advantage of this form over that of Statement (2.2) is that it is an equation, and thus can be worked with using the formal rules of arithmetic or algebra.

In this example we have a generalization that is *plausible* because it is verified in all the instances we have checked. But to assert it as a *factual property* of all natural numbers, we must *prove* it. And our proof cannot consist of checking *every* number; we would be faced with an unending task. We must have a general means of proof to match the generality of what we claim to be true.

There are a variety of arguments available to demonstrate the universal truth of our statement about adding odd numbers, varying from pictorial and intuitive arguments to those of considerable formality. For example, a proof known to the Greeks no later than the second century B.C. uses a pictorial approach. (See Figure 2.1.) This figure shows a square array consisting of 25 dots, 5 dots on a side. The separating lines emphasize the fact that this array can be thought of as "growing" from one dot by the addition of 3 dots in a "bent row," 5 dots in the next "bent row," 7 dots in the next, and 9 in the last. We could add the next "bent row," which would contain 11 dots, and have a 6-by-6 array. In general, it should be clear that each successive "bent row" of dots will add on the next odd number, and the result will always be a square. In other words,

The sum of the first *n* odd numbers is the *square* of *n*.

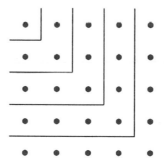

Figure 2.1 *Sums of successive odd numbers.*

An alternative proof relying on arithmetic and algebraic notation is also available. Suppose we did not already know the sum of the first *n* odd numbers and chose to represent this sum by some letter, say *s*. Then

$$s = 1 + 3 + 5 + 7 + \ldots + (2n - 1)$$

It follows that

$$s = (2n - 1) + (2n - 3) + (2n - 5) + \ldots + 5 + 3 + 1$$

(Don't let this step throw you off. We have merely written the same sum in reverse order. Of course, we are writing in a few of the numbers implied, but not written, in the first step [found by decreasing $(2n - 1)$ by 2, by 4, etc., following the pattern of odd numbers], but nothing is changed.) If we now add the two equations, combining corresponding terms of each on the right, we obtain

$$2s = 2n + 2n + 2n + 2n + \ldots + 2n + 2n$$

On the right, there are n numbers [Why?], and thus we can write

$$2s = n \times 2n$$

or

$$s = n \times n = n^2$$

and we are finished.

EXERCISES 2.1

1. What is the 100th even number?

2. What is the 235th even number?

3. What is the 235th odd number?

4. What is the 994th odd number?

5. What is the ordinal position of the even number 578? (*Ordinal position* is the position in the sequence; e.g., 12 is the 6th even number, so it is in ordinal position 6.)

6. What is the ordinal position of the odd number 4321?

7. What is the sum of the first 100 odd numbers?

8. What is the sum of the first 235 odd numbers?

9. What is the sum of all odd numbers from 1 to 1987, inclusive?

10. What is the sum of all odd numbers from 1 to 4321, inclusive?

11. What is the sum of all odd numbers from 1989 to 4321, inclusive?

12. What is the sum of all odd numbers from 135 to 579, inclusive?

Exercises 13–16 refer to the sequence of numbers that starts:

$$3, 6, 9, 12, 15, 18, \ldots$$

13. What is the next number in this sequence?

14. What is the 10th number in this sequence?

15. What is the 200th number in this sequence?

16. Determine an expression for the nth number in this sequence.

Exercises 17–20 refer to the sequence of numbers that starts:

$$1, 4, 7, 10, 13, 16, \ldots$$

17. What is the next number in this sequence?

18. What is the 10th number in this sequence?

19. What is the 200th number in this sequence?

20. Determine an expression for the nth number in this sequence.

Exercises 21–24 refer to the sequence of numbers that starts:

$$10, 17, 24, 31, 38, 45, \ldots$$

21. What is the next number in this sequence?

22. What is the 10th number in this sequence?

23. What is the 200th number in this sequence?

24. Determine an expression for the nth number in this sequence.

Exercises 25–30 refer to the sequence of numbers that starts:

$$1, 3, 6, 10, 15, 21, \ldots$$

These numbers were called **triangle numbers** by the Greeks. Thus, the first triangle number is 1, the second is 3, the third is 6, and so on.

25. Why are these numbers called triangle numbers? (*Hint*: Look at the arrangement of the 10 pins in a bowling alley or of the 15 balls on a pool table.)

26. What is the seventh triangle number?

27. Determine an expression for the nth triangle number. (*Hint*: Note that the nth triangle number is equal to

$$1 + 2 + 3 + \ldots + (n - 1) + n$$

and follow the procedure used in the second proof of (2.3).)

28. What is the 100th triangle number?

29. A clock chimes the correct number of times on the hour, and chimes once on the half hour. How many times does it chime in a twenty-four-hour period?

30. What is the result when a pair of consecutive triangle numbers is added? Formulate a general statement. Prove your generalization, if possible.

WRITING EXERCISES

1. There are many situations in which a sequence of numbers occurs. In some cases the sequence is "natural" or "intrinsic," while in others it is arbitrary or contrived, no more appropriate to the situation than many other possible sequences. Describe one particular instance of *each* of these two types of sequences as clearly as you can. For the first kind, explain how and why the sequence is natural; for the second, explain how and why it is arbitrary.

2. Exercises 25–30 deal with *triangle numbers*; you are already familiar with *square numbers*. How would you generalize from these two kinds of numbers to define **figurate numbers**? Which regular polygons[2] of plane geometry correspond naturally to number patterns and which do not? Identify the first few numbers of each type you consider, and explain why they correspond to the polygon with which you identify them.

3. Research and report on what the Greeks considered to be figurate numbers. Give specific examples besides triangle and square numbers.

4. How might you generalize the idea of *square number* to *rectangle number*? Explain the pros and cons of doing this.

5. How might you generalize the idea of *triangle number* to
(a) *right triangle number*;
(b) *isosceles triangle number*;
(c) *scalene triangle number*.
Explain the pros and cons of doing this.

6. The 3-dimensional analogue of *square number* is *cube or cubic number*. What is a 3-dimensional analogue of *triangle number*? Explain how you would define such a kind of number and why it is an appropriate analogue. Include in your description the first five such numbers.

---

[2]A *regular* polygon is one with all sides congruent and all angles congruent.

2.2   Divisibility

If we consider the four basic arithmetic operations as applied to the set of natural numbers, we find some simple, but significant, differences. For example, the *sum* or *product* of any two natural numbers is always a natural number, but this is not the case, in general, for the *difference* or *quotient* of two natural numbers. For some choices of x and y, $y-x$ is a natural number; for other choices, it is not. For example: $5-3$ equals the natural number 2, but $3-5$ is not a natural number. If, for a particular x and y, $y-x$ happens to be a natural number, then that indicates some special relationship must hold between x and y. That relationship, of course, is that x **is less than** y (or equivalently, y **is greater than** x), which is denoted by the familiar symbolism

$$x < y \quad \text{or} \quad y > x$$

Thus, because $5-3$ is a natural number, we write $3 < 5$, and say, "3 is less than 5."

A completely analogous situation holds for division. Sometimes $y \div x$ is a natural number; sometimes it is not. When it is, there is a special relationship that holds between x and y.

DEFINITION   If $y \div x$ is a natural number, then x **divides** y (or x is a **divisor** of y, or x is a **factor** of y, or y is a **multiple** of x, or y is **divisible by** x), which is denoted by $x \mid y$.

Just as we denote the negation of "equals" by putting a slash mark through the symbol (e.g., $5 \neq 8$, $2 + 2 \neq 3$), we denote the negation of "divides" with a slash mark through the divisibility symbol. Thus, all of the following are true statements:

$$2 \mid 6, \quad 8 \mid 8, \quad 9 \nmid 2, \quad 12 \nmid 9$$

Of course, with some numbers we may have to do some arithmetic to determine whether one divides the other.[3] For instance, it is not immediately obvious whether 37 divides 703. This is settled by dividing 703 by 37, obtaining the quotient 19. Note that the statement "$37 \mid 703$" does not contain the information that 19 is the quotient of 703 divided by 37. It only states that there is *some* (natural number) quotient.

---

[3]The same is true of the relation "less than." For example, it is not immediately obvious whether 5^7 is less than 7^5, or, beyond the set of natural numbers, whether $\frac{12}{19}$ is less than $\frac{103}{163}$.

It is worth noting that $x \mid y$ is equivalent to

$$\text{For some natural number } z, \quad x \cdot z = y$$

This follows, of course, from the familiar connection between multiplication and division. This "multiplication form" of the definition, however, exhibits the symmetric role of x and z, and verifies that z is also a factor of y. These two general principles are restated for emphasis in the following list of statements, along with several others that should be equally clear after a bit of reflection.

(2.4)	$x \mid y$ if and only if [4] there is some natural number z such that $x \cdot z = y$.
(2.5)	If $x \mid y$, then $\dfrac{y}{x}$ is a natural number, and $\dfrac{y}{x} \mid y$.
(2.6)	$1 \mid y$.
(2.7)	$y \mid y$.
(2.8)	If $x \mid y$ and $y \mid z$, then $x \mid z$.
(2.9)	If $x \mid y$ and $x \mid z$, then $x \mid (y + z)$.
(2.10)	If $x \mid y$, then $x \leq y$.

EXERCISES 2.2

For Exercises 1–16, determine whether the given statement is *true* or *false*.

1. $9 \mid 99$
2. $7 \nmid 59$
3. $12 \mid 6$
4. $8 \mid 300$
5. $17 \nmid 102$
6. $37 \nmid 444$
7. $5 \mid 24{,}680$
8. $12 \mid 153{,}624$
9. $2 \nmid 123{,}456{,}789$
10. $3 \mid 123{,}456{,}789$
11. $2^2 \mid 2^3$
12. $3^8 \nmid 3^{11}$
13. $5^7 \mid 10^7$
14. $5^7 \mid 7^5$

15. If $x \leq y$, then $x \mid y$.
16. If $x \mid y$ and $y \mid x$, then $x = y$.

17. Suppose that b and c are two numbers with the property that $b \mid c$ and $b \mid (c + 1)$. What is b?

18. Use the notation of divisibility to define an *even number*. (That is, complete the statement, "n is an *even number* if and only if ....")

---

[4] If you are unfamiliar with this phrase, you can find an explanation of it in Appendix Section A.3.

19. Use Property (2.9) to explain why the sum of two even numbers is even. (See Exercise 18.)

20. Use Property (2.8) to explain why the product of an even number and any number is even. (See Exercise 18.)

21. Explain why the product of any two consecutive numbers is even. (See Exercise 20.)

22. If $5 \mid k$, what is the next number larger than k that is a multiple of 5?

23. Construct a table with six columns, as follows: in the first column, list the natural numbers from 1 to 30; head the column "n." In the second column, next to each number in the first, enter all the divisors of that number; head this column "divisors." In the third column, tally the number of divisors entered in the second column; head this column "$D(n)$." (For example, the entries corresponding to 18 in the second and third columns would be "1, 2, 3, 6, 9, 18" and "6," respectively.) Leave the last three columns blank. They will be used for Exercises 1 and 2 of Section 2.5. (This table is used as a source of data for the Topics for Papers at the end of the chapter.)

WRITING EXERCISES

1. The use of "$\mid$" to stand for divisibility and of letters to stand for arbitrary numbers of a system typifies the mathematical use of symbols as a convenient shorthand.

   (a) Write each of the statements (2.4) through (2.10) completely in words, without using *any* symbolic notation, not even letters to represent numbers. (Be careful to quantify the statements properly. That is, if a statement is intended to be a property of *all* numbers, make sure that your translation reflects that fact, etc.)

   (b) Discuss the pros and cons of using symbols in presenting mathematical ideas.

2. The text identifies an analogy between subtraction and *less than*, on the one hand, and division and divisibility, on the other. Elaborate on this analogy by showing how any four of the divisibility statements (2.4) through (2.10) translate into analogous statements about *less than*. Are there any of these seven statements that *cannot* be translated by analogy? Justify your answer.

2.3   Counting Divisors

In Exercise 23 of Section 2.2, you were asked to tabulate the number of divisors of each integer from 1 to 30. Here we shall pursue that notion. For brevity, we adopt the following notational definition:

NOTATION   The number of divisors of n will be denoted by $D(n)$. (This is read "D of n.")

For example, there are 4 divisors of 10—namely, 1, 2, 5, and 10—thus, we write $D(10) = 4$. Similarly, because there are 9 divisors of 36—namely, 1, 2, 3, 4, 6, 9, 12, 18, and 36—we write $D(36) = 9$.

Divisibility by 5. From Calandri's Arithmetica, *Florence, 1491.*

Our goal in this section is to develop the ability to determine $D(n)$ for any number n.[5] There is, of course, the "brute force" method, whereby we test every number less than or equal to n, recording (or at least counting) those that divide n. [Property (2.10) points out that it would be fruitless to test numbers larger than n, because none can possibly be divisors of n.] This is precisely the method suggested in forming the table of Exercise 23 in Section 2.2. Unfortunately, this method is not very practical for large numbers. Imagine, for instance, testing 10,000 numbers! Of course with the aid of a computer this could be done in a reasonable amount of time, but even with such assistance, testing *every* number less than or equal to n is highly inefficient.

---

[5] Recall that throughout this chapter "number" means "natural number," unless otherwise specified.

A great deal of time and effort can be saved by observing the import of Property (2.5). This says, for example, that because $40 \mid 10,000$, then $\frac{10,000}{40}$ (which equals 250) is also a divisor of 10,000. Put another way, when we test 40 as a divisor of 10,000 and obtain 250 as a quotient, not only do we establish that 40 is a divisor, but also that 250 is a divisor. In other words, divisors occur in pairs. Thus, in testing for divisors of 10,000, we find pairs: 1 and 10,000, 2 and 5000, 4 and 2500, and so on. The real advantage of this pairing process is seen in the observation of which numbers are *not* divisors. For example, when we test 3 and find that it is not a divisor of 10,000, we also establish that there are no divisors of 10,000 between 2500 and 5000. For if there were, there would be a "companion" divisor between 2 and 4—which is clearly false.

To find all the divisors of 10,000, then, we need only test "halfway," locating pairs of divisors. Note that halfway does not mean up to 5000; that would be halfway if we were *adding* pairs of numbers to obtain 10,000. For pairs of numbers whose product is 10,000, halfway is 100—the *square root* of 10,000. Each divisor less than 100 must pair up with a divisor greater than 100, and vice versa, so that when we have found all divisors from 1 to 100, with their paired divisors from 100 to 10,000, we have thereby found all the divisors of 10,000. In general:

> To find all the divisors of a number, say n, test all the numbers up to $\sqrt{n}$, identifying pairs of divisors.

For most values of n, $\sqrt{n}$ is not a natural number, of course, but the reasoning is just the same in either case. In practice, if we test for divisibility by successive integers starting with 1, we merely keep going until we obtain a quotient less than or equal to the divisor being tested. At that point we know we have passed $\sqrt{n}$ and can stop testing. It is not necessary to know or compute $\sqrt{n}$.

Example 2.1   **Problem:** What is $D(200)$?

Solution: To find all the divisors of 200, we divide by 1, 2, 3, and so on, obtaining a quotient and remainder in each case. The results of this arithmetic are recorded in Table 2.1: Each time we obtain a remainder of 0, we have found *two* divisors of 200. There is no need to test numbers larger than 14, because the quotient when 200 is divided by 14 is 14 (with a nonzero remainder). Testing numbers larger than 14 will produce quotients less than 14, and we already know which of these are divisors. Thus, the divisors of 200 are: 1 and 200, 2 and 100, 4 and 50, 5 and 40, 8 and 25, 10 and 20. Therefore, $D(200) = 12$. (Incidentally, we have also found that $\sqrt{200}$ is between 14 and 15.)   □

Test Divisor	Quotient	Remainder
1	200	0
2	100	0
3	66	2
4	50	0
5	40	0
6	33	2
7	28	4
8	25	0
9	22	2
10	20	0
11	18	2
12	16	8
13	15	5
14	14	4

Table 2.1   Testing for divisors of 200.

This method is significantly better than our first idea of testing *all* numbers less than or equal to n. And it can be marginally improved by noting that we need not test for divisibility by any number that is a multiple of one known not to be a divisor. (In Example 2.1, testing 6, 9, and 12 was unnecessary, given that 3 is not a divisor.) But it is not the best we can do. Look at the table that you were asked to construct in Exercise 23 of Section 2.2. Reorganize that information by asking which numbers have exactly one divisor, which have exactly two divisors, and so on. The resulting information is found in Table 2.2.

k	Numbers with exactly k divisors
1	1
2	2, 3, 5, 7, 11, 13, 17, 19, 23, 29, ...
3	4, 9, 25, ...
4	6, 8, 10, 14, 15, 21, 22, 26, 27, ...
5	16, ...
6	12, 18, 20, 28, ...
7	
8	24, 30, ...

Table 2.2   Numbers classified by the number of their divisors.

We do not have a lot to go on, but some observations can be made. Clearly, 1 is unique in having only one divisor. The numbers with exactly two divisors are probably familiar to you. They play a very important role in number theory.

DEFINITION   A **prime number** is a number that has exactly two divisors.

An equivalent statement is:

(2.11) | A number is prime if and only if it cannot be written as the product of two smaller numbers.

Note that we have not explained why certain numbers have exactly two divisors. We have only established a label for such numbers. There are many fascinating theorems and conjectures involving prime numbers. We will see some of these as we proceed, but we will not digress at this point from our pursuit of a formula or process for finding $D(n)$.

Looking now in Table 2.2 at numbers with exactly three divisors, we find 4, 9, 25 (and others, if we were to look beyond 30). A bit of reflection shows that these numbers are just the squares of the prime numbers. If we test this idea a bit, we find, for example, that

the only divisors of 7^2 (or 49) are 1, 7, and 49;
the only divisors of 11^2 (or 121) are 1, 11, and 121

In general, it should be clear that if a number n is the square of a prime p (that is, if $n = p^2$), it will have the three distinct divisors 1, p, and p^2. But does that guarantee that $D(n) = 3$? In other words, for the squares of some primes, might there not be more than these three divisors? To answer this question, we need a bit more terminology and one more basic fact about numbers.

DEFINITION   The number n, or any representation of n as a product of two or more numbers is called a **factorization** of n. If all factors in a factorization are prime, it is called a **prime factorization**.

Example 2.2   Here are all the factorizations of 36:

$$36 \qquad 3 \times 4 \times 3 \qquad 6^2$$
$$1 \times 36 \quad 1 \times 2 \times 3 \times 6 \quad 2^2 \times 3^2$$
$$4 \times 9 \quad 2 \times 3 \times 2 \times 3 \quad 1 \times 2 \times 2 \times 1 \times 3 \times 3$$

Of these, $2 \times 3 \times 2 \times 3$ and $2^2 \times 3^2$ are prime factorizations of 36. Note that $1 \times 2 \times 2 \times 1 \times 3 \times 3$ is *not* a prime factorization because 1 is not a prime number. Note also that 36 is a factorization of 36.   □

Observe that every number has factorizations, and every number except 1 has prime factorizations. The latter fact is true because we can omit 1 from a factorization of any number other than 1, and we can write nonprime numbers as products of smaller numbers, which in turn are either prime or are products of smaller numbers. Eventually, the process of breaking numbers into products must terminate in a product of primes. This is the easy part of the proof of a basic fact about numbers:

> **The Fundamental Theorem of Arithmetic:** Every number greater than 1 has a prime factorization that is unique aside from the order in which the primes are written.

The proof of the "uniqueness" part of this theorem is somewhat involved and will not be given here. (See the references at the end of the chapter.) The theorem states that every number greater than 1 has a *unique collection* of primes associated with it. (Note that these primes may not all be distinct from each other.) The product of these primes is the number, and no other collection of primes multiplied together will produce that number. Thus, every number other than 1 is either prime or is the product of a unique collection of two or more (possibly repeated) primes. This prompts another definition:

DEFINITION A number is a **composite number** if it is the product of two or more (possibly repeated) primes.

For example, 6 $(= 2 \times 3)$ and 9 $(= 3 \times 3)$ are composite numbers. Notice that 1 does not fit this definition, and thus it is neither prime nor composite. In conjunction with the definition of prime and Property (2.11) we can characterize composite numbers as follows:

(2.12) | A number is composite if and only if it has three or more divisors.

Simply put, a prime number is so called because it is "primal" relative to factorization; that is, it cannot be factored as the product of smaller numbers. A composite number is composed (by multiplication) of smaller numbers, and ultimately of (a unique collection of) prime numbers. An interesting analogy can be drawn between numbers and chemical compounds, with primes corresponding to atoms and composites to compounds. The analogy is far from perfect, as there are an infinite number of primes that can be combined in an infinite number of ways.

The Fundamental Theorem provides the key to the question posed earlier about divisors of the square of a prime, and indeed about $D(n)$ in general, by yielding the following characterization of divisibility:

(2.13)   If $m > 1$, then $m \mid n$ if and only if each prime factor of m is also a prime factor of n.

The proof of this fact is fairly easy if we utilize Property (2.4) and the Fundamental Theorem. If $m \mid n$, then there is some number k such that $m \cdot k = n$. If we replace each of the numbers m, k, and n by its prime factorization, the uniqueness of the prime factorization of n establishes that the same primes must occur on either side of the equals sign. Thus, every prime in the factorization of m is also in the factorization of n. Conversely, if the primes in the factorization of m are all in the factorization of n, then the product of the "excess" primes, if any, in the factorization of n form a number k such that $m \cdot k = n$. Therefore, $m \mid n$. (If there are no "excess" primes, then $m = n$ and $k = 1$.)

For example, the prime factorizations of 15 and 180 are

$$3 \times 5 \quad \text{and} \quad 2 \times 2 \times 3 \times 3 \times 5$$

respectively, implying that $15 \mid 180$. (The "excess" primes in 180, relative to 15, are 2, 2, and 3, whose product equals 12, and $12 \times 15 = 180$.) Similarly, 8, which equals $2 \times 2 \times 2$, has one more factor of 2 than does 180; thus, this theorem tells us that $8 \nmid 180$.

Property (2.13), applied to the square of any prime, $n = p^2$, shows that the only divisors other than 1 are p and p^2. There are only three divisors, so $D(n) = 3$. In general, Property (2.13) tells us that the number of divisors (other than 1) of a number n is the number of distinct subcollections of primes that can be formed from the collection of primes in the prime factorization of n. We can even fit the divisor 1 into this description by observing that 1, the only number with no prime factors, corresponds to the "collection" of primes with nothing in it.

Example 2.3   **Problem:** Find all divisors of 144.

Solution: We begin by factoring 144, say as

$$12 \times 12 = 3 \times 4 \times 3 \times 4 = 3 \times 2 \times 2 \times 3 \times 2 \times 2 = 2^4 \times 3^2$$

We then list all the distinct collections of primes that can be formed from these four factors of 2 and two factors of 3. This is done in Table 2.3, where the divisor 1 is also listed.

Without multiplying out each of these, it is easy to count that there are 15; so $D(144) = 15$.   □

Early Illustration of Calculations on a Blackboard.
From Böschensteyn's Rechenbüchlin, *Ausburg, 1514.*

1	2	2^2	2^3	2^4
3	2×3	$2^2 \times 3$	$2^3 \times 3$	$2^4 \times 3$
3^2	2×3^2	$2^2 \times 3^2$	$2^3 \times 3^2$	$2^4 \times 3^2$

Table 2.3 *All the divisors of 144.*

A closer inspection of the pattern in Table 2.3 shows that we actually did more than necessary to compute $D(144)$. There are 15 numbers in the table because it is composed of 3 rows of 5 columns each. The first row contains the divisors of 144 with no factors of 3; the second row contains divisors with one factor of 3; the third row contains divisors with two factors of 3. The columns are analogously described in terms of the number of factors of 2. Thus, the numbers of rows and columns in the table necessarily correspond to the number of choices of factors of 3 and the number of choices of factors of 2, respectively. But these numbers can be determined *without forming the table*!

Example 2.4   **Problem**: What is $D(25,000)$?

Solution: The prime factorization of 25,000 is $2^3 \times 5^5$. If we formed a table listing all the divisors, it would be 4-by-6 in size because there are 4 divisors of 2^3—1, 2, 2^2, and 2^3—and there are 6 divisors of 5^5—1, 5, 5^2, 5^3, 5^4, and 5^5). Therefore, there are 24 divisors of 25,000. Symbolically,

$$D(25,000) = D(2^3 \times 5^5) = D(2^3) \times D(5^5) = 4 \times 6 = 24 \qquad \square$$

The procedure developed in the preceding two examples generalizes to numbers whose prime factorization involves three or more distinct primes.

Example 2.5   **Problem**: What is $D(7056)$?

Solution: The prime factorization of 7056 is $2^4 \times 3^2 \times 7^2$. As was seen in Example 2.3, there are 15 divisors of $2^4 \times 3^2$. We can multiply each of these 15 divisors by 1 or 7 or 7^2 (the three divisors of 7^2) to produce 45 distinct divisors of 7056, and thus

$$
\begin{aligned}
D(7056) &= D(2^4 \times 3^2 \times 7^2) \\
&= D(2^4) \times D(3^2) \times D(7^2) \\
&= 5 \times 3 \times 3 \\
&= 45 \qquad \qquad \square
\end{aligned}
$$

Our process for determining $D(n)$, then, for $n > 1$, is as follows:

Step 1.   Determine the prime factorization of n.

Step 2.   Collect repeated primes and write the prime factorization in exponential form.

Step 3.   Compute "D" of every power of each distinct prime by adding 1 to each exponent.

Step 4.   Multiply the resulting numbers obtained in *Step 3*.

The simple calculation in *Step 3* determines the number of choices of factors of each prime; adding 1 accounts for the choice of using no factors of that prime. *Step 4* computes the total number of combinations.

Example 2.6   **Problem**: What is $D(9828)$?

Solution: The prime factorization of 9828 is $2^2 \times 3^3 \times 7 \times 13$. Adding 1 to each exponent, we obtain 3, 4, 2, and 2, respectively. (Don't forget that 7 and 13 have unwritten exponents of 1.) Thus, $D(9828) = 3 \times 4 \times 2 \times 2 = 48$.
$\square$

The problem of determining $D(n)$ is thus ultimately reduced to finding the prime factorization of n and doing some trivial arithmetic. The Fundamental Theorem guarantees a unique prime factorization, so that any method that repeatedly factors n (and avoids the trivial factor of 1) will ultimately produce the prime factorization. Arithmetic skill, a hand calculator, and common sense can all be used. One way is to factor a composite number and any composite factors of it repeatedly into two smaller numbers, as in Example 2.3. A more organized process is to factor all the 2s, then all the 3s, then all the 5s, and so on.

Example 2.7   **Problem:** Determine the prime factorization of 5280.

Solution:

$$
\begin{array}{r|l}
2 & 5280 \\
2 & 2640 \\
2 & 1320 \\
2 & 660 \\
2 & 330 \\
3 & 165 \\
5 & 55 \\
  & 11
\end{array}
$$

Thus, $5280 = 2^5 \times 3 \times 5 \times 11$.   □

Example 2.8   **Problem:** Determine the prime factorization of 126,000.

Solution: Here, we can save some time by recognizing that each terminal zero represents a factor of 10, or 2×5. Thus,

$$
\begin{array}{r|l}
2^3 \times 5^3 & 126000 \\
2 & 126 \\
3 & 63 \\
3 & 21 \\
  & 7
\end{array}
$$

Consequently, $126{,}000 = 2^4 \times 3^2 \times 5^3 \times 7$.   □

Example 2.9   **Problem:** Determine the prime factorization of 20,889.

Solution: 20,889 is not divisible by 2.

$$
\begin{array}{r|l}
3 & 20889 \\
3 & 6963 \\
  & 2321
\end{array}
$$

2321 is not divisible by 3, 5, or 7, but

$$
\begin{array}{r|l}
11 & 2321 \\
   & 211
\end{array}
$$

211 cannot be divisible by 2, 3, 5, or 7. (Otherwise, 2321 would be; see

Property (2.8)). It is not divisible by 11, and thus we try the next prime, 13. It is not divisible by 13, so we try 17. Now, 17 is not a factor, but $2321 \div 17$ equals 12, with remainder 7; the fact that the quotient obtained is less than 17 tells us that we need go no further [Why?]; 211 is a prime number. So $20{,}889 = 3^2 \times 11 \times 211$.   □

EXERCISES 2.3

Determine the prime factorization for each of the following twelve numbers. Express your answer in exponential form.

1. 91   **2.** 120   **3.** 299

4. 313   **5.** 347   **6.** 496

7. 836   **8.** 1878   **9.** 4173

10. 100,100   **11.** 299,299   **12.** 1,000,000

13–24. Determine $D(n)$ for each of the numbers in Exercises 1–12, and identify which of these numbers are prime and which ones are composite.

25. If p is a prime, what is $D(p^3)$? What is $D(p^4)$? In general, what is $D(p^k)$?

26. Show that if $n > 1$ and if n has no prime divisor less than or equal to $\sqrt{n}$, then n must be prime. (See Examples 2.1 and 2.9.)

27. Prove or disprove: $D(n)$ is odd if and only if n is the square of some number.

28. Prove or disprove: If p is prime, then $p^2 - 2$ is prime.

In Exercises 29–34, you are asked to reason backward from information about $D(n)$ to infer something about n. For example, if $D(n) = 6$, then the form of the prime factorization of n is either p^5 or $p^2 \times q$, where p and q are distinct primes.

29. What are the possible forms of the prime factorization of n if $D(n) = 10$?

30. What are the possible forms of the prime factorization of n if $D(n) = 11$?

31. What are the possible forms of the prime factorization of n if $D(n) = 12$?

32. What is the smallest number that has exactly 12 divisors?

33. What is the smallest number that has exactly 18 divisors?

34. Determine all numbers less than 200 with exactly 12 divisors.

35. What is the prime factorization of 1001? Use your answer to explain the "parlor trick" of Section 2.1. (See page 19.)

36. The prime factorization of 10,001 is 73×137. Design a "parlor trick" similar to that of Section 2.1 that uses this fact. (See page 19.)

37. (a) What is the prime factorization of 10,101?

(b) Use your answer from (a) to describe a "parlor trick" like that of Section 2.1. (See Exercise 35.)

WRITING EXERCISES

1. The text mentions an analogy between prime and composite numbers in mathematics, and atoms and compounds in chemistry.

(a) Explore this analogy in some detail, discussing how it is apt and how it is inappropriate. (In particular, what in chemistry is analogous to the Fundamental Theorem of Arithmetic in mathematics?)

(b) Identify and discuss an analogy in some other subject area that mirrors, at least partially, the relationship between prime numbers and composite numbers.

2. Each of the statements given at the end of this question purports to be a definition of *prime number*. However, all but one of them are inaccurate (grammatically, syntactically, or logically). Identify the correct statement, and justify how the one you picked describes exactly the same thing as the text definition. Then choose any three incorrect definitions and explain why they are incorrect.

    (a) A prime number is a number divisible by itself and one.

    (b) A prime number is a number that cannot be written as the product of two other numbers.

    (c) A prime number is a number that is divisible only by itself and one.

    (d) A prime number is a number that is divisible by just one number smaller than itself.

    (e) A prime number is one like 2, 3, 5, 7, etc.

    (f) A prime number is a number with two different factors.

    (g) A prime number is a number that has no factor that is larger than one and smaller than itself.

    (h) A prime number is when you try to divide it by something besides one or itself but you can't.

3. Explain in your own words how the Fundamental Theorem of Arithmetic justifies the correctness of the four-step process for determining $D(n)$. How do we know that every factor of n is counted? ...that no numbers are counted which are not factors? ...that no factor is mistakenly counted twice?

2.4   Summing Divisors

We now turn our attention to the problem of *adding* all the divisors of a given number. As with the problem of determining the number of divisors, our practical challenge is not how to get an answer, but how to do it efficiently. As in the previous section, we introduce a notational definition for the sake of brevity:

NOTATION   We denote the sum of all the divisors of n by $S(n)$. For instance,

$$S(10) = 1 + 2 + 5 + 10 = 18$$

and

$$S(24) = 1 + 2 + 3 + 4 + 6 + 8 + 12 + 24 = 60$$

Much of the work necessary to determine $S(n)$ is already behind us. In fact, the development of our process of determining $D(n)$ identified that each combination of primes chosen as a subset of the prime factorization of n produces a divisor of n. Thus, we already have an efficient method of listing all the divisors of a number.

Example 2.10   **Problem:** List all divisors of 675.

   Solution: The prime factorization of 675 is $3^3 \times 5^2$. Thus, its divisors are

$$\begin{array}{llll} 1 & 3 & 3^2 & 3^3 \\ 5 & 3 \times 5 & 3^2 \times 5 & 3^3 \times 5 \\ 5^2 & 3 \times 5^2 & 3^2 \times 5^2 & 3^3 \times 5^2 \end{array} \qquad \square$$

We could now go on to compute each of the divisors listed in Example 2.10, add them up, and obtain $S(675)$. Let us do this in several carefully chosen steps. The twelve divisors of 675 are listed in four columns of three. Let us add each column, and then add up the subtotals. The sum of the numbers in the first column is $1 + 5 + 25$, which equals 31. We can now use the internal pattern of the list of divisors to help us. Because each number in the second column is exactly 3 times the corresponding number in the first column, the sum of these three numbers must be 3 times 31. Similarly, the sum of the numbers in the third column must be $3^2 \times 31$ and the sum of those in the fourth must be $3^3 \times 31$. Thus, the grand total of all twelve numbers is

$$31 + (3 \times 31) + (9 \times 31) + (27 \times 31)$$

which, in turn, equals

$$(1 + 3 + 9 + 27) \times 31 = 40 \times 31 = 1240$$

Notice that 31, the sum of the first-column numbers, is the sum of the divisors of 5^2, and 40, the sum of the multipliers, is the sum of the divisors of 3^3. In short,

$$S(3^3 \times 5^2) = S(3^3) \times S(5^2)$$

Example 2.11   **Problem:** Determine $S(288)$.

Solution: The prime factorization of 288 is $2^5 \times 3^2$. If we were to list the eighteen divisors in three rows of six columns, as in the previous example, the first column would contain 1, 3, and 3^2, whose sum is 13. And the six multipliers of 13 would be 1, 2, 2^2, 2^3, 2^4, and 2^5, whose sum is 63. Thus,

$$S(288) = S(2^5) \times S(3^3) = 63 \times 13 = 819 \qquad \square$$

Notice that in Example 2.11 it is not necessary to determine or list most of the divisors of 288. The problem reduces to finding $S(2^5)$ and $S(3^2)$ and multiplying the results. This pattern holds true for any number. Thus:

To compute $S(n)$:

Step 1. Determine the prime factorization of n.

Step 2. Collect repeated primes and write the prime factorization in exponential form.

Step 3. For each power of a prime, p^k, in this factorization, compute $S(p^k)$.

Step 4 Multiply the resulting numbers.

(Notice the strong parallel between this process and that discussed in Section 2.3.)

Example 2.12 **Problem:** Determine $S(1800)$.

Solution: The prime factorization of 1800 is $2^3 \times 3^2 \times 5^2$.

$$S(2^3) = 1 + 2 + 4 + 8 = 15$$
$$S(3^2) = 1 + 3 + 9 = 13$$
$$S(5^2) = 1 + 5 + 25 = 31$$

Therefore,

$$S(1800) = 15 \times 13 \times 31 = 6045$$ □

Thus, the problem of computing $S(n)$ is reduced to factoring and computing $S(p^k)$, where p is a prime. The former has been treated earlier. For the latter, the divisors of p^k are trivial to note:

$$1, \ p, \ p^2, \ p^3, \ldots, \ p^{k-1}, \ p^k$$

If k is small, the simplest computation is just to add the few numbers. For large k, the following formula may be used:

(2.14) $\quad$ If $n > 1$, then $1 + n + n^2 + \ldots + n^{k-1} + n^k = \dfrac{n^{k+1} - 1}{n - 1}$

Example 2.13 **Problem:** Determine $S(11^2)$.

Solution: $S(11^2) = 1 + 11 + 11^2 = 1 + 11 + 121 = 133$ □

Example 2.14 **Problem:** Determine $S(3^{10})$.

Solution: $S(3^{10}) = 1 + 3 + 3^2 + \ldots + 3^{10}$. By (2.14), the expression on the right side equals

$$\frac{3^{11} - 1}{3 - 1} = \frac{3^{11} - 1}{2} = 88{,}573 \qquad \square$$

Note that the arithmetic involved in the last step is not much easier than that involved in adding the eleven numbers. (It is easier with a hand calculator that has an exponent, or "y^x," key.) For some of our purposes, though, it will be useful to be able to represent the desired sum as a fraction, even if we don't do the implied arithmetic.

EXERCISES 2.4

For Exercises 1–6, list, in factored form, all the divisors of the given numbers.

1. 91   2. 120   3. 288
4. 405   5. 496   6. 756

For each of the numbers in Exercises 7–18, determine $S(n)$. (You have already determined the prime factorizations in Exercises 1–12 of Section 2.3.)

7. 91   8. 120   9. 299
10. 313   11. 347   12. 496
13. 836   14. 1878   15. 4173
16. 100,100   17. 299,299   18. 1,000,000

19. What is $S(2^5)$?

20. What is $S(3^5)$?

21. What is $S(2^5 \times 3^5)$?

22. What is $1 + 6 + 6^2 + 6^3 + 6^4 + 6^5$? Is this $S(6^5)$? (See the previous problem.)

23. Prove Statement (2.14). (*Hint*: Let $1 + n + n^2 + \ldots + n^{k-1} + n^k = z$. Determine a comparable equation for $n \times z$. Combine these two by subtraction and solve for z.)

24. Suppose you had a job that paid 1 cent the first day and that doubled in pay each successive day. How much money would you earn in 30 days at this job?

WRITING EXERCISES

1. Describe how at least three of the dozen problem-solving tactics of Section 1.2 can be used to help in the solution of Exercise 23. (Even if you can't complete the proof for Exercise 23, you should be able to do this writing exercise.)

2. In what way does the four-step process for computing $S(n)$ depend on the Fundamental Theorem of Arithmetic? Explain.

2.5   Proper Divisors

Numbers can be classified in many ways. They are either even or odd; they may be prime or composite; they may have exactly 12 divisors. A classification introduced by the ancient Greeks, one that will be of interest to us, depends on the sum of the divisors of a number. Greek mathematicians thought of divisors slightly differently from the way presented here; they considered them to be "parts" of that number, and understandably, did not consider a number to be part of itself. To match up with this classification, we introduce the following definition.

DEFINITION   A number x is a **proper divisor** of y provided that $x \mid y$ and $x < y$.

Note that y is then the only "improper" divisor of y; all other divisors are proper. We can now approach the Greeks' classification determined by sums of proper divisors.

NOTATION   We denote the sum of all proper divisors of a number n by $P(n)$.

Example 2.15   The proper divisors of 15 are 1, 3, and 5, so it follows that
$$P(15) = 1 + 3 + 5 = 9 \qquad \qquad \square$$

Example 2.16   **Problem**: Determine $P(1800)$.

Solution: In Example 2.12 we noted that $1800 = 2^3 \times 3^2 \times 5^2$, so $D(1800) = 36$. Thus, 1800 has 35 proper divisors. It would be tedious to determine these 35 numbers and then add them; fortunately, there is no need to do so. We found before that $S(1800) = (15)(13)(31) = 6045$. Because this is the sum of all 36 divisors, and because 1800 is the lone "improper" divisor included in the sum, $P(1800) = 6045 - 1800 = 4245$. $\qquad \square$

In general:

(2.15)
$$\boxed{P(n) = S(n) - n}$$

We can thus use the techniques of Section 2.4 to compute $S(n)$, and then subtract n to determine $P(n)$. Example 2.15 indicates that it is not always necessary to compute $S(n)$ first, however. It would be a waste of time when there are only a few divisors. Use common sense to select your method.

Finally, we can observe an interesting distinction in the results of Examples 2.15 and 2.16. It is not surprising that $P(1800)$ is much larger than $P(15)$. What is noteworthy is that $P(1800)$ is larger than 1800, but $P(15)$ is smaller than 15. Thus, we can categorize the two numbers differently, as the Greeks did.

DEFINITION   If If $P(n) > n$, then n is **abundant**.
If $P(n) < n$, then n is **deficient**.
If $P(n) = n$, then n is **perfect**.

The simple connection between $P(n)$ and $S(n)$ suggests the following simple characterizations:

(2.16) n is deficient if and only if $S(n) < 2n$.

(2.17) n is abundant if and only if $S(n) > 2n$.

(2.18) n is perfect if and only if $S(n) = 2n$.

Thus, 15 is deficient and 1800 is abundant. As the exercises show, there are infinitely many numbers that are abundant and infinitely many that are deficient. Questions about perfect numbers will occupy us for the next two sections.

EXERCISES 2.5

1. Determine $S(n)$ and $P(n)$ for each number from 1 to 30. Using the table you constructed for Exercise 23 of Section 2.2, record these values in the fourth and fifth columns, under the headings "$S(n)$" and "$P(n)$," respectively.

2. Classify each number from 1 to 30 as *deficient*, *abundant* or *perfect*. Use the last column of the table you constructed for Exercise 23 of Section 2.2 to record these answers, under the heading "type."

For each of the numbers in Exercises 3–14, determine $P(n)$. (You have already determined $S(n)$ for these numbers in Exercises 7–18 of Section 2.4.)

3. 91	**4.** 120	**5.** 299
6. 313	**7.** 347	**8.** 496
9. 836	**10.** 1878	**11.** 4173
12. 100,100	**13.** 299,299	**14.** 1,000,000

15–26.   Classify each number in Exercises 3–14 as *deficient*, *abundant* or *perfect*.

27. Show that any prime number is deficient. (Notice that this implies that there are infinitely many deficient numbers.)

28. Show that any multiple of 12 is abundant. (*Hint*: Some, but not necessarily all, of the proper divisors of $12x$ are x, $2x$, $3x$, $4x$, and $6x$.) Notice that this implies that there are infinitely many abundant numbers. Why?

29. Show that every multiple of an abundant number is abundant. (See the preceding exercise.)

30. Show that 2^k is deficient for every value of k.

31. Give an example of an odd abundant number, if possible; if not possible, explain why not.

WRITING EXERCISES

1. In what ways are the words "abundant," "deficient," and "perfect" appropriate labels for the concepts they represent? In what ways are they inappropriate? Can you suggest an alternate set of labels for these types of numbers that is more descriptive or more appropriate? If so, give an argument to support your suggestion.

2. Describe how at least three of the dozen problem-solving tactics of Section 1.2 can be used to help in the solution of Exercise 30. (Even if you can't complete Exercise 30, you should be able to do this writing exercise.)

2.6   Even Perfect Numbers

In doing the exercises of Section 2.5, you will have encountered several perfect numbers:

$$6 \qquad 28 \qquad 496$$

This is not a lot, but we should expect perfect numbers to be rare. After all, there are many more ways for $P(n)$ to be unequal to n than to be equal to it. Our goal is to identify the characteristic properties of perfect numbers. Are the three perfect numbers cited here the only ones? If there are more, how many more? Is there a formula that will determine every perfect number, just as the formula $\frac{n(n+1)}{2}$ determines every triangle number? (See Exercise 27 of Section 2.1.)

In pursuit of answers to these questions, all we have to go on are the three perfect numbers identified so far. What do they have in common? Certainly, they are all even numbers; thus, we might restrict our attention to even numbers for the present. The distribution of these three perfect numbers strongly suggests that we have missed some. The numbers 6 and 28 are fairly close together, but there is a long way to 496. We could test every number between 28 and 496. However, before launching into such a sizable task, we should attempt some intelligent guesswork, perhaps along the following line. Because 6 and 28 differ by 22, we might guess that another perfect number could occur at about 22 more than 28. Following this idea, let us test the number 50, as well as a few even numbers in the neighborhood of 50.

Example 2.17   **Problem:** Determine whether any of the numbers 46, 48, 50, 52, or 54 is perfect.

Counting Board. Medieval woodcut.

Solution: $P(46) = 26$, $P(48) = 76$, $P(50) = 43$, $P(52) = 46$, and $P(54) = 66$. Thus, 46, 50, and 52 are deficient, whereas 48 and 54 are abundant; none of them are perfect.   □

The idea of an additive pattern doesn't seem to work, so let's try a multiplicative pattern. Note that 28 is between 4 and 5 times 6, and 496 is between 17 and 18 times 28. We can look for a potential "missing" perfect number about 4 times 28 and expect 496 to be about 4 times that. Because $4 \times 28 = 112$ and $\frac{1}{4}$ of 496 equals 124, we might reasonably look at even numbers in the range of 112 to 124.

Example 2.18   **Problem:** Determine whether any even number from 112 to 124 is perfect.

Solution: $P(112) = 136$, $P(114) = 126$, $P(116) = 94$, $P(118) = 62$, $P(120) = 240$, $P(122) = 64$, and $P(124) = 100$. Thus, 116, 118, 122, and 124 are deficient; 112, 114, and 120 are abundant; none of them are perfect.

$\square$

Neither of these ideas has worked. Perhaps we did not look far enough. Perhaps we have been looking in the wrong place. If we reflect on the things we have seen in the earlier sections, we must note that determining $D(n)$, $S(n)$, and $P(n)$ depends on first finding the prime factorization of n. In fact, the Fundamental Theorem of Arithmetic provides the proof that the prime factorization of a number determines all the properties of the number that are definable in terms of its divisors. This suggests that we should search for the key to perfect numbers in their prime factorizations.

Table 2.4 shows a very strong pattern — in fact, several patterns. We have a perfect number with the factor 2 appearing once, one with the factor 2 appearing twice, and one with the factor 2 appearing four times. And each perfect number has a single odd prime factor, with the magnitude of that prime increasing as the number of 2s increases. Perhaps we should look for a perfect number with the factor 2 appearing three times (one may be missing from the list) and then one with 2 appearing five times, six times, and so on. Or we might look for a perfect number with five or more 2s in its prime factorization, trying some pattern that starts with 1, 2, 4, the three powers of 2 observed so far.

n	Prime Factorization
6	2×3
28	$2^2 \times 7$
496	$2^4 \times 31$

Table 2.4   The prime factorizations of three perfect numbers.

To pursue the first idea, we should examine numbers of the form $2^3 \times p$, where p is some odd prime between 7 and 31. The information about these numbers is given in Table 2.5. Although we do not find another perfect number among these, we do see some interesting information there. When multiplied by 2^3, smaller odd primes produce abundant numbers; larger primes produce deficient numbers. If we consider the numerical difference n minus $P(n)$, we find it closest to zero (the difference for a perfect number) when the value of the prime is 13 or 17, and we see that this difference is farther away from zero as we pick odd primes farther away in either direction.

p	$n = 2^3 \times p$	$P(n)$	Classification	$n - P(n)$
11	88	92	abundant	-4
13	104	106	abundant	-2
17	136	134	deficient	2
19	152	148	deficient	4
23	184	176	deficient	8
29	232	218	deficient	14

Table 2.5   Testing the primes between
7 and 31.

The differences recorded in the last column of Table 2.5 suggest that we should try 15, the number halfway between 13 and 17, and hope for a difference of $n - P(n) = 0$, which will mean that n is a perfect number. However, as we saw in Example 2.18,

$$P(2^3 \times 15) = P(120) = 240$$

and thus 120 is "very" abundant. This is to be expected, of course, because 15 is composite, and 120 has an abundance of divisors, 16, whereas all the numbers in Table 2.5 have only 8 divisors. What is called for is a *prime* number halfway between 13 and 17. Of course, no such number exists; but if it did, we would have the perfect makings of another perfect number. For if 15 were prime, there would be only eight divisors of 120, namely 1, 2, 4, 8, 15, 30, 60, and 120, and the sum of the proper divisors would be 120.

Another way to put all this is: Each perfect number in Table 2.4 consists of a power of 2 times an odd number that happens to occur in "the right place" and is prime; but there happens to be no *prime* number in the right place to match up with 2^3. Observe that 3, 7, and 31 are primes, and are factors of perfect numbers. Also note that 15 is an odd number that neatly fits into the gap between 7 and 31, but it is composite and fails to yield a perfect number.

We have not yet found another perfect number, but we have made excellent progress in that direction. We now know "good" odd numbers to be 3, 7, 15, and 31. It is not too difficult to see a pattern in these numbers. Each is twice the previous number, plus 1. Another way to describe the pattern is to note that each number is respectively 1 less than 4, 8, 16, and 32. Thus,

$$3 = 2^2 - 1, \ 7 = 2^3 - 1, \ 15 = 2^4 - 1, \text{ and } \ 31 = 2^5 - 1$$

The former observation is useful for successively computing terms of the sequence; the latter description is preferable for identifying a formula for a typical term — namely, $2^k - 1$. The next number in this sequence of odd numbers is 63, which should be multiplied by 2^5. Because 63 is composite, we might expect a situation similar to that encountered with $2^3 \times 15$. This is precisely the case, as you are asked to show in the exercises.

If we go to the next step in this progression, we find

$$(2 \times 63) + 1 = 127$$

which is a prime number. And when we examine

$$2^6 \times 127 = 8128$$

we find that

$$
\begin{aligned}
S(8128) &= S(2^6) \times S(127) \\
&= (2^7 - 1) \times 128 \\
&= 127 \times 128 \\
&= 16{,}256
\end{aligned}
$$

so we have

$$P(8128) = 16{,}256 - 8128 = 8128$$

Thus, 8128 is a perfect number!

We should now have enough information to formulate a precise conjecture and, hopefully, to prove it.

(2.19) | If $2^k - 1$ is prime, then $(2^{k-1})(2^k - 1)$ is perfect. |

The proof of the conjecture is merely a generalization of the computation done above for 8128. We will calculate $S(n)$, where $n = (2^{k-1})(2^k - 1)$:

$$S(n) = S(2^{k-1}) \times S(2^k - 1)$$

as we saw in Section 2.4. But

$$S(2^{k-1}) = 2^k - 1$$

by (2.14), and

$$S(2^k - 1) = (2^k - 1) + 1 = 2^k$$

because $2^k - 1$ is prime. Therefore,

$$S(n) = (2^k - 1)(2^k)$$

This is almost the prime factorization of n (with the factors interchanged). The difference is a factor of 2^k, rather than of 2^{k-1}. Therefore,

$$S(n) = (2^k - 1)(2^k) = (2^k - 1)(2^{k-1})(2) = 2n$$

and, by the characterization in Statement (2.18), n is perfect. We have proved Statement (2.19) and can label it a theorem.

This theorem was known to the ancient Greeks and was recorded in Euclid's *Elements*. Perfect numbers of this form are thus known as "Euclidean (perfect) numbers."

DEFINITION   A number of the form $(2^{k-1})(2^k - 1)$ is called a **Euclidean number**, and is denoted E_k.

Thus, 6 is a Euclidean number because

$$6 = 2 \times 3 = 2^1 \times (2^2 - 1) = (2^{2-1})(2^2 - 1)$$

Specifically, $6 = E_2$. Similarly,

$$E_3 = 4 \times 7 = 28, \quad E_4 = 8 \times 15 = 120$$

and so on. Using this notation, our theorem could be reworded:

(2.20) | If $2^k - 1$ is prime, the Euclidean number E_k is perfect.

Euclidean perfect numbers are equal to a power of 2 times a single odd prime. Might there not be even perfect numbers with two or more odd prime factors? Surprisingly, the answer is no, as the following statement indicates.

(2.21) | If n is an even perfect number, it is of the form $(2^{k-1})(2^k - 1)$, where $2^k - 1$ is prime.

In other words, if n is an even perfect number, it is Euclidean, with a single odd prime factor. The proof of this statement is considerably more involved than the others presented here, and thus it is omitted.

EXERCISES 2.6

1. Verify that $2^5 \times 63$ is abundant.

2. Show that $2^5 \times p$ is abundant if p is any odd prime less than 63, and is deficient if p is any prime greater than 63. (See Table 2.5.)

3. Prove or disprove: If an even number has two or more odd prime factors, it is abundant.

4. Prove or disprove: The sum of the reciprocals of the divisors of a perfect number is 2. (The reciprocal of x is $\frac{1}{x}$.)

5. Prove or disprove: Every Euclidean number is a triangle number. (See Exercise 27 of Section 2.1.)

6. Show that any multiple of a perfect number other than the perfect number itself is abundant. (*Hint*: See Exercise 28 of Section 2.5.)

WRITING EXERCISES

1. Discuss how at least four of the problem-solving techniques of Section 1.2 are used in developing the ideas of this section. Identify specific passages in the text to illustrate your discussion.

2. (a) Write Statement (2.19) completely in words, without using *any* symbolic notation, not even letters to represent numbers.

   (b) Write the proof of Statement (2.19) completely in words, without using *any* symbolic notation, not even letters to represent numbers.

   (c) Which parts (if any) of the statement and proof of Statement (2.19) are easier without symbols? Which parts (if any) are harder? Which parts (if any) are impossible? Justify your answers briefly.

   (d) Use Parts (a) and/or (b) to illustrate a general discussion of the pros and cons of using symbols in presenting mathematical ideas.

2.7   Mersenne Primes

In Section 2.6 we found that every odd prime number of the form $2^k - 1$ is a factor of a Euclidean perfect number and also that every even perfect number is Euclidean and has exactly one odd prime factor, which is of the form $2^k - 1$. The search for additional even perfect numbers, then, boils down to an examination of numbers of the form $2^k - 1$, to determine which are prime.

DEFINITION   A number of the form $2^k - 1$ is called a Mersenne number[6] and is denoted by M_k.

In terms of this notation, we can rephrase Statement (2.20) as follows:

---

[6] Marin Mersenne (1588–1648) was a Franciscan Friar and amateur mathematician. He spent a great deal of time corresponding about a variety of matters, including mathematics, with prominent intellectuals of his day, fostering communication among them. His interest in number theory is honored by the numbers that bear his name.

(2.22)

If M_k is prime, then E_k is perfect.

(Note that Statement (2.21) implies that the converse[7] of this statement is also true.)

To determine whether or not M_k is prime, we could always resort to the definition of prime numbers. Given sufficient time, M_k can be tested (for a specific value of k) to determine if it is prime. But, as we have said so often before, there ought to be a better way. We organize what we have already observed in Table 2.6.

k	M_k	
1	1	
2	3	prime
3	7	prime
4	15	composite
5	31	prime
6	63	composite
7	127	prime

Table 2.6   The first seven Mersenne numbers.

$M_8 = 255$ is clearly composite, which suggests the possibility that the Mersenne numbers are alternately prime and composite, after an initial anomaly. Unfortunately, this hypothesis is disproved by $M_9 = 511 = 7 \times 73$, which is composite.

A different conjecture, with no noticeable exception, is readily suggested by comparing the subscript k with the classification of M_k. So far, M_k is prime when k equals 2, 3, 5, or 7 and is composite when k equals 4, 6, 8, or 9. The pattern is striking in its simplicity, and we formulate it as follows:

(2.23)   *Conjecture*: M_k is prime if k is prime, and composite if k is composite.

We even have $M_1 = 1$, the subscript matching the value, in the special case of the only number that is neither prime nor composite. This suggests that M_{10} should be composite, M_{11} prime, M_{12} composite, and so on. Note that there are two parts to the Conjecture (2.23); let us separate them, taking the "composite" part first.

We can obtain a lead to a more precise connection between k and M_k by observing some of the divisors of the composite occurrences of M_k as listed

---

[7]The term *converse* is explained in Appendix Section A.3.

in Table 2.7. We note in this table that a Mersenne prime is always among the prime factors. Specifically,

$$3 \quad (= M_2) \quad \text{is a factor of } M_4, M_6, M_8, \text{ and } M_{10};$$
$$7 \quad (= M_3) \quad \text{is a factor of } M_6 \text{ and } M_9;$$
$$31 \quad (= M_5) \quad \text{is a factor of } M_{10}.$$

Again, there is a striking correlation among the Mersenne numbers and their subscripts, suggesting the following:

(2.24)

$$\boxed{\text{If } c \mid d, \text{ then } M_c \mid M_d.}$$

This turns out to be fairly easy to prove, using Formula (2.14) and some laws of exponents. We consider the fraction

$$\frac{M_d}{M_c} = \frac{2^d - 1}{2^c - 1}$$

Because $c \mid d$, we know there is some integer b such that $d = b \cdot c$ — so our fraction becomes

$$\frac{M_d}{M_c} = \frac{2^{bc} - 1}{2^c - 1} = \frac{(2^c)^b - 1}{2^c - 1}$$

By Formula (2.14), this fraction is the "answer" to the "problem"

$$1 + 2^c + (2^c)^2 + (2^c)^3 + \ldots + (2c)^{b-1}$$

Thus, $\frac{M_d}{M_c}$, which equals the sum of a series of whole numbers, must itself be a whole number, implying that $M_c \mid M_d$.

k	M_k	Factorization of M_k
4	15	3×5
6	63	$3^2 \times 7$
8	255	$3 \times 5 \times 17$
9	511	7×73
10	1023	$3 \times 11 \times 31$

Table 2.7 Factorizations of Mersenne numbers.

As a simple corollary, if d is composite, it has a divisor c such that $1 < c < d$. Then

$$1 < M_c < M_d \quad \text{and} \quad M_c \mid M_d$$

establishing that M_d is composite. This is the proof of

(2.25)

> If k is composite, then M_k is composite.

Thus, we have proved half of the Conjecture (2.23). The other half, asserting that M_k is prime when k is prime, turns out to be false! It is untrue for $k = 11$, and for many other prime values of k. Although 11 is prime, M_{11} is composite, as you are asked to verify in the exercises.

This is one of the classic examples in mathematics of a persuasive pattern turning out to be misleading. Many mathematicians believed in the truth of the conjecture about Mersenne primes, and as recently as the beginning of this century printed works appeared with the erroneous assertion that M_{11} is prime.[8] The incorrectness of such a persuasive conjecture, based on several verified examples, is a paramount example of why mathematics insists on rigorous proofs for every assertion, even the seemingly obvious ones. Simplicity of form, successful prediction of examples, and majority belief can all be wrong.

Where do we stand in our search for Mersenne primes, which are literally the key factors for even perfect numbers? Statement (2.25) narrows the search to Mersenne numbers with prime subscripts, but each such number must be tested. Although some additional narrowing can be effected by the form of the prime, it is basically true that, to date, no pattern has been found to characterize Mersenne primes. Special algorithms for testing the primeness of large numbers and high-speed computers have extended the list of known Mersenne primes (see Table 2.8), but no general characterization of Mersenne primes has been proved.

As you can see from Table 2.8, the occurrence of Mersenne primes is sporadic, and the frequency of these primes among small prime values of k is deceptive as a guide to the general "pattern." The magnitude of these numbers, after the first few, is truly monumental. The thirty-first (and largest known to date) perfect number contains 130,100 digits. This number, if printed on a single line in ordinary newsprint with the requisite 43,366 commas, would require a line of print over 1200 feet long, the length of four football fields!

---

[8] This is so despite the fact that M_{11} was shown to be composite in 1603 by Cataldi.

k	M_k (Number of Digits)	Discoverer and Date
2	3	unknown (before 300 B.C.)
3	7	" "
5	31	" "
7	127	" "
13	8191	Regius (1536)
17	131,071	Cataldi (1603)
19	524,287	Euler (1772)
31	2,147,483,647	" "
61	(19)	Seelhoff and Pervusin (1886)
89	(27)	Powers and Cunningham (1914)
107	(33)	Uhler (1952)
127	(39)	" "
521	(157)	Robinson (1952)
607	(183)	" "
1279	(386)	" "
2203	(664)	" "
2281	(687)	" "
3217	(969)	Lucas and Lehmer (1962)
4253	(1281)	" "
4423	(1332)	" "
9689	(2917)	" "
9941	(2993)	" "
11,213	(3376)	Gillies (1964)
19,937	(6002)	Tuckerman (1971)
21,701	(6533)	Nickel and Noll (1978)
23,209	(6987)	Noll (1979)
44,497	(13,395)	Slowinski (1979)
86,243	(25,962)	" (1982)
110,503	(33,265)	Colquitt and Welsh (1988)
132,049	(39,751)	Slowinski (1983)
216,091	(65,050)	" (1985)

Table 2.8   The known Mersenne primes.

To round out our discussion of perfect numbers, we consider now the question of odd perfect numbers. Briefly, nothing definitive is known about odd perfect numbers. More precisely, no examples of such numbers are known, and theoretical methods have shown that any possible odd perfect numbers must be extremely large. (It can be proved, for example, that an odd perfect number must have at least six distinct prime factors; and it has been shown that there are no odd perfect numbers less than 100,000,000,000,000,000,000.) Many mathematicians believe that no odd perfect numbers exist; so far, however, this conjecture has defied proof.

EXERCISES 2.7

In Exercises 1–8, state whether the following are prime or composite.

1. M_{11} 2. M_{12} 3. M_{19}

4. M_{25} 5. M_{29} 6. M_{39}

7. M_{89} 8. M_{121}

In Exercises 9–16, state whether the following are perfect or not.

9. E_{11} 10. E_{12} 11. E_{19}

12. E_{25} 13. E_{29} 14. E_{39}

15. E_{89} 16. E_{121}

17. What number should be multiplied by M_{89} to produce a perfect number?

18. What number should be multiplied by $(2^{2203} - 1)$ to produce a perfect number?

19. What odd number should be multiplied by 2^{12} to produce a perfect number?

20. What number can be multiplied by 2^{10} to produce a perfect number?

In Exercises 21–26, find one prime divisor of each of the following.

21. M_{11} 22. M_{12} 23. M_{19}

24. M_{25} 25. M_{39} 26. M_{121}

27. Determine two distinct prime factors of M_{11}.

28. Determine three distinct prime factors of M_{21}.

29. Determine three distinct prime factors of M_{20}.

30. What is the prime factorization of M_{15} (which equals 32,767)?

WRITING EXERCISES

1. Look up Marin Mersenne in a book on the history of mathematics and also in a book on the history of music; then write a short biographical sketch of him. (Be sure to list the source(s) you used in writing the paper, and be careful not to simply copy or paraphrase the material from your sources.)

2. Who are Nickel and Noll? (See Table 2.8.) Research this question by consulting appropriate library sources and write a paragraph or two summarizing what you find.

3. If you were setting out to search for an odd perfect number, where and how would you start? Look back at Section 1.2 for help in answering this question.

LINK: 2.8

Number Theory and Cryptography

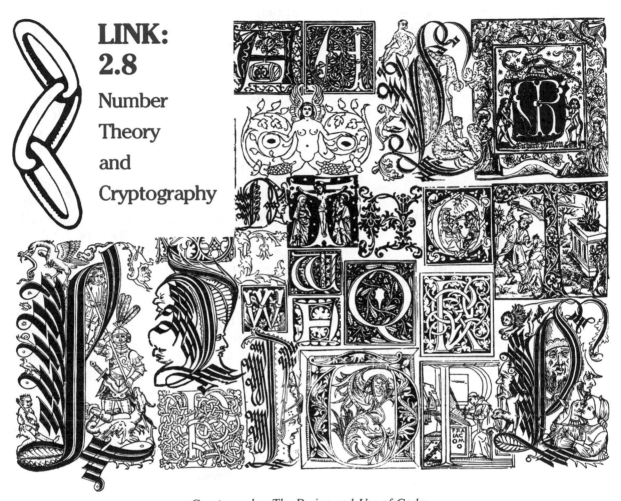

Cryptography: The Design and Use of Codes.

We have strongly suggested that number theory is an example of "pure" mathematics, and this is so. But there are numerous applications of this field within and outside of mathematics. From the pencil-and-paper algorithms of arithmetic to the design of computer software, one finds number theory. A rather surprising use of some basic properties of numbers is found in the area of cryptography — design and use of codes — and cryptanalysis — the science of code breaking.

The desire for secrecy in communication and the countereffort to discover others' secrets are probably as old as man himself. And from the earliest stages of those secret communications, number patterns have been used to systematize the coding and to aid in the code breaking. For example, Julius Caesar used a simple code in his military messages, substituting for each letter the one occurring three letters earlier in the alphabet. Of course, this simplest of coding methods is also simple to figure out.

As societies grew in complexity, so did their cryptography and cryptanalysis. By the Renaissance cryptography had reached the level of a science. In the 16th century Blaise de Vigenere formalized the so-called "polyalphabetic" system of coding. In this system, one uses several of the 26 cyclic rearrangements of the alphabet, such as

$$U V W X Y Z A B C \ldots R S T$$

in a substitution code. The choice and pattern of substitute alphabets is determined by a key word or sequence of numbers, hopefully known only to the sender and receiver of the message. Later generations have used noncyclic rearrangements of the alphabet to tap the millions of such variations, again maintaining utility of these by use of longer and longer key sequences of letters or numbers.

Each escalation of complexity in cryptography has called forth an equal complexity in cryptanalysis, drawing not only on number theory, but on probability, statistics, abstract algebra, and graph theory. And until recently, virtually every code and key used by the cryptographer has been unravelled and discovered by the cryptanalyst.

In today's world of high-speed communication, banks, corporations, law-enforcement agencies, and so on need to transmit confidential information over public phone lines or airwaves to a large number of other, similar institutions. This would be feasible with coded messages if each different pair of correspondents shared a different code key for their messages. Of course, the keys themselves could not be transmitted openly, and would require personal meetings or safe couriers, and the maintenance of accurate and secure records of which correspondent matched which key would be a bookkeeper's nightmare and a security agent's undoing.

To meet these communication problems, cryptographers have developed "public-key" systems. In such a system both the method of coding and a "one-way" key for each potential user are public information. In one such system, the RSA (developed by Ronald Rivest, Adi Shamir, and Leonard Adleman of M.I.T.), a user selects a pair of large (say 100-digit) prime numbers. After calculating the product of these two primes and the number of natural numbers that have no prime factors in common with this product (a number we shall call a), the user chooses (arbitrarily) another number that has no prime factors in common with a. This number and the *product* of the two large primes are made public. Anyone wishing to send this individual a coded message uses this public key, following a known encoding process. In order to decode the message, it is essential to be able to compute the original prime factors of the product key number. Without this factorization, even someone who knows both the coded message and the encoding process would not be able to break the code. To date, no one has found a practical method for factoring such large composite numbers. Although, as we saw in this chapter, any number can be factored uniquely into primes, this is a theoretical fact, not a practical method for handling large (200-digit) numbers. Any procedures currently known, even with the largest and fastest computers, would require many thousands of years to factor such a number!

In 1982, a ripple of concern ran through the National Security Agency, the nation's code-watching authority, when two Europeans, Hendrik Lenstra and Henri Cohen, devised a practical computer method for identifying large prime numbers of arbitrary type (as opposed to primes of special types, such as Mersenne primes.) The fear was that similar methods might solve the factorization problem for large composite numbers. So far, this has not proved to be the case, and some feel that the RSA public-key crypto system will continue to be unbreakable.

Whatever the ultimate fate of the factorization and the current coding system, cryptography and its use of both simple and increasingly sophisticated mathematics seems destined to be with us for the foreseeable future.

Topics for Papers – Chapter 2

General Instructions: Consider the reader of your paper to be an anonymous student in your class — someone who knows the material of Chapter 2 fairly well, but does not know anything about your particular topic. Introduce your topic clearly and take your reader step by step through an orderly development of your ideas to your desired conclusion.

1. *Note*: This describes a "family" of topics; the particular one you work on should be chosen with the concurrence of your instructor.

   Examine the six-column table you constructed for Exercises 2.2.23, 2.5.1, and 2.5.2; look for some interesting property that a few, but not too many, of the numbers between 1 and 30 have.

   Example: Suppose your favorite number is 18. Looking at the data for 18, you see that $D(18) = 6$, $S(18) = 39$, and $P(18) = 21$. Now, all of these numbers are divisible by 3, and 18 is the only number on the list that has this property! There must be others, one would think. What would they look like? How can they be found? Is there a formula that describes all of them? or some of them?

Once you have found one or two properties like this, get your instructor's approval to use one of them as a topic. (If you can't find any such property, ask your instructor to help you.) The type of number described by the topic you choose will be yours to name and investigate, *using as many of the problem-solving techniques of Chapter 1 as you find helpful.* Your paper is to be a description of this investigation, reporting what specific questions you asked yourself, describing in detail the particular problem-solving tactics you pursued, showing where they led, etc. (In this type of investigation, even a blind alley is something to report, provided you can describe the alley.) The paper should begin with a clear definition of your kind of number, and it should end with some questions that seem to be promising avenues for further investigation. If you generate more numerical data, explain how it was done. Be as creative as you like in looking for patterns and making conjectures, but try to prove your conjectures or at least supply some evidence or argument to support them.

Remember: The main theme of the paper is *how you went about investigating your kind of number.* The results of your investigation, while interesting to report, are less important than the procedures you used to find them.

2. Research one or more introductory texts on number theory for some of the following topics:

- semiperfect numbers;
- pseudo-perfect numbers;
- multiply perfect numbers (doubly perfect, triply perfect, etc.);
- weird numbers;
- amicable pairs.

Write a descriptive paper that explains in detail one or two of these topics. For each topic you discuss, your first paragraph should include a definition of the type of number in question, and the last paragraph should include some questions that seem to be promising avenues for further investigation of the topic.

For Further Reading

1. Armendariz, Efraim P., and Stephen J. McAdam. *Elementary Number Theory.* New York: Macmillan Publishing Co., Inc., 1980.

2. Beck, A., M. Bleicher, and D. Crowe. *Excursions Into Mathematics.* New York: Worth Publishers, Inc., 1969.

3. Devlin, Keith. *Microchip Mathematics: Number Theory for Computer Users*. Cheshire, England: Shiva Publishing Limited, 1984.

4. Dunham, William. *Journey Through Genius: The Great Theorems of Mathematics*. New York: John Wiley and Sons, Inc., 1990, Chapter 3.

5. Gardner, Martin. "Mathematical Games: A short treatise on the useless elegance of perfect numbers and amicable pairs." *Scientific American*, **218** (3): 121-126, 1968.

6. Hellman, Martin E. "The Mathematics of Public-Key Cryptography." *Scientific American*, **241** (2): 146-157, 1979.

7. Kahn, David. *The Code Breakers*. New York: The Macmillan Co., Inc., 1967.

8. Slowinski, D. "Searching for the 27th Mersenne Prime." *Journal of Recreational Mathematics*, **11** (4): 258-267, 1979.

Euclid Greeting Students at the Outer Gate of the Circle of Knowledge.

CHAPTER

MATHEMATICS OF FORM: GEOMETRY

3.1 What Is Geometry?

The key to geometry lies in its history. The word itself describes its own beginnings — "geo-metry" means "earth measurement." The mathematical ancestors of modern geometers were the land surveyors of ancient Egypt, whose job was to reestablish boundaries that had been washed away by the periodic flooding of the Nile. They, along with the builders of Egypt and Babylon and the navigators of the coastal-trading civilizations along the Mediterranean, were among the first people to use the mathematical properties of lines, angles, and circles in systematic ways to produce useful results. However, just as we use a power saw to help us build a house without having to know the electrical principles that make it work, so those ancient craftsmen used their geometric tools without probing for any unifying theory to explain their reliability.

Geometry as a mathematical discipline began with Thales, a wealthy Greek merchant of the 6th Century B.C. He is considered to be the first Greek philosopher, as well as the father of geometry as a deductive study. The significance of Thales' work is that he began the search for unifying rational explanations of reality, thus moving away from the previous reliance on religion and mythology to explain the phenomena of nature. His search for some underlying unity in geometric ideas led him to investigate how some geometric statements could be derived logically from others. The statements themselves were well-known, but the process of linking them by logic was new.

The Pythagoreans and other Greek thinkers continued the logical development of geometric principles, which culminated in Euclid's *Elements*.

63

Ancient Egyptian Surveyors. Geometric methods were developed to reestablish property boundaries obliterated by the Nile's annual flooding.

Building on some three centuries of work, Euclid organized and extended all that was known about mathematics in the Greece of 300 B.C. Although the thirteen books of the *Elements* contain considerable material on arithmetic and number theory, Euclid paid special attention to geometry. His goal was to systematize the various observable relationships among spatial figures, which he, like Plato, Aristotle, and the other Greek philosophers, regarded as ideal representations of physical things.

Euclid listed a small number of basic statements that appeared to capture the essential properties of points, lines, angles, etc.; he then proceeded to derive the rest of geometry from these basic statements by means of a series of carefully proved propositions. Euclid's work in geometry was so comprehensive and clear that his *Elements* became the universally accepted source for the study of plane geometry from his time on. Even the geometry studied in high school today is essentially an adaptation of Euclid's *Elements*.

Euclid's work was so good, in fact, that it took two thousand years to find flaws in his system. Not until the nineteenth century did mathematicians begin to suspect that there might be other ways of looking at geometry. They gradually realized that the "space" we live in need not conform to Euclid's assumptions, and that some of its properties actually can be explained more easily from a different viewpoint, that is, in another geometric system. The question that flows immediately from this realization is the motivating theme of this chapter:

If Euclidean geometry is not THE geometry, then what *is* geometry?

This question is almost as difficult as "What is mathematics?" and replies to it seem nearly as numerous as geometers themselves. Rather than giving you a glib, shallow definition, we try to capture in this chapter a sense of the historical evolution of the subject, employing Euclid's answer to the question as the basis for discussing more recent viewpoints. In the course of doing this we describe briefly several different kinds of geometry. After you have seen several geometries "in operation" we shall be able to draw some conclusions about geometry in general, and also about the relationship between modern mathematics and the world we live in.

EXERCISES 3.1

WRITING EXERCISES

1. (a) Find three English words besides *geometry* that begin with "geo-" and indicate how their meanings involve "earth."

   (b) Find three English words besides *geometry* that contain "metr-" and indicate how their meanings involve "measure."

2. Suppose you want to measure and mark off a 100-by-200-foot rectangular shore lot at the edge of a lake in such a way that one short side of the rectangle runs along the (straight) lake shore. The only measuring tool you have is a 50-foot cloth tape measure. Describe how you would do it and what principles of plane geometry your method would depend on.

3.2   Euclidean Geometry

Euclidean geometry is essentially the same as the geometry you studied in high school, so our discussion of it will presuppose some familiarity with its basic language. However, several difficulties that often arise in the treatment of this material need to be pointed out, especially insofar as they are due in large measure to Euclid himself.

Recall that Euclid's goal was to derive as many geometric statements as possible by strictly logical argument from a small number of initial statements assumed to be "obviously" true. The derived statements he called theorems; some initial statements were called axioms, others postulates. The axioms were general statements about quantity and logic, such as

The whole is equal to the sum of its parts.

The postulates were specific geometric ideas that were assumed to be true. Euclid based all of his plane geometry on the following five postulates:

E1. A straight line can be drawn from any point to any other point.

E2. A finite straight line can be produced continuously in a straight line.

E3. A circle may be described with any center and distance.

E4. All right angles are equal to one another.

E5. Through a given point can be drawn exactly one line parallel to a given line. (*Note*: E5 is an alternate and equivalent form of Euclid's fifth postulate; it is named Playfair's Postulate, after the British mathematician John Playfair (1748-1819) who developed it.)[1]

Euclid's conception of points and lines as things that can be drawn is apparent even in his postulates, and "proof by picture" crept unnoticed into many of his deductive arguments. For instance, there is nothing in Euclid's definitions or postulates to guarantee that a straight line passing through the center of a circle must actually have a point in common with the circle, and yet many of his proofs depend on the location of such an intersection point.

"But," you say, "it is obvious that a line cannot pass from the inside to the outside of a circle without crossing it at some point, because there are no gaps in a circle; it is continuous."

A moment's reflection should convince you that you are deriving this property of circles from the way we visualize them, not from what Euclid explicitly postulated. Euclid tacitly assumed many similar ideas, such as

A line which intersects one side of a triangle but does not pass through any vertex of the triangle must intersect one of the other sides.

and

Given any three distinct points on the same line, one of them is between the other two.

These statements are "obvious" only if we think of geometry in terms of pictures; they cannot be deduced from Euclid's axioms and postulates and must be stated as postulates if they are to be used.

In order to circumvent such logical difficulties, David Hilbert, in the early part of this century, renovated the foundations of Euclidean geometry by selecting a full set of assumptions that accurately describes geometry as Euclid envisioned it. In Hilbert's system all the things that "ought to work"

---

[1]A complete list of Euclid's original definitions, axioms, and postulates may be found on pages 38–39 of [4] in the list *For Further Reading* at the end of this chapter.

in Euclidean geometry are established from his set of explicit axioms.[2] From this point on, we ignore the historical distinction between Euclid's geometry and Hilbert's, treating this study as a single type of geometry called **Euclidean** to distinguish it from the other geometric systems of modern mathematics.

Despite what has just been said, a large part of the legitimate study of Euclidean geometry throughout its history has consisted in drawing pictures — pictures of a very special kind, called **compass-and-straightedge constructions**. This is one of the great classical "games" of mathematics, perhaps the greatest in terms of depth of investigation and number of players. It is played in the Euclidean plane (an "ideal" flat surface), and its rules are Euclid's axioms (postulates) E1, E2, and E3. The construction tools are called **Euclidean tools**; they are chosen to reflect accurately the scope of those axioms. A straightedge is used for drawing lines between points and for extending these lines, and a compass is used for drawing circles.

The axioms do not provide for any measurement of length, so neither tool is allowed to have a memory; the straightedge cannot be marked in any way, nor can the compass preserve a fixed opening after its points have been lifted from the plane. This latter restriction may surprise you because the compasses used in high school geometry classes can preserve length, and that seems to be "cheating." The dishonesty is only an illusion, however; it can be shown that the distance-preserving compass is no more powerful than the collapsing one. (See Exercise 2.)

Each compass-and-straightedge construction involves a specific geometric object with one or more special properties, and the goal is to prove that this object can be constructed using only the Euclidean tools. Notice that the process is deductive by nature; the construction need not be carried out, but merely proved possible. Thus, the entire construction "game" could be played without actually drawing a picture, although this is seldom done because visual representations are very helpful to the imagination and memory of the player.

Example 3.1   **Problem:** Given a line segment $\overline{AB}$ and a point P not on the segment, construct a segment $\overline{PQ}$ equal to $\overline{AB}$. (*Note*: Strictly speaking, the last part of the problem should read "$\overline{PQ}$ *congruent* to $\overline{AB}$," in the sense that the two figures have the same size and shape but they do not necessarily contain the same points. However, it has been traditional to apply the word *equal* to congruent line segments and congruent angles. We follow this practice, relying on common sense to maintain the distinction.

---

[2]Modern mathematics no longer distinguishes between the terms *axiom* and *postulate*; they are considered synonyms that refer to any statements assumed to be true without proof.

"The Measure of Any Height in a High Place." Woodcut from Tanner's Manual, London, 1587.

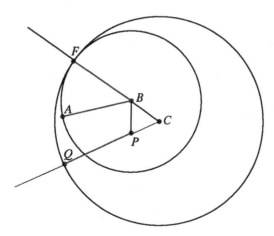

Figure 3.1 $\overline{BF} = \overline{AB} = \overline{PQ}$

Proof: (See Figure 3.1.) Starting with the given line segment and point, draw segment $\overline{BP}$. Then construct the equilateral triangle BPC with $\overline{BP}$ as one side. (Exercise 1 asks you to justify this construction.) Next, construct a circle with center B and radius $\overline{BA}$, and extend $\overline{CB}$ beyond B to intersect this circle at point F. (The existence of such a point is one of those pictorially obvious items that slips through the logical cracks in Euclid's postulates.) Now construct a circle with center C and radius $\overline{CF}$, and extend $\overline{CP}$ beyond P to intersect this circle at point Q. Then $\overline{CF} = \overline{CQ}$ (Why?) and $\overline{CB} = \overline{CP}$ (Why?) Subtracting the latter from the former, we conclude that $\overline{PQ} = \overline{BF}$. But $\overline{BF} = \overline{BA}$, so $\overline{PQ} = \overline{BA}$, as desired.   □

The construction (theorem) in Example 3.1, once proved, becomes a useful tool for further constructions, such as copying or bisecting a given line segment or angle, constructing a line perpendicular to a given line, or constructing squares and rectangles of given dimensions. From such humble beginnings, the game proceeds to more complex constructions, such as a line tangent to a circle, a regular 15-sided figure inscribed in a circle, or a square equal in area to a given parabolic segment.

Euclidean geometry is by no means confined to examining circles and straight lines. The ancient Greeks made extensive studies of various types of curves. Notable examples of such figures are the **conic sections**, first thoroughly treated by Apollonius about 225 B.C. These are curves representing the intersection of a plane with a (double) cone. The cone is obtained by rotating two intersecting lines about an axis that bisects their angle of intersection. The cone itself is the collection of all points in space that the rotating lines

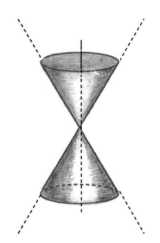

Figure 3.2   A double cone.

"pass through," as shown in Figure 3.2. There are four essentially different ways in which a plane that does not contain the intersection point of the cone's generating lines can intersect that cone. They are shown in Figure 3.3. The four kinds of figures that result are called **circles, ellipses, parabolas,** and **hyperbolas.** One of the truly remarkable facts of mathematics is that, once Descartes (in 1637) had connected geometry with algebra by means of his coordinate system and mathematicians began to study the geometric representation of algebraic equations, it was discovered that every conic section can be represented by a quadratic (2nd-degree) equation in two variables!

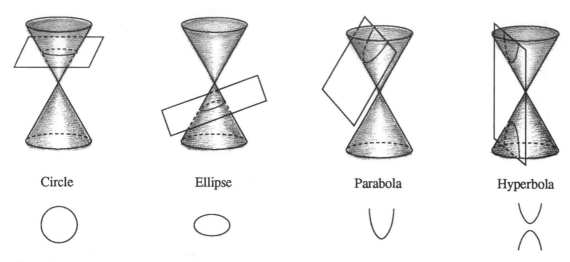

Circle   Ellipse   Parabola   Hyperbola

Figure 3.3 *The four conic sections.*

Geometry can be regarded in one sense as the theory of comparison of shapes, with different norms of comparison yielding different geometries. In Euclidean geometry, shapes or figures are compared in terms of either **congruence** or **similarity**. It is often said that two figures are congruent if one can be made to coincide with the other by a "rigid motion." This is a bit misleading, since there really is no movement involved in the notion of congruence, but the general idea is correct. The formal mathematical definition of congruence captures the idea of superimposing one figure on the other, thus checking to see that both shape and size are preserved. Other kinds of geometry also deal with congruence of figures, but only Euclidean geometry studies similarity as a concept distinct from congruence. The formal definition of similarity provides for a distinction between size and shape, and figures are called "similar" if they have the same shape but not necessarily the same size.

Example 3.2 Any two circles are similar, but to be congruent they must have the same radius. Any two equilateral triangles are similar, as are any two squares; but to be congruent two equilateral triangles or two squares must have sides of the same length. By contrast, two rectangles are not similar unless their corresponding sides have proportional lengths. □

We close this section with a proof of one of the most important statements in all of mathematics—the Pythagorean Theorem. Since it is both a numerical statement and a geometric one, this theorem applies to a remarkably wide variety of mathematical topics. From among the almost

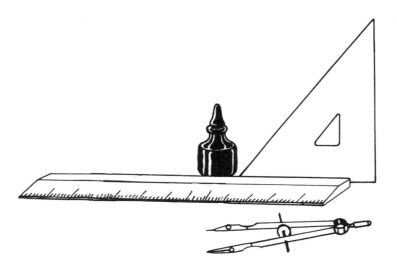

Compass-and-Straightedge Constructions.

incredible variety of proofs for this theorem, one attributed to the 19th-century German geometer Jacob Steiner stands out because of its simplicity and elegance. It depends solely on classical constructive methods, using no numerical or algebraic arguments at all. Thus, it is a fitting example of classical Euclidean geometry "in action."

> **The Pythagorean Theorem**: The square of the hypotenuse of a right triangle is equal to the sum of the squares of its legs.[3]

(All terms of this theorem can be interpreted in a strictly geometric sense. Thus, the square of a line segment is a square whose side is equal in length to the line segment, and the sum of two squares will be considered equal to another square if the first two squares can be "cut into pieces and rearranged" to form the third. These intuitive ideas can be made rigorous, but a pictorial approach enhances the elegance of the argument and best suits our purpose. The justifications for most steps of the construction should be clear, so they have been omitted for the sake of brevity.)

---

[3] Recall that the hypotenuse of a right triangle is the side opposite the right angle.

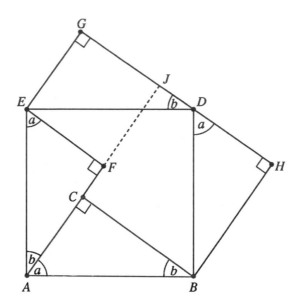

Figure 3.4   *Area EFJG + Area CBHJ = Area BHGEFC = Area ABDE*

Proof

Starting with right triangle ABC, construct the square $ABDE$ as shown in Figure 3.4. Extend line AC, and locate a point F on that line such that $AF = BC$. Now,

$$\angle ABC = \angle EAF$$

because they both are complements of the same angle, and $AB = EA$ (Why?), so

$$\triangle ABC \text{ is congruent to } \triangle EAF.$$

Construct two more copies of this triangle outside the square, on sides BD and DE. Because $\angle GDE$ and $\angle HDB$ are complementary, GDH is a straight line. Observe that $EFJG$ is a square that has one leg of the original triangle ABC as a side, and $CBHJ$ is a square that has the other leg of triangle ABC as a side. Their sum is represented by the irregular hexagon (6-sided figure) $BHGEFC$. If we "move" triangles BHD and DGE to their congruent images, triangles AFE and ABC, leaving the irregular pentagon (5-sided figure) $BDEFC$ fixed, it is easy to see that the hexagonal sum of the two smaller squares has been rearranged to form the original large square. This completes the proof of the theorem.

EXERCISES 3.2

In the following exercises you may use any of the familiar Euclidean theorems about congruent triangles, as well as the visually "obvious" properties of figures mentioned in the text. You may also use the result of any exercise in figuring out ones that follow it.

In Exercises 1–8, show how a (collapsing) compass and (unmarked) straightedge can be used to make each of the constructions called for.

1. Given a segment $\overline{AB}$, construct an equilateral triangle with $\overline{AB}$ as one side.

2. Given a segment $\overline{AB}$, a line l that does not contain $\overline{AB}$, and a point P on l, construct a point Q on l such that $\overline{PQ} = \overline{AB}$. (This construction shows that a collapsing compass and an unmarked straightedge can be used to transfer length.)

3. Given an angle, construct its bisector. (*Note*: An *angle bisector* is a segment that has one end at the vertex of the angle and divides the angle into two equal adjacent angles.)

4. Given a segment, find its midpoint.

5. Given a segment and a point on it, construct a perpendicular segment with that point as one endpoint.

6. Given a line and a point not on it, construct a perpendicular segment with that point as one endpoint.

7. Given a segment, construct a square with that segment as one side.

8. Given two segments, construct a rectangle with adjacent sides equal to the two segments.

9. Give a reason to justify each step in the proof of the Pythagorean Theorem given in this section.

WRITING EXERCISES

1. A careful study of how Euclid used his third axiom would suggest the following somewhat physical image for that axiom: "Any line segment can be held fixed at one end while the other end rotates, but the segment as a whole cannot be moved." Explain in your own words the difference between a collapsing compass and a noncollapsing compass, and how the image described here corresponds to the former.

2. Criticize the following "proof" of the statement that there are two perpendiculars from a point to a line: *Proof*: (See Figure 3.5.) Consider two intersecting circles with centers O and O', and call one of their intersection points A. Draw the diameter of each circle from A, and let B and C be the other endpoints of those diameters. Draw $\overline{BC}$, intersecting the two circles at points D and E. Draw $\overline{AD}$ and $\overline{AE}$. Both $\angle ADB$ and $\angle AEC$ are right angles because each is inscribed in a semicircle. Therefore, there are two perpendiculars from the point A to the line $\overleftrightarrow{BC}$.

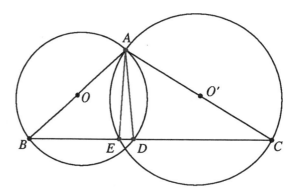

Figure 3.5   Are there two perpendiculars from point A?

3.3   Axiom Systems

The organization of information into axiomatic systems can be traced back to the Greeks, who as early as 600 B.C. began to study the logical connections among mathematical facts. About 300 B.C., Euclid organized most of the known mathematics of his time so that virtually all statements were proved from a small collection of definitions and axioms, and thus the **axiomatic method** was born. Today the axiomatic method is the distinctive structure of mathematics (and much of science). A staggering amount of material has been added to mathematical knowledge in the more than 22 centuries since Euclid. Nevertheless, no mathematical claim is accepted unless it can be proved from basic axioms, and whenever a new mathematical theory emerges, mathematicians follow Euclid's pattern in working to identify an axiomatic basis for it. Such is the influence of the axiomatic method.

Simply put, an **axiom system** has four essential components — *defined terms*, *undefined terms*, *axioms*, and *theorems*. As we describe each of these in turn, it should also become clear how they fit together.

When we define terms in mathematics, we are merely assigning name tags to ideas. A **definition** is the statement of a single, unambiguous idea that the word, phrase, or symbol being defined represents from that point on — no more and no less. The single, unambiguous idea is called the **characteristic property** of the definition; it is a condition such that, given any object whatsoever,

(1) we can determine whether or not that object satisfies the condition, and

(2) the term being defined is used to label everything that satisfies the condition and is not used to label anything else.

Example 3.3   The statement "A square is a four-sided figure" is accurate, but it is not a good definition of a *square* (in its usual sense) because there are four-sided figures that we do not want to call squares. We might use "A square is a rectangle with four equal sides" to define *square* as the term is normally understood, provided we have already defined *rectangle*.   □

"Defining" a word simply by giving a synonym (another word that means the same thing), a method common in some dictionaries, is either meaningless or useless from a mathematical viewpoint. If we do not know what the synonym means, we have no way of applying it as a characteristic property; if we do know what the synonym means, then that word is a perfectly good name for the idea and there is no need to confuse the issue by supplying another. Moreover, that practice can lead to a problem of "circularity." For instance, if we define *number* as *quantity*, define *quantity* as *amount*, and define *amount* as *number*, we have stated that there are three different labels for the same idea, but we haven't said what the idea is! In general, then a good definition must not be **circular**; that is, its defining condition must not use the term itself or use terms which are themselves defined using the term being defined.

Example 3.4   "An **odd number** is a whole number that is not even" and "An **even number** is a whole number that is not odd," taken together, are a circular pair of definitions because the characteristic property of each one depends on the other. Either definition taken by itself is fine, provided that the other term has been defined independently.   □

The requirement that definitions cannot be circular means that some terms must be **undefined**. We cannot have an infinite regression of terms being defined by more basic terms that are themselves defined by more basic terms, and so on, so there must be some "most basic" terms that are not themselves defined at all. The fact that a term is undefined does not mean it is meaningless. For example, the term *set* is undefined in formal mathematics. The formal meaning of *set* is determined by the way that word is used within the underlying assumptions of its mathematical system, the axioms of set theory. In other words, the undefined terms of a mathematical system acquire their meanings by context.

Just as we cannot expect every term to be defined, so we cannot expect every mathematical statement to be proved from previously developed statements. We must have some statements to start with, statements that are assumed without proof. These statements are called **axioms** (or sometimes **postulates**). In an ideal mathematical system, all other statements are derived from the axioms by strict logical proof. Statements that are proved from the axioms are called **theorems**.

Euclid recognized the need for undefined terms and axioms in his treatment of geometry. He regarded geometry as a description of the physical world, and he tried to systematize that description by putting it on a deductive foundation. He defined all the technical terms he used (such as *point*, *line*, and *plane*), but in doing so he used other words such as *part*, *length*, *equal*, etc., without defining them. Euclid also distinguished between axioms and postulates. His axioms were statements of common ideas he regarded as obvious, such as

The whole is greater than any of its parts

and

If equals are subtracted from equals, the remainders are equal

whereas his postulates were statements dealing explicitly with geometry, such as

All right angles are equal

and

A straight line can be drawn from any point to any point

These latter statements he regarded as true observations about the physical world, rather than as arbitrary assumptions. In that sense, Euclid's geometry typifies what is now known as "material axiomatics." A **material axiomatic system** considers its undefined terms to have meanings derived from reality, and its underlying axioms to be statements that are "obviously" true in the real world.

In contrast to material axiomatics is the modern deductive approach to mathematical systems called "formal axiomatics." A **formal axiomatic system** begins by assuming that its undefined terms have no meaning at all and behave much like algebraic symbols. Using these terms, a few statements are devised and, although these statements have no necessary connection with truth or reality outside the system, they are assumed to be "true" within the system itself. These assumed statements are the *axioms* or *postulates*. (These two words are interchangeable in formal axiomatics.) From there on, all new words used in the system are defined from the undefined terms and all other statements (theorems) are proved from the axioms. The mathematics constructed in this way is called **pure mathematics**

Example 3.5   **Undefined terms**: squirrel, tree, climb

Axioms: **1.** There are exactly three squirrels.
   2. Every squirrel climbs at least two trees.
   3. No tree is climbed by more than two squirrels.

If we consider this to be a formal axiom system, its terms have no meanings other than what they acquire from their use in the axioms. Hence, this system is exactly the same as:

Undefined terms: X, Y, relates to

Axioms: **1.** There are exactly three X's.
   2. Every X relates to at least two Y's.
   3. No Y relates to more than two X's.   □

In a highly formal axiomatic development, not only are there undefined and defined terms and axioms, but also there are explicit rules of logic, usually quite restricted. Such systems often are highly symbolic, rather than verbal, because symbols are less likely to invite errors caused by unjustified assumptions about the "real meaning" of a term or axiom. Words, even in mathematics, can be deceiving.

In a less formal development, rules of logic are not specified, but instead, we allow ourselves a sort of "undefined" logic — the formalized (and sometimes symbolized) mathematical logic that is merely a sharpened form of common sense. This is the logic used in everyday discourse and practiced in arithmetic, algebra, science, (occasionally politics), and, especially, geometry. Even in developments of this sort, a considerable spread of precision and formalism is found. In some treatments, nothing is accepted as "understood" about the system except what is contained in the definitions and axioms. Even a slight variation, such as in the tense or mood of a verb, is not allowed unless formally introduced by a definition or language rule. In other treatments familiar variations of grammar are accepted without question, and results of other familiar parts of mathematics are used without formal justification.

We shall adopt this last, most informal mode for the rest of the chapter. It is most likely the style you saw in your high-school geometry course, and it is the one used in the preceding two sections. There is an obvious comfort in allowing ourselves to use logical and grammatical "common sense," but we must be aware of the danger of this approach. We can grow so casual that we believe we have proved a theorem, when, in fact, we have been duped by a persuasive, but flawed, argument. To help you avoid such difficulties, you might find it useful to review the basic principles of logic summarized in Appendix A.

Exercises 3.3

In the following exercises: (a) determine whether the given statement accurately defines the given word or phrase; (b) if not, identify whether the statement is circular or is not characteristic; (c) if the statement is not characteristic, give an example (verbally or by means of a figure) that fits the description but is not the word or phrase described, or vice versa.

1. Two angles are supplementary if each is the supplement of the other.

2. The angle bisector of $\angle ABC$ is a ray with B as its endpoint.

3. An acute triangle is a triangle with no obtuse angles.

4. A right triangle is a triangle containing a right angle.

5. A right triangle is a triangle containing two acute angles.

6. A right triangle is a triangle whose sides have length 3, 4, and 5, respectively.

7. A right triangle is a triangle with one pair of perpendicular sides.

8. An obtuse triangle is a triangle that is obtuse.

9. An isosceles triangle is a triangle in which one side is not congruent to the other two.

10. A parallelogram is a figure with opposite sides parallel.

11. A parallelogram is a quadrilateral with opposite sides congruent.

12. A rectangle is a quadrilateral that is rectangular.

13. A rectangle is a figure with all angles congruent to each other.

WRITING EXERCISES

1. Is it possible to have *no* defined terms in an axiom system? For example, in geometry we could avoid using the term *quadrilateral* by using the phrase "four-sided figure" whenever we needed to refer to one. Can this be done for all the defined terms of geometry? If so, what are the advantages and disadvantages?

2. Relying on what you know about real squirrels and real trees, which of the assertions about them in Example 3.5 are true? In light of your response, write a short paper on the pros and cons of choosing abstract terms, as in the second version of that axiom system, versus terms with some meaning in everyday language.

3.4   Models

All the emphasis on formality and the absence of meaning in the previous section might lead you to believe either that pure mathematics is a useless game or that mathematicians are somehow psychic or phenomenally lucky in what they choose to play with. That is not the case at all. Mathematicians do not pick arbitrary symbols and statements at random and just happen to come across useful systems most of the time; they often have in mind a concrete system they are trying to describe, just as Euclid did. The essential difference is that, unlike Euclid, modern mathematicians recognize that the concrete system they are describing does not *predetermine* the abstract system they are formulating. The axioms are not statements that are true by nature; they are formal sentences that are considered true in the system merely by agreement. The abstract system is, in general, subject to many interpretations, of which its prototype is only one.

DEFINITION   An **interpretation** of an axiom system is any assignment of specific meanings to the undefined terms of that system. If an axiom becomes a true statement when its undefined terms are interpreted in a specific way, then we say that the interpretation **satisfies** the axiom. A **model** for an axiom system is an interpretation that satisfies all the axioms of the system.

Thus, the word *model* is used in mathematics in much the same way as in ordinary English. Consider, for example, an architect's model of a proposed building. In this instance, the model builder often is working from the blueprints of the proposed building. Different model builders might con-

Dürer. Institutiones Geometricae.

struct different, but equally accurate, models from the same set of blueprints; the choice of materials, scale of the model, landscaping, etc., would not be prescribed by the blueprints, so considerable variance might appear in the models. This kind of modeling is mirrored in the mathematical use of the word. Our blueprints are the axioms of a system. Just as an architect's model makes the design ideas more concrete and visible, so too, a model of an axiom system makes its ideas more concrete. And, because there usually are many things *not* specified by the axioms, a considerable variety of models may be possible. **Applied mathematics** can be described as the study of models (physical, biological, social, etc.) of abstract mathematical systems.

Let us look at an example of how an axiom system is taken from its original setting, formalized, then modeled. In this instance, the artistic problem of picturing objects realistically led to an axiom system for a special kind of geometry, thereby providing a fine example of the interplay between reality and abstraction. The story begins more than five hundred years ago.

As Europe passed from Middle-Age night to Renaissance morning, man's awareness of the world around him yawned, stretched, and began to search for its slippers. While scientists and philosophers explored the physical world, artists searched for ways to mirror this reality on paper and canvas.

One of their problems was that of dimension — how to portray depth on a flat surface. The artists of the 15th century realized that this problem was primarily geometric, so they began a mathematical investigation of the properties of spatial figures as the eye sees them. The problem was approached by considering the surface of a picture to be a window through which the artist views the object to be painted or drawn. As the lines of vision from the object converge to the point where the eye is viewing the scene, the picture captures a cross section of the figure formed by them. (See Figure 3.6.) This simple principle became the basis for the artistic theory of perspective, pioneered by Leonardo Da Vinci and the German artist Albrecht Dürer.

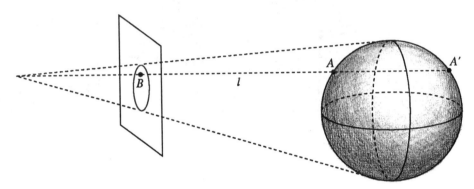

Figure 3.6   *The lines of vision from the sphere to the eye form a cross section on the plane.*

Abstracting this approach from the artistic complications of color, shading, motion, and the like, we obtain a relatively simple geometric process called *projection* — a way of making the points of an object correspond to the points of its representation on a planar surface.

One well-known illustration of perspective is the image of railroad tracks stretching into the distance. Although we know they are the same distance apart both near and far away, they appear to be meeting at a point just beyond the horizon. Indeed, if they are painted, the artist must treat the track lines on the canvas plane as lines that meet somewhere "just beyond" the picture. Thus, for the artist, a realistic theory of representing parallel lines requires that those lines be regarded as lines that intersect. This requirement, along with a few other basic assumptions about points and lines, became the basis for an important type of geometric system with applications far beyond the artistic questions that spawned it. That system is called **projective geometry.** An axiom system for the projective geometry of a single plane is as follows:

P1. There exist at least one point and one line.

P2. If A and B are distinct points, then exactly one line passes through both of them.

P3. If l and m are distinct lines, then exactly one point is common to both of them.

P4. There are at least three points on any line.

P5. Not all points are on the same line.

Now, if we consider *point, line*, and the various ways of expressing *on* (including "pass through," etc.) as undefined terms, this axiom system can be restated as follows:

Undefined terms: X, Y, related to

Axioms: P1. There exist at least one X and one Y.

P2. If a and b are distinct X's, then exactly one Y is related to both of them.

P3. If c and d are distinct Y's, then exactly one X is related to both of them.

P4. At least three X's are related to any Y.

P5. Not all X's are related to the same Y.

In this form, not only is it unclear that we are talking about geometry, but even the relationships among the various terms of the system becomes obscure. The only information that has been retained by the terms of this formal system comes from the syntax (grammatical form) of the original statements. Specifically, X and Y are understood as nouns of some sort, and "related (to)" is understood as some sort of connective verb form that can be used with these nouns.

Let us build a model using brass rings for the X's, wires for the Y's, and "attached" for *related to*. Take seven brass rings (labeled A, B, C, D, E, F, G) and seven wires (labeled 1, 2, 3, 4, 5, 6, 7), and attach them as shown in Figure 3.7.

It is not hard to see that this interpretation is a model for axiom system **P1–P5**:

P1. There exist at least one brass ring and one wire.

P2. Any two distinct brass rings are attached to exactly one wire. (Check some examples.)

P3. Any two distinct wires are attached to exactly one brass ring. (Check some examples.)

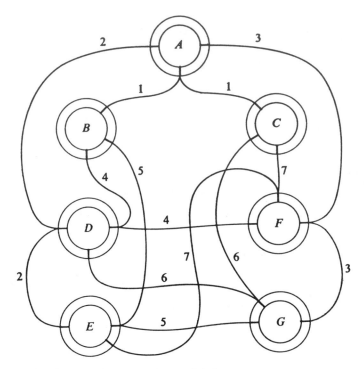

Figure 3.7 *A ring-and-wire model for
axiom system P1–P5.*

P4. Three brass rings are attached to each wire.

P5. Not all brass rings are attached to the same wire.

Of course, this little collection of rings and wires is not the only model of
that axiom system. Many other models exist, including the artists' infinite
projective plane. You will be asked to find yet another model in Exercise 3.

It is also worth observing that, like many models, the ring-and-wire
contraption shown in Figure 3.7 can be used as a model for other axiom
systems, as well. Consider, for example, the following axiom system, which
describes some of the basic properties of the Euclidean plane:

Undefined terms: point, line, on

Axioms: A1. There is at least one point.
   A2. Every point is on at least two lines.
   A3. Every line is on at least two points.
   A4. Given any two points, there is exactly one line on
   both of them.

If we interpret *point* as "brass ring," *line* as "wire," and *on* as "attached," then Figure 3.7 is a model for this axiom system.

Example 3.6   Another (different) model for axiom system **A1–A4** can be constructed as follows:

> Let *point* mean "one of the numbers 1, 2, or 3";
> let *line* mean "one of the sets {1, 2}, {1, 3}, or {2, 3}";
> let *on* mean "contains" or "is contained in," as appropriate.

It is easy to check that all four axioms are true in this case. Notice that this model differs from the ring-and-wire model in some essential ways. Not only are the objects involved apparently different, but the models actually have some different properties. For instance, this model only has 3 *points* and 3 *lines*, whereas the ring-and-wire model has 7 *points* and 7 *lines*. Nevertheless, both models satisfy all axioms of the system.   □

The axiomatic approach to mathematics has one vitally important benefit:

> Any statement that can be proved from the axioms of a system
> *must be true in every model* of the system.

That is, a single proof of a theorem in an abstract axiom system guarantees the truth of that statement as it is interpreted in all models of that axiom system, regardless of how different the various interpretations appear to be. Besides being a valuable observation about applied mathematics, this fact will be useful in understanding the ideas of the next section.

Example 3.7   Prove the following theorem for axiom system **A1–A4**:

Theorem: Given a line, there is a point not on that line.

Proof: Let L be the given line.
There is a point on L (by **A3**); label it p.
There is (at least) one other line, M, on p (by **A2**).
Then there is (at least) one other point, q, on M (by **A3**).
This point cannot be on L, for if it were, M and L would be
the same (by **A4**).

Thus, no matter how odd or unusual the meanings of *point*, *line*, and *on* in any model of axiom system **A1–A4**, we are guaranteed that, for each *line* of that model, there is a *point* not *on* it.   □

EXERCISES 3.4

1. Show that the ring-and-wire contraption shown in Figure 3.7 is a model for axiom system **A1–A4** by verifying that each of the axioms is true. (Check all possibilities.)

2. Construct an axiom system containing three axioms; then find two different models for it.

3. Find a model for axiom system **P1–P5** that differs from the ring-and-wire model given in this section and also from the standard projective-plane model. Verify that your model actually satisfies the axioms; then describe how it differs from the other two models.

For Exercises 4–8, find a model for the given axiom system:

4. **Undefined terms**: pencil, word, write
   Axioms:
   (a) There are exactly three pencils.
   (b) All pencils write the same word.
   (c) There is exactly one word that no pencil writes.

5. **Undefined terms**: cat, mouse, catch
   Axioms:
   (a) All cats catch mice.
   (b) Some mice do not catch cats.
   (c) There are at least two cats.

6. **Undefined terms**: point, line, on
   Axioms:
   (a) There are at least four points.
   (b) Every point is on two lines.
   (c) Some line is not on any point.

7. **Undefined terms**: spade, heart, trump
   Axioms:
   (a) There is at least one heart.
   (b) For any two distinct spades, there is exactly one heart that trumps them.
   (c) Each heart trumps at least two distinct spades.

(d) Given any heart, there is at least one spade it does not trump.

8. **Undefined terms**: X, Y, related
   Axioms:
   (a) There are exactly five X's.
   (b) At least one Y is related to each X.
   (c) No X is related to every Y.
   (d) Every X is related to at least two Y's.

9. Verify that the theorem proved in Example 3.7 is satisfied in both models for axiom system **A1–A4** given in this section.

10. Prove the following theorem for axiom system **A1–A4**:

    There are at least three lines.

11. Consider the following statement in axiom system **A1–A4**:

    Every line is on at least three points.

    (a) Is this statement verified in the model of Figure 3.7?

    (b) Is this statement verified in the model of Example 3.6?

    (c) Is this statement a theorem of the axiom system?

12. In general, how can you show that a given statement is *not* a theorem of a given axiom system? (*Hint*: See Exercise 11.)

WRITING EXERCISES

1. Discuss how the mathematical usage of the word "model" is analogous to its nonmathematical usage.

2. Does changing the undefined terms in an axiom system — as was done in this section in

the axioms for projective geometry — have any effect on the existence or variety of models? Justify your answer by appropriate argument or examples.

3. If there is a model that faithfully represents each of two different axiom systems, does this indicate that the two abstract systems are equivalent, aside from purely linguistic differences? Justify your answer by appropriate argument or examples.

4. Create your own example of a formal axiom system using "snark" as one of your undefined terms. Design your system so that no axiom explicitly states that there are an infinite number of snarks, but any model of the system must actually interpret "snarks" as an infinite set. Expalin why this is so.

3.5   Consistency and Independence

Consider the following axiom system:

Undefined terms: *A*, *B*, is

Axioms: **1.** *A* is *B*.
          2. *A* is not *B*.

This clearly is an example of an abstract mathematical system because it is a collection of statements about some undefined terms; yet our common sense tells us that something is wrong. Axiom 2 is the logical opposite of Axiom 1; therefore, regardless of the interpretations we assign to *A* and *B*, one of the axioms must be true and the other false. However, because they are *axioms* of our system, Axioms 1 and 2 must be true within the system. In other words, this axiom system is useless because there can be no interpretation that will preserve the truth of the axioms, so there are no models. In general, any system containing contradictory axioms is of no practical value at all.

Let us pursue this matter a little further. Suppose the axioms of a system are not apparently contradictory to each other but, at some point in the development of the system, two contradictory statements can be proved by correct deductive reasoning. What can be said about the axiom system? The assumption that the axioms are all true and the deductive arguments are valid leads inescapably to the fact that two contradictory statements are both true. Thus, the contradiction was at least latently present in the original system, even though it may have been disguised by the phrasing of the axioms. As before, this means that there is no possible model of it. Such a system is said to be "inconsistent." In slightly more formal terms:

DEFINITION   The **negation** of a statement is a statement whose truth value is always opposite to that of the original statement. [See Appendix A for examples and further discussion.]

DEFINITION   An axiom system is **inconsistent** if it is possible to prove both a statement and its negation from the axioms. A system is **consistent** if it is not inconsistent.

As we have already remarked, any system that has a model cannot involve a contradiction within itself. This fact provides us with a convenient way of testing the consistency of an axiom system — all we have to do is find a model for it. For example, the existence of the ring-and-wire model for the projective-geometry axiom system **P1–P5**, described in Section 3.4, guarantees that the geometry built on this axiom system will be free of contradiction, no matter how complicated its theorems become! This test of consistency is based mainly on physical models, but all or part of an abstract system that has already been proven consistent may also serve as a model to verify the consistency of another system. For instance, since Euclidean plane geometry has been proven consistent, any plane geometric figure can serve as a legitimate model for another axiom system.

Example 3.8   The squirrels-and-trees axiom system given in Example 3.5 (in Section 3.3) can be proved consistent by using a triangle as a model. If we interpret the *squirrels* as the three vertices, the *trees* as the three sides, and *climb* as "on," then it is clear that all three axioms are satisfied:

1. There are exactly three vertices.
2. Every vertex is on (exactly) two sides.
3. No side is on more than two vertices.   □

Sometimes it is helpful to make a model of a new system in terms of an older, more familiar, system, even though the latter has not itself been proven consistent. In this case, the consistency of the new system depends on that of the older one; thus, we call the new system **relatively consistent** with respect to the old one. Any model of a system using our usual arithmetic is an example of a relative consistency proof, because our arithmetic is a simple, workable, but *not* provably consistent system.

Although consistency is all that is really necessary to insure the logical reliability of an axiom system, the natural desire for simple efficiency leads us to another consideration. In specifying axioms for a system, we surely would avoid stating the same axiom twice because repeating a statement adds no new information to the system. Similarly, we wish to avoid giving as an axiom any statement that can be proved from the other axioms of the system because such a statement also adds no new information to the system. The logical dependence of such a statement on the other axioms can be characterized by the fact that, if it were replaced as an axiom by its negation, the resulting system would contain a contradiction. This observation leads to a useful definition.

DEFINITION   An axiom A in a consistent axiom system is said to be **independent** if the axiom system formed by replacing A with its negation is also consistent. An axiom is **dependent** if it is not independent. An axiom system is **independent** if each of its axioms is independent.

Recall that the negation of a statement is true if and only if the original statement is false. Thus, to prove that an axiom (in a consistent system) is independent, we can again use a model; but this time it must be a model for which the axiom in question is *false* while all the other axioms are true. This model insures the consistency of the "new" axiom system, and it necessarily will have to be different from the model used to prove the consistency of the original system.

Example 3.9   Consider the abstract form of the projective-geometry axiom system, **P1**–**P5**, as it was given in Section 3.4. To show that Axiom **P4** is independent, we must construct a model in which there are fewer than three X's related to some Y, but all the rest of the axioms are true statements. A triangle can be used for such a model, with its vertices as X's, its sides as Y's, and *is related to* interpreted as "is on" (or "connects," when grammatically more comfortable). In this model **P1**, **P2**, **P3**, and **P5** are all true:

P1. There exist at least one vertex and one side.

P2. If a and b are distinct vertices, then exactly one side connects both of them.

P3. If c and d are distinct sides, then exactly one vertex is on both of them.

P5. Not all vertices are on the same side.

Moreover, **P4** is false because no side has three vertices on it. Thus, **P4** is independent of the other axioms.   □

To prove an entire axiom *system* independent, we need to show that each axiom is independent. This requires a different model for each axiom, so that each one, in turn, can be modeled as false while the others are modeled as true. These models, together with a model for which all the axioms are true, form a proof of the consistency and independence of an axiom system.

Thus, to prove the independence of the axiom system **P1–P5**, we would need four more models. Some of these are left as exercises.

There is a logical trick that sometimes comes in handy for constructing models. It involves axioms that are "conditional statements"— that is, statements that can be put in the form

If p, then q

In order for such a statement to be false, its hypothesis p must be true *and* its conclusion q must be false. (See Appendix Section A.3.) Thus, for example, the statement

If I get home in time, then I'll call.

is false only if you *do* get home in time *and* you *don't* call. The trick is this:

> If you construct a model in which the hypothesis of an axiom is never true, then that axiom *can never be false* in this model.

Thus, we can show a conditional axiom is true either by showing that both its hypothesis and its conclusion are true *or* by showing that its hypothesis is false. In the latter case we say the statement is **vacuously satisfied**.

Example 3.10   We can prove that Axiom **P1** of the projective-geometry system of Section 3.4 is independent as follows: Using the single-element set $\{\#\}$ as a model, interpret the symbol "#" to be the only X, let there be no Y's, and let *related to* mean "connected to." Axiom **P1** is false in this case because there are no Y's. **P5** is true because the only X in this model is not related to any Y. The other axioms are vacuously satisfied for the following reasons:

P2: Because there are not two distinct X's, this hypothesis is false, so the statement must be true.

P3: Because there are no Y's at all, this hypothesis is false.

P4: This axiom can be rephrased as "If there is a Y, then at least three X's must be related to it." Because there are no Y's, this hypothesis is always false.   □

EXERCISES 3.5

Exercises 1–3 refer to the following axiom system:

Undefined terms: letter, envelope, contain

Axioms:

(a) There are at least two envelopes.
(b) Each envelope contains exactly three letters.
(c) No letter is contained in all the envelopes.

1. Prove that the system is consistent.

2. Write the negation of each axiom.

3. Prove that Axiom (b) is independent.

4. Prove that the following axiom system is inconsistent:

Undefined terms: point, line, on
Axioms:
(a) Any two points are on exactly one line.
(b) There are exactly four points.
(c) Every line is on exactly two points.
(d) There are exactly five lines.

Exercises 5–7 refer to the following axiom system:
Undefined terms: X, Y, related
Axioms:
(a) There are at least two X's and two Y's.
(b) Each pair of X's is related to exactly one Y.
(c) Each Y is related to at least one X.

5. Is the system consistent? Prove your answer.

6. Is Axiom (b) independent? Prove your answer.

7. Is Axiom (c) independent? Prove your answer.

Exercises 8–12 refer to the following axiom system:
Undefined terms: box, crate, in
Axioms:
(a) There are exactly four boxes.
(b) There is at least one crate.
(c) Every box is in at least two crates.
(d) Not all boxes are in the same crate.

8. Prove that the system is consistent.

9. Prove that one of the axioms is dependent on the others.

10. Prove that three of the four axioms are independent.

11. Supply a fifth axiom that makes the system inconsistent. Prove your answer.

12. Supply a fifth axiom that is independent of the other four. Prove your answer.

Exercises 13–15 refer to the following axiom system:
Undefined terms: point, line, on

Axioms:
(a) There are at least two lines that are on equally many points.
(b) There are at least two lines that are not on the same point.
(c) Each line is on at least two points.
(d) Not all lines contain the same number of points.
(e) There are at least four points.

13. Prove that the system is consistent.

14. Prove that Axiom (d) is independent.

15. One of the axioms is dependent on the others. Which one is it, and why?

Exercises 16–19 refer to the following axiom system:
Undefined terms: student, book, read
Axioms:
(a) There is at least one student.
(b) Some students read at least two books.
(c) At least one student does not read any book.
(d) Every book is read by at least two students.

16. Prove that the system is consistent.

17. Is the statement
     There are at least three students.
    dependent or independent? Justify your answer.

18. Is the statement
     There are exactly three students.
    dependent or independent? Justify your answer.

19. Is the statement
     If there is only one student, then there are at least ten books.
    true or false in this system? Why?

20. Complete the proof that the projective-geometry axiom system **P1–P5** of Section 3.4 is independent by finding three more models to establish the independence of Axioms **P2**, **P3**, and **P5**.

WRITING EXERCISES

1. Explain why the following definition of *dependent axiom* is equivalent to the definition given in the text:

   "An axiom A in an axiom system S is **dependent** if and only if it is a theorem in the axiom system consisting of all the axioms of S other than A."

2. (a) Explain why the deletion of a dependent axiom from an axiom system has no effect on what theorems can be proved or on what models the system has.

   (b) If an axiom system has two dependent axioms, does it follow that *both* could be deleted with no substantive effect on the system? Explain your answer.

3.6   Non-Euclidean Geometries

From its earliest days, the axiomatic approach to mathematics has been driven by a desire to base the topics it treats upon the simplest possible foundation. Long before formal axiomatics raised the question of consistency, mathematicians' concern for the independence of axiom systems led them to ferret out unnecessary assumptions wherever they could be found. The problem of finding such redundancies often involved great difficulties. Mathematicians had to rely on their "logical instinct" for some intuitive hint that an axiom or postulate might be dependent, and only then would the task of proving that statement from the other assumptions begin.

Almost from the very moment Euclid proposed his five postulates for geometry, that logical instinct began to stir doubts about the fifth of these statements, often called the **parallel postulate**:

> Through a given point not on a line, there is exactly one line parallel to the given line.

(You might find it helpful in reading this section to review all five of Euclid's postulates, stated as **E1–E5** at the beginning of Section 3.2.)

Even Euclid himself seemed to be bothered by the apparent dependence of this postulate; he avoided using it until the proof of his twenty-ninth theorem. The belief that the parallel postulate could be proved from the other four spread rapidly, and over the centuries many scholars established their mathematical reputations by constructing "proofs" of it in which the flaws were so subtle as to escape notice during their lifetimes. Even these flaws were uncovered eventually, however, and the independence of the parallel postulate remained an open question through the end of the seventeenth century.

Railroad Tracks Stretching into the Distance.

In the early 1700s, an Italian teacher and scholar named Girolamo Saccheri made the first noteworthy attempt to attack the problem indirectly. He proposed to negate the parallel postulate and then to probe the resulting geometric tangle until he found a contradiction. The existence of that contradiction would establish the dependence of the postulate. (This is the approach discussed in the preceding section.) Now, the negation of the parallel postulate has two parts:

> Through a given point,
> *i.* there are no lines parallel to a given line, or
> *ii.* there are at least two lines parallel to a given line.

Using the assumption that Euclid's second postulate requires straight lines to be infinitely long, Saccheri found a contradiction resulting from the first part. The second case was much more stubborn, and to get his thread of logic to snag, he had to knot it himself. After skillfully proving many valuable results, he ended by forcing a weak and vague conclusion about lines that merge at infinity, which he managed to twist into a contradiction. It apparently convinced almost no one, and even Saccheri himself was sufficiently skeptical to attempt another solution. However, his second effort was no more successful than the first.

Saccheri's work had little influence on the mathematical thought of his day. Apparently everyone was so convinced of the dependence of the parallel postulate that no attention was paid to the alternative possibility. This alternative would imply that the negation of the parallel postulate combined with Euclid's postulates **E1–E4** are the basis for a new, but logically consistent, geometry that is essentially different from the system Euclid described. Such a situation was not, it seems, within the realm of speculation for the scholars of the seventeenth century.

Almost a hundred years later Saccheri's indirect approach was revived independently and almost simultaneously by four men. Shortly after 1800, the great German mathematician Carl Friedrich Gauss was the first to recognize that the replacement of the parallel postulate by part *ii* of its negation resulted in the birth of a new geometry, but he did not publish his findings. It is speculated that his reluctance stemmed from the dominance of Immanuel Kant's philosophy within the European intellectual community of that day. Kant's theory of metaphysics was based in part on the assertion that the human perception of space is necessarily Euclidean, and hence the assertion of an equally valid geometry with different properties would bring Gauss into direct conflict with Kant. Perhaps even Gauss did not want to pit his reputation as Germany's greatest mathematician against that of its greatest philosopher.

The first such publication appeared in 1829, written by a Russian mathematician, Nicolai Lobachevsky, who devoted much of his life to developing this type of geometry. Part of his work was anticipated by Janos Bolyai, a young Hungarian army officer, but Bolyai did not publish his results until 1832. Bolyai was also interested in developing what he called "absolute" geometry, a system based on Euclid's first four postulates alone. A geome-

Nicolai Lobachevsky. He postulated that more than one line can be constructed parallel to a given line.

try based on these four postulates and part i of the negation of the parallel postulate appeared in 1854, when Bernhard Riemann (also of Germany) proved that Saccheri's contradiction in this case could be avoided because the postulate that a finite straight line could be produced continuously (**E2**) need not imply the infinitude of the resulting line.

The geometries developed by Lobachevsky and Riemann are called **non-Euclidean geometries**. They differ from Euclidean geometry only in their rejection of the parallel postulate. There are two types of non-Euclidean geometry, corresponding to the two parts of the negation of **E5**:

- the non-Euclidean geometry that allows no line through a given point parallel to a given line is called **Riemannian geometry**;

- the type that requires more than one line through a given point parallel to a given line is called **Lobachevskian geometry**.

Many different models are possible for each of these axiom systems. We shall confine our study to an informal discussion of one model for each geometry.

To visualize a model of Riemannian geometry, consider the *plane* to be a sphere. *Points* are positions on the sphere and *lines* are great circles (that is, circles that divide the sphere into two equal parts). Because the shortest path between any two points on a sphere is an arc of a great circle through those points, it is natural for such circles on the sphere to be analogous to straight lines in the Euclidean plane. This model is called **Riemannian spherical geometry** (or **double elliptic geometry**). It is a fairly good interpretation of Euclid's first four postulates (with just a little refinement needed), but the parallel postulate does not hold, because any two great circles must intersect in exactly two diametrically opposite points, Q_1 and Q_2. Thus, given a point P not on a line l, any line through P must intersect l, as in Figure 3.8.

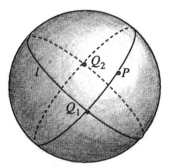

Figure 3.8   *Any two distinct great circles must intersect at two points.*

The plane for our model of Lobachevskian geometry is a little harder to construct. Imagine a little child walking along, pulling a toy attached to a string. If the child makes an abrupt left turn, the toy will trail behind, not making a sharp corner, but slowly curving around until it is almost directly behind once more. The curved path that the toy travels is called a **tractrix**, and it is used to generate the surface we want. We take two opposite copies of a tractrix, as shown in Figure 3.9(a), and rotate this double curve about the line AB, as in Figure 3.9(b). The resulting surface is called a **pseudosphere**; it looks somewhat like two trumpet bells joined together, with the narrow part of each one tapering gradually as it becomes infinitely long. This is the *plane* of the model; *points* are locations on this pseudosphere, and *lines* are "straight" paths on it, with straightness again characterized by the minimum distance between points.

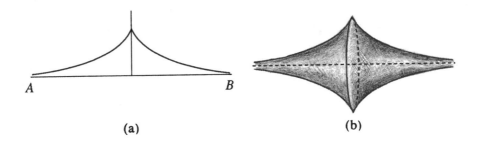

(a)

(b)

Figure 3.9   The formation of a pseudosphere.

It is difficult to draw or visualize pictures on the pseudosphere, but this surface has the property that through a point not on a line there are many parallels to the given line. Figure 3.10 illustrates how this would look on a flat surface; if that picture were "wrapped around" the pseudosphere, all the lines in it would be straight. (Notice that parallel lines in this geometry may not be everywhere equidistant from each other.)

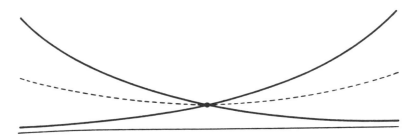

Figure 3.10   Several parallels to a line through a point on a pseudosphere.

To see how these two young geometries compare with their Euclidean ancestor, we return to the work of Saccheri. His approach to the problem of parallels provides a result that typifies the differences among the three systems. For simplicity, we shall use pictures taken from our models of these geometries, with the plane, sphere, and pseudosphere as the basic surfaces. Nevertheless, it should be noted that the results described do not depend on the models, but can be proved directly from the axiomatic foundations of their respective systems.

Saccheri began his investigation by constructing an *isosceles birectangular quadrilateral*; that is, he formed a four-sided figure with two opposite sides of equal length and two right angles. This figure, also called a **Saccheri quadrilateral**, serves as our starting point.

In any of these geometries we may consider a line segment $\overline{AB}$, construct equal perpendicular segments $\overline{AD}$ and $\overline{BC}$ at each of its endpoints, and join them by the line segment $\overline{CD}$. The figures thus formed on the three surfaces are quite different from each other, as shown in Figure 3.11. The planar figure is a rectangle with $\overline{CD} = \overline{AB}$ and right angles at C and D. On the sphere, however, $\overline{CD}$ is shorter than $\overline{AB}$, and the angles at C and D are obtuse. (The size of an angle may be visualized by considering tangents to its sides at the vertex.) On the pseudosphere, $\overline{CD}$ is longer than $\overline{AB}$, and the angles at C and D are acute.

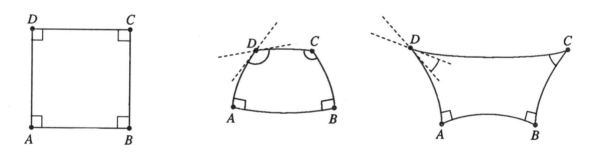

Figure 3.11 *Saccheri quadrilaterals on the plane, sphere, and pseudosphere.*

We can use this information to prove a comparative theorem:

Theorem

The sum of the angles of a triangle is equal to two right angles in Euclidean geometry, is greater than two right angles in Riemannian geometry, and is less than two right angles in Lobachevskian geometry.

[The steps of this proof are independent of the parallel postulate, so they apply to all three geometries. Figure 3.12 illustrates the construction. This figure, *which is not necessarily "flat,"* will look approximately the same if it is drawn on a Euclidean plane or on a small portion of a large sphere or a small pseudosphere.]

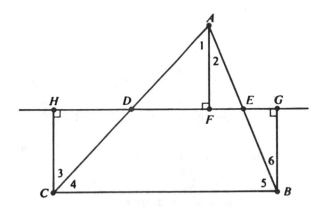

Figure 3.12   *The sum of the angles of a triangle.*

Proof

Let ABC be a given triangle, and let D and E be the respective midpoints of sides $\overline{AC}$ and $\overline{AB}$. Draw the line through D and E. Drop perpendiculars $\overline{AF}$, $\overline{BG}$, and $\overline{CH}$ from the three vertices of the original triangle to line $\overleftrightarrow{DE}$. In each geometry it can be shown that triangles ADF and CDH are congruent and that triangles AFE and BGE are congruent. It follows that

$$\angle 1 = \angle 3 \text{ and } \angle 2 = \angle 6$$

Thus, the sum of the angles of triangle ABC, which is

$$\angle 1 + \angle 2 + \angle 4 + \angle 5$$

is equal to

$$\angle 3 + \angle 6 + \angle 4 + \angle 5$$

which can be regrouped in the form

$$(\angle 3 + \angle 4) + (\angle 5 + \angle 6)$$

But, because $\overline{BG} = \overline{AF} = \overline{CH}$, by congruence, $HGBC$ is a Saccheri quadrilateral. This implies that

$$\angle 3 + \angle 4 \text{ and } \angle 5 + \angle 6$$

are right angles, obtuse angles, or acute angles, depending upon the geometry they are in, and hence the desired result is proved.

EXERCISES 3.6

Exercises 1–6 refer to Riemannian geometry.

We define the **excess of a triangle** ABC as the sum of the interior angles of the triangle minus two right angles; we denote this quantity by **exc**(ABC). Similarly, we define the **excess of a quadrilateral** $ABCD$ as the sum of the interior angles of the quadrilateral minus four right angles; we denote this by **exc**$(ABCD)$.

1. Show that congruent triangles have the same excess.

2. Let $PQRS$ be a quadrilateral with diagonal $\overline{PR}$. Prove that
$$\text{exc}(PQRS) = \text{exc}(PQR) + \text{exc}(PSR)$$

3. Referring to Figure 3.12, show that
$$\text{exc}(ABC) = \text{exc}(BCHG)$$

4. Let D be a point on side $\overline{BC}$ of a triangle ABC. Show that
$$\text{exc}(ABC) = \text{exc}(ABD) + \text{exc}(ACD)$$

5. For any triangle, show that there must exist another triangle whose excess is at most half that of the given triangle. (See Exercise 4.)

6. Show that there must exist a triangle whose excess is as small as we like. (See Exercise 5.) What is the angle sum of such a triangle?

Exercises 7–12 refer to Lobachevskian geometry.

We define the **defect of a triangle** ABC as two right angles minus the sum of the interior angles of the triangle; we denote this quantity by **def**(ABC). Similarly, we define the **defect of a quadrilateral** $ABCD$ as four right angles minus the sum of the interior angles of the quadrilateral; we denote this by **def**$(ABCD)$.

7. Show that congruent triangles have the same defect.

8. Let $PQRS$ be a quadrilateral with diagonal $\overline{PR}$. Prove that
$$\text{def}(PQRS) = \text{def}(PQR) + \text{def}(PSR)$$

9. Referring to Figure 3.12, show that
$$\text{def}(ABC) = \text{def}(BCHG)$$

10. Let D be a point on side $\overline{BC}$ of a triangle ABC. Show that
$$\text{def}(ABC) = \text{def}(ABD) + \text{def}(ACD)$$

11. For any triangle show that there must exist another triangle whose defect is at most half that of the given triangle. (See Exercise 10.)

12. Show that there must exist a triangle whose defect is as small as we like. (See Exercise 11.) What is the angle sum of such a triangle?

Exercises 13 and 14 refer to Riemannian geometry. Refer also to the spherical model described in the text. (See Figure 3.8.)

13. In the model:

(a) What kind of figure represents a line?

(b) What kind of figure represents a circle?

(c) What is the distinction between these two?

14. Consider a circle in the Riemannian plane (or its representation in the model). Let R be the ratio of the circumference to the diameter.

    (a) Is R less than, equal to, or greater than π?

    (b) Is the value of R the same for all circles in the Riemannian plane?

    (c) How small could R be?

    (d) How large could R be?

15. Answer the questions of Exercise 14 for the Lobachevskian plane. (You will probably find it easier to reason by analogy, rather than to use the model of the pseudosphere.)

WRITING EXERCISES

1. Could there be other non-Euclidean geometries besides the Riemannian and Lobachevskian geometries? Explain your answer.

2. Defend the thesis:
   Non-Euclidean geometry could have been developed before Euclidean geometry.

3. Defend the thesis:
   Non-Euclidean geometry could not have been developed before Euclidean geometry.

3.7   Abstract Mathematics and the Real World

The emergence of the non-Euclidean geometries provoked a major shift in Western scientific thought. What started as an attempt to show the dependence of Euclid's fifth postulate, presumably an exercise internal to mathematics, ended in a dramatic change in the concept of what mathematics is, what "the truth" about the real world is, and how the two relate to each other.

To begin to appreciate this shift, let us follow up on the concepts introduced in Section 3.6. We saw there that it is logically consistent to deny Euclid's parallel postulate and to assert in its place either that there are many parallels to a given line (Lobachevskian geometry) or that there are none (Riemannian geometry). Certainly, both of these alternatives seem contrary to our intuition. This intuition, bolstered by our high school geometry experience, might easily persuade us that the geometries of Lobachevsky and of Riemann are merely intellectual curiosities. "After all," we might say, "the three geometries — Euclidean, Riemannian, and Lobachevskian — are mutually contradictory, so one of them must be true and the other two false. *Obviously*, the real world follows the properties described by Euclid."

But on what basis do we assert that the geometry of the real world is Euclidean? As we saw earlier, a geometry is a formal system whose basic terms have no meaning. In dealing with geometries, it is the form and internal consistency of the systems that concerns mathematicians, not their connection with the real world. Statements about points and lines could just as well be about rings and wires, or sets and numbers, or X's and Y's. In the language of this chapter, then, the question of the correct geometric description of the real world is a question of which of the geometries is satisfied by the real world as a model and, so far, at least, we have little more than Euclid's word to justify calling reality a model for his geometry, and his alone.

"This is just a semantic issue," you say. "In the terminology of models, can we not assert that when the terms *point*, *line*, and *on* are interpreted in the real world, we find that Euclidean geometry is satisfied and not the non-Euclidean geometries?"

It's not that simple. First, we must be precise about what in the real world is the interpretation of a point and a line and what we mean by *on*; only then can we verify the truth of Euclid's axioms with that interpretation. Suppose, for example, we interpret *point* as a speck of graphite made by pressing a pencil to paper, and *line* as a streak of the same stuff laid down by drawing along the edge of a ruler. Then how do we test the parallel property? Given a streak and a speck, how can we assert that there is one and only one streak through that speck which does not meet the given streak, no matter how far the streaks are extended? Surely, on any piece of paper we can make two or more streaks through the given speck, neither of which intersect the original streak. Does that mean the geometry of the real world is Lobachevskian?

Perhaps we were a bit hasty in our choice of interpretations. You might opt for some description such as "a location in space" for *point*. But now we are speaking of an abstract concept created in our minds, not something that exists in the physical universe. Thus, whatever we conclude about such "points" may well describe the geometry we are *thinking* about, but it has no necessary connection with the geometry of the universe around us.

There is another, more serious, problem with making a physical interpretation. No matter what we choose for it, we are also doomed to failure in trying to distinguish among the three geometries by looking directly at parallelism. The distinction, at least between Euclidean and Lobachevskian geometry, *requires* that we be able to extend lines infinitely in order to decide whether or not certain pairs of lines really *never* meet. However, in the real world we have access only to a finite portion of the universe, so we would never be able to know with certainty whether or not (any) lines were parallel.

This problem suggests that we seek a test that avoids the direct confrontation with the differing parallel axioms, and instead, attempts to distinguish on the basis of some derived property. An obvious and promising candidate is the angle sum of a triangle. The theorem at the end of Section 3.6 says that the angle sum of a triangle varies with the type of geometry we are in. Therefore, if we construct a physical model of a triangle and measure its angles carefully, then the sum of those angles should tell us which geometry the physical triangle is in. That is, the physical world will be Lobachevskian, Euclidean, or Riemannian depending on whether the angle sum is less than, equal to, or greater than two right angles, respectively.

We again face the problem of interpretation, but, bypassing that for the moment, suppose we carefully draw a triangle with a very sharp pencil and measure the angles with a protractor. As you know, particularly if you have done any lab work in physics or chemistry, when two different people (or even the same person two different times) measure something, the results will vary slightly, no matter how much care is taken. That is why measurements in physical experiments are normally done several times, and the average value is used, often with some expression of the range of possible error. Thus, any careful triangle construction and angle measurement can be expected to produce a result like

$$179.95 \text{ degrees} \pm 0.15 \text{ degrees}$$

Unfortunately, this says that the "true" angle sum might be less than, equal to, or greater than two right angles (180°), so we have not distinguished among the geometries.

Perhaps, having studied carefully the exercises for Section 3.6, you have some hint of why this problem arises and how it might be fixed. In looking at the concept of "excess" in Riemannian geometry and of "defect" in Lobachevskian geometry, you may have noted that "small" triangles in either geometry have angle sums that are very close to two right angles. In fact, all the properties of these two geometries are approximately Euclidean for "small" portions of space. Thus, even if we have reasonable physical interpretations for point and line, as long as we are dealing with a "small" part of the real world, the shortcomings of physical measurement will make the distinction among geometries impossible. The obvious solution, then, is to work with a "big" triangle. We might, for example, use thin poles in a large field, measuring angles with a surveyor's transit. Or we might try three mountain tops. Or we might think really big and use two widely separated locations on the earth and a reflector on the moon. (Note that in all these situations we are interpreting *line* as the path taken by a ray of light.) Gauss himself tried this experiment in the early part of the nineteenth century. Using three mountain peaks in Germany (Brocken, Hohehagen, and Inselsberg), he built a signal fire and set up mirrors to form a large triangle of reflected light rays. The sides of the triangle were 69, 85, and 197 km (about 43, 53, and 123 miles), and his angle-sum measurement was

*Gauss' Experiment. Using three mountain peaks, he built a signal
fire and set of mirrors to form a triangle of reflected light rays.*

$$180° + 14.85 \text{ seconds}$$

However, this excess was far smaller than the margin of error for that experiment, so the results proved nothing.

Could we argue that the mere fact that all these results are very close to 180° (i.e., within the limits of error of measurement) demonstrates the Euclidean nature of the real world? Unfortunately, we cannot! After all, compared to the size of the physical universe, even just the part of it we know, distances between mountains — or between planets in the solar system — are incredibly small. We are still working with little triangles. It is not surprising that, until relatively recently, Euclidean geometry was the only one recognized, and no wonder that we have built up such a strong bias toward it. But we should take care not to mistake its pragmatic value or that strong bias as proof that Euclidean geometry is "true." As long as there is physical measurement involved, some error must be expected and taken into account. And as long as there is some error, it is impossible to conclude such an experiment in favor of Euclidean geometry.

Interestingly enough, it is conceivable that careful measurement *might* support one or the other of the non-Euclidean options. For instance, a triangle sum of 181° ±0.5° could only be true for a Riemannian triangle. Some experiments have shown just such a discrepency from Euclidean predictions, though the experiment was not on the angle sum of a triangle. Specifically, an astronomical experiment done some thirty years ago showed that light from a distant star followed a path that was *not* a Euclidean line. One way to explain the phenomenon is to say that the geometry of space really is Euclidean, but the light was following a curved path. Another explanation is that the light went in a straight line, but the geometry is Riemannian. The "truth" is that *either explanation is acceptable*; you may pick either, depending on whether your bias is stronger for Euclidean geometry or for the straightness of light paths. But it *is* a matter of bias, or of convenience, not one of "truth."

In other words, asking about the "true" geometry of the physical universe is pursuing an unanswerable and misleading question. The universe is whatever way it is, and geometry is a mathematical concept, created by human mathematicians. A reasonable question might be which geometry is most useful to apply in some particular physical situation, with certain appropriate interpretations of the geometric terms. Clearly, for most measuring, building, travelling, etc., where we are dealing with a tiny portion of physical space, Euclidean geometry is the simplest to use. It works. For large-scale measurements (galactic size and up), many scientists argue for the use of Riemannian geometry as the simplest way to explain and predict phenomena. It works. In still other situations, Lobachevskian geometry is the type that works, or works best. The type of geometry used in a scientific investigation, then, is determined simply by which one best fits the observable data, not by a claim to truth *a priori*.

The so-called "big bang" theory of the origin of the universe provides a striking modern example of this approach. Building on Einstein's general theory of relativity, Alexander Friedmann, a Russian physicist-mathematician, concluded in 1922 that the universe must be expanding. His theoretical findings were verified in the late 1920s by the observations of American astronomer Edwin Hubble and others. Subsequent scientific investigations over the next four decades pointed to the likelihood that the universe as we know it began with a mighty cosmic explosion of matter and energy, and that the universe has been expanding "outward" from that initial burst. Assuming that this is the case, we are faced with three distinct models of the expansion of the universe, all of which fit the big-bang theory:

- The expansion of the universe is slowed by the gravitational attraction between galaxies to the extent that it will eventually stop. Then the universe will begin to contract, pulling back into a single infinitely dense point at some distant future time (the "big crunch").

- The expansion is proceeding so rapidly that gravity can never stop it. The universe will continue to expand indefinitely, getting bigger and bigger, at a rate that will eventually become nearly constant (due to the retarding effect of gravity).

- The gravitational attraction between galaxies will eventually counter-balance the expansive force of the "big bang" almost exactly. That is, the expansion of the universe will slow almost to a halt, continuing just enough to avoid reversal, gradually approaching (and never exceeding) some maximum finite size.

The most convenient geometry for describing the universe depends on which of these models you accept. The first one is a Riemannian universe, the second one is Lobachevskian, and the last is Euclidean. Which one is the "real" universe? We don't know. To date, there is no conclusive scientific evidence to eliminate any of these three models. Thus, we might be in a Riemannian or a Lobachevskian universe, rather than a Euclidean one. But, even if we are, we can still *choose* to use the Euclidean axioms for our geometry because they provide a convenient approximation of the behavior of our small, known part of the large, unknown universe.[4]

Now perhaps you can begin to see the far-reaching effects of the development of the non-Euclidean geometries. At the beginning of the 19th century, it was generally accepted that truth about reality was to be found through logical analysis of observed data. The so-called "scientific method" typifies this approach to truth, as it proceeds from observation to hypothesis to experimental verification to "law." The origins of Euclidean geometry are firmly rooted in such observation of the real world, and no area of human knowledge has received a more thorough logical analysis. Yet, paradoxically, the final result of all that analysis is the realization that "the truth" about the geometry of the real world does not really exist.

EXERCISES 3.7

WRITING EXERCISES

1. When *you* say that something is true, what do you mean? Has anything you have seen in this chapter led you to rethink your own use of the word "true"? If so, what? If not, defend your use of the word as being consistent with the ideas presented in this section.

2. (a) In light of this section, and recalling that the literal meaning of *geometry* is "earth measurement," comment on the propriety of using the term *geometry* to describe the Euclidean and non-Euclidean systems.

   (b) Is it more appropriate to apply the term *geometry* to the Euclidean system than to the Riemannian or Lobachevskian systems? Why or why not?

----

[4]For a fuller discussion of this topic, see Chapter 3 of [5] in the list *For Further Reading* at the end of this chapter.

LINK: 3.8

Geometry and Society

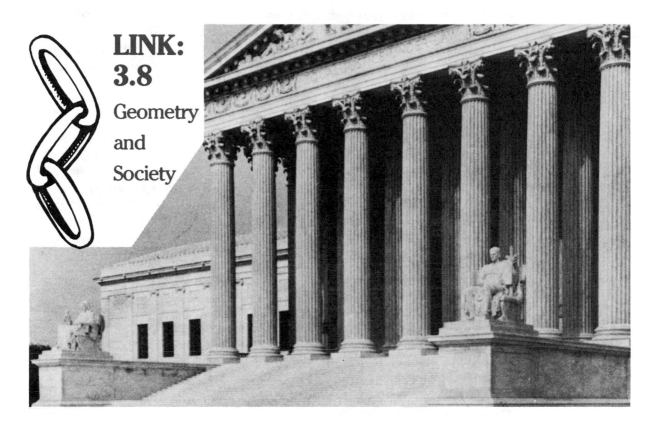

The Supreme Court Building, Washington, D.C.

In Section 3.7 we saw a truly ironic twist in the history of mathematics. Geometry, one of man's oldest formal studies, a study whose very name — "earth measurement" — describes its concrete, practical roots, has now evolved to the point that technically it is no longer a study of the earth or of anything physical at all, but rather is an abstract study of consistency and internal logical relationships.

Actually, this deliberately overstates the case a bit. Even today, Euclidean geometry is constantly being applied to the real world, with dependable results. Moreover, in recent years the non-Euclidean geometries have been put to use in studying the realities of our universe — Riemannian geometry has been employed in theoretical physics, and other non-Euclidean geometries have been developed and used in examining the cosmos. Nevertheless, the work of the past 150 years has established a fundamental separation between the subject itself and the reality that spawned it. Today it is understood that geometries are created and studied in the abstract world of pure mathematics, and then applied to real-world problems "from the outside" as a tool to be used, just as a hammer used to build a house is not part of the house itself. The recognition of that separation has thereby given this abstract tool a much broader applicability to human affairs because geometry is no longer confined to the spatial limitations of the physical universe.

Even before the 19th century's formal recognition of geometry as an abstract study, the "geometric method" was being used to bring the order and persuasion of logic to ideas outside the realm of traditional mathematics and physical science. We close this chapter with two examples of nonmathematical works that clearly trace their form to the axiomatic method crystallized so many centuries ago by Euclid.

One of the most striking examples of the axiomatic form in a nonmathematical setting appears in a major work of the 17th-century Dutch philosopher Benedict de Spinoza. Although Jewish by birth, Spinoza's independence of thought resulted in his excommunication from Judaism at the age of 23. He devoted most of his life to the study of philosophy, supporting himself (modestly) as a lens maker. Spinoza's *Ethics*, published in 1677 (the year of his death), is constructed in five parts, each consisting of a list of definitions, a set of axioms, and a number of propositions derived from the axioms. For example, "Part One, Of God," defines eight technical terms, including *cause of itself*, *substance*, *attribute*, and *eternity*. The text then gives seven axioms, such as

A true idea must agree with that of which it is the idea.

Thirty-six propositions are then deduced from that beginning, including the necessary existence of God.

It is not our purpose to examine Spinoza's philosophy, but rather, to point out a classic example of the pervasive nature of the axiomatic form. The fact that the form of the *Ethics* is obviously borrowed from Euclid's *Elements* suggests more than just the use of a good literary or expository style. It was Spinoza's intention to do in philosophy what Euclid had done in geometry — to express a set of basic statements (the axioms) that the reader should find so compelling as to be undeniable, and to follow them with a step-by- step logical development of inescapable consequences.

Spinoza's use of the axiomatic form is obvious. Other uses outside of mathematics are less obvious, but no less exemplary. For example, in the late eighteenth century, a group of men gathered together to fashion a brief political document. Although they did not use the word *axiom* in stating their premises, they surely thought of them that way.

> We hold these truths to be self-evident, that all men are created equal; that they are endowed by their Creator with certain unalienable rights; that among these are life, liberty, and the pursuit of happiness. That, to secure these rights, governments are instituted among men, deriving their just powers from the consent of the governed.

Euclid may or may not have agreed with these assertions, but he certainly would have understood their form. These are the axioms from which one can logically derive necessary propositions. Notice that several distinct models could fit these axioms, such as a confederation of states, a republic, or a democracy. But a monarchical form of government is *not* a model allowed by these axioms. The only logical conclusion for the framers of this axiom system was that their society, the United States of America, must become free and independent states.

One might argue that any human society — family, club, nation, world — determines its operational rules, however formal or informal, on the basis of fundamental principles, stated or understood. These are the axioms and theorems, and the resulting society is merely a model (perhaps one of many possible) of that axiom system. In some instances the principles are written down, as in the Declaration of Independence or the Constitution of the United States; in others they are not. Sometimes those fundamentals seem to change and evolve (and perhaps they do) with the agreement of the society. Some social change results from the discovery that certain rules are not axiomatic, but are only accidental properties of one model. In this political society called the United States we have a group of individuals whose sole, explicit charge is to safeguard our axiom system, the Constitution, by confirming modifications of our model that conform to the axioms and discarding those that do not. This group is called the Supreme Court.

We could give other less lofty examples, or we could study some of the logical consequences of these examples. Either of those paths leads to interesting places, but we need not travel any further. Our point is already made: Whatever role geometry has played or will play in mathematics and science, its mark has been indelibly left on human society in the form of the axiomatic method.

Topics for Papers – Chapter 3

1. The purpose of this project is to help you understand the role of axiom systems as a bridge between the "real world" and the world of abstract

science. Write a paper about axiom systems, constructed according to the following outline:

(a) Describe informally a real-life situation you know — sport, game, organization, political structure, dorm, class, or anything else — that involves rules or procedures.

(b) Describe formally *some aspect* of that situation by an axiom system consisting of 3 or 4 basic terms and 3 or 4 axioms. (Do yourself a favor — keep them simple!)

(c) Rework your axiom system, if necessary, to insure that
   i. the system is consistent, and
   ii. all the axioms are independent.
   Prove these two assertions.

(d) State at least one theorem and prove it from the axioms.

(e) Treating the basic terms of your axiom system as undefined, find an interpretation for the system that is different than the one you started with. Confirm the difference by stating a property of the basic terms in the original situation that is not shared by the basic terms in your new interpretation.

2. Research one of the following historical figures and write a detailed report on his attempt to prove Euclid's fifth postulate:

> Poseidonios (c. 135–c. 51 B.C.)
> Proclus (410–485)
> John Wallis (1616–1703)
> Johann Heinrich Lambert (1728–1777)
> John Playfair (1748–1819)
> Adrien-Marie Legendre (1752–1833)

Be sure to provide a bibliography that lists any sources you used in writing the paper, and be careful that you do not simply copy or paraphrase the material from your sources. Any material not explicitly quoted should have been mentally digested by you, so that your own words describe what really are your own ideas when you write them down.

3. Write a fictional short story wherein the central theme is the non-Euclidean geometry of the universe in which the characters live. The story may or may not be directly about geometry or mathematics, and may or may not touch upon the possible existence of other types of geometry, at your discretion.

For Further Reading

1. Bonola, Roberto. *Non-Euclidean Geometry: A Critical and Historical Study of Its Development*, transl. by H. S. Carslaw, 1911. Reprint. New York: Dover Publications, 1955.

2. Davis, Philip J., and Reuben Hersh. *The Mathematical Experience.* Boston: Birkhäuser, 1981.

3. Dunham, William. *Journey Through Genius: The Great Theorems of Mathematics.* New York: John Wiley and Sons, Inc., 1990, Chapters 1 and 2.

4. Eves, Howard, and Carroll V. Newsom. *An Introduction to the Foundations and Fundamental Concepts of Mathematics*, Rev. Ed. New York: Holt, Rinehart and Winston, 1965.

5. Hawking, Stephen. *A Brief History of Time.* New York: Bantam Books, 1988.

6. Kline, Morris. *Mathematical Thought from Ancient to Modern Times.* New York: Oxford University Press, 1972.

7. Lieber, H. G., and L. R. Lieber. *Non-Euclidean Geometry.* Lancaster, PA: The Science Press Printing Co., 1931.

8. Newman, James R., ed. *The World of Mathematics*, Vols. 1–4. New York: Simon & Schuster, Inc., 1956.

9. Prenowitz, Walter, and Meyer Jordan. *Basic Concepts of Geometry*, 1965. Reprint. New York: Ardsley House, Publishers, Inc., 1989.

10. Trudeau, Richard. *The Non-Euclidean Revolution.* Boston: Birkhäuser, 1987.

CHAPTER

MATHEMATICS OF CHANCE: PROBABILITY AND STATISTICS

4.1 The Gamblers

Much of mathematics is concerned primarily with deducing valid conclusions from premises that are assumed to be true. This process is logically satisfying, but any attempt to apply it to the world we live in is hampered by a major obstacle: There is no way of verifying with certainty most assumptions that deal with reality. The twin branches of mathematics called *probability* and *statistics* face this problem squarely and try to bridge the gap between hypothesis and fact. They consider uncertain situations, examine the conclusions that can be drawn from the possible alternatives, and make quantitative judgments about the reliability of those conclusions.

The mathematical theory of probability began in 1654 when the Chevalier de Méré, a wealthy French nobleman with a taste for gambling, proposed to the mathematician Blaise Pascal a problem involving the distribution of stakes in an unfinished game of chance. This was the famous "problem of points," in which a simple gambling game between two players is ended before either wins. The question is how to divide the prize money if the partial scores of the players are known, and its answer is based on the likelihood of each player winning the game. Pascal communicated the question to Pierre de Fermat, another prominent French mathematician, and from their correspondence a new field of mathematics emerged. Both Pascal and Fermat arrived at the same answer to the problem, but their methods of solution were different. They generalized the problem and its solution, then extended their investigations to other games of chance. Their discussions aroused considerable interest in the European scientific community, and soon

Blaise Pascal. One of the founders of the theory of probability.

Creait: Smithsonian Institution

other scholars took up the challenge of analyzing gambling games.[1]

As Pascal, Fermat, and others in Europe were investigating games of chance, researchers in England were hard at work studying a slightly more respectable, but equally costly, form of gambling. The many hazards of life in the 16th and 17th centuries had established an eager market for a remarkable new service called "life insurance." The groups of people who backed these first life-insurance policies must be ranked among the most daring of gamblers, for they often risked large sums of money on a single turn of fate. In order to understand this life-and-money game of chance better, people began to investigate death records. The first important mortality tables were compiled by Edmund Halley in 1693 as a basis for his study of annuities. This second type of probabilistic investigation is known as *statistical probability* or *statistics*.

The first book devoted entirely to probability and statistics was Jacob Bernoulli's *Ars Conjectandi*, published in 1713. In this book Bernoulli anticipated by more than 200 years the practical importance of the subjects by suggesting their application to questions of government, law, economics, and morality. As the study of statistics has developed since that time, it

---

[1] For further information about Pascal and Fermat, see Appendix Section B.6.

Games of Chance. The study of probability originated out of problems concerned with gambling. Pictured are games of chance in various ages around the world. From upper left: Card Party, Early 16th Century, Western Europe. Gambling with a Revolving Pointer, China. Three-handed Game of Monte, Mexico. At the right: Gambling Sticks of the Haida Tribe, Queen Charlotte Islands.

has provided the means by which probability theory is applied not only to the fields suggested by Bernoulli, but to education, the social sciences, and many other areas as well.

Probability and **statistics** are two distinct, but closely related, fields of the mathematics of "uncertainty"; in fact, they are devoted to investigating opposite sides of the same fundamental problem. The question asked of probability is:

> Given a *known collection* of objects, what can be said about the characteristics of an *unknown sample* of that collection?

The "problem of points" is such a question; the set of all possible outcomes of the gambling game is known, but the particular outcome of an unfinished game is not. Statistics attempts to answer the converse question:

> Given a *known sample* of an *unknown collection* of objects, what can be said about the characteristics of the collection as a whole?

For instance, the statistical question of death rates may be phrased: "Knowing the life span of each person in a relatively small group, what can be said about the life spans of people in general?"

The principles of elementary probability are easy extensions of a "commonsense" approach to chance situations. A little experience with flipping a coin or playing a simple card game should provide the insight needed for the fundamentals of the theory presented in this chapter. Statistical investigations rely heavily on the principles of probability theory, and elementary statistical concepts also flow easily from common-sense notions. However, the powerful advanced techniques in both probability and statistics require an understanding of calculus and are beyond the scope of this book; we present only a general discussion of these areas.

EXERCISES 4.1

WRITING EXERCISES

1. What is *gambling*? Is the insurance business really gambling? If so, who are the gamblers? If not, why not?

2. What is *gambling*? Is investing in the stock market really gambling? If so, who are the gamblers? If not, why not?

3. Describe an aspect of or situation in one of the following areas that you think involves probabilistic and/or statistical thinking (beyond mere data gathering) in some significant way:

   (a) government
   (b) law
   (c) morality
   (d) psychology
   (e) sociology

Are YOU ÆTNA-IZED?

Over 100,000 Accident Claims Paid by the Ætna Life Insurance Company

Railroad, Street car and Steamboat	31,634	$2,613,145.98
Falls on pavement, stairs, etc	20,624	1,475,359.86
Horse and carriage	8,705	751,565.66
Automobile	2,362	519,949.76
Fire Arms	628	392,751.23
Bathing or drowning	737	288,932.39
Burns or scalds	3,257	273,456.69
Bicycle	2,835	192,489.50
Septic wounds (blood poisoning)	2,021	177,811.91
Athletic sports	3,201	154,482.82
Falling of heavy weights	2,957	149,407.97
Cuts with edged tools or glass	5,030	139,035.36
Fingers crushed	4,503	120,144.73
Eye injuries	2,281	108,541.76
Elevator	320	89,420.29
Machinery	1,653	70,903.84
Assaults	434	52,998.86
Toes crushed	1,577	50,418.01
Stepping on nails or glass	1,668	45,647.52
House accidents (contact with furniture)	483	45,285.66
Hands lacerated on hooks, nails, etc	1,465	41,162.50
Bites by dogs or insects	783	31,973.20
Tripped over mats or rugs	237	19,514.54
Splinters in hands or feet	508	18,284.70
Motor boats	147	10,505.47
Fingers caught in electric fans	147	6,892.06
Miscellaneous accidents	7,803	476,365.40
Total	**108,000**	**$8,316,447.67**

Insurance. The investigation of death records for insurance purposes was another early impetus to the study of probability and statistics. Pictured here is an insurance ad that appeared in Life Magazine in 1911.

4.2   The Language of Sets

The brief discussion of probability and statistics in Section 4.1 shows that both areas are fundamentally concerned with collections of objects. In mathematics (of any sort), collections of objects are usually called *sets*. The language of sets is basic to virtually all of modern mathematics. The importance of set-theoretic language comes from the fact that it applies to so many different areas and thereby acts as a unifying force within the subject. A significant application occurs in this chapter. In our study of probability, we restate the problems in set-theoretic language. This restatement makes it easier to explain new ideas by using precise terminology and notation. As a bonus, the use of set language here may help you to notice analogies between parts of probability theory and other mathematical areas.

If you are familiar with the language of sets, then this section may be skimmed quickly in order that you see the notation we use in this book. However, if you haven't seen much set-theoretic language before or if what you have seen is a little rusty in your memory, we invite you to study this section with some care. It contains a succinct summary of all the basic set theory needed for understanding the rest of this chapter.

Any collection of objects is called a **set**, and the objects themselves are called **elements** of that set. We use the term **collection** as a synonym for "set." (Notice that we do *not* have a formal definition of set here, because of the obvious circularity between the terms *set* and *collection*. It is presumed that at least one of these two words has a familiar, common-sense meaning. A similar observation can be made about the terms *element* and *object*.) The standard notation used to write sets and elements includes the following conventions:

- Sets are usually denoted by capital English letters, such as A or B or S, and elements by small letters, such as a, b, c, etc. We assume, unless otherwise specified, that different small letters denote different elements.

- The symbol "$\in$" stands for the phrase *is an element of* (or *is in*). Thus, "$a \in S$" is read "a is an element of the set S."

- Negation is denoted by a slash mark through the symbol being negated. For example, we write "b is not an element of S" as "$b \notin S$." (This is similar to the elementary-algebra custom of writing "1 does not equal 2" as "$1 \neq 2$.")

- If a set is described by listing its elements, that list is enclosed in braces. For example, the set consisting of the first four letters of the alphabet is written $\{a, b, c, d\}$.

Sometimes it is inconvenient or impossible to list all the elements in a set. In that event there are two alternatives. One of them only makes sense if there is an obvious sequential pattern to the set. Then an ellipsis (...) may be used to indicate that the pattern is continued, either to a specified stopping point or indefinitely. Thus, the set of all letters of our alphabet is {a, b, c, ..., z}, and the set of all even whole numbers is {0, 2, 4, 6, ...}.

The other, more general alternative is to specify some property or properties that characterize the elements included in the set. (By "characterize" we mean that they describe all the elements in the set and nothing else.) In this way it becomes possible to "build" the set from its description. In such cases we use a standard form of notation, called (not surprisingly) **set-builder notation**:

$$\{x \mid x \text{ has a certain property}\}$$

This is read "the set of all x such that x has a certain property." The vertical bar stands for "such that" in this context, and the variable x (or whatever expression is used in its place) represents the general form of a single typical element of the set.

Example 4.1   The set of all digits can be denoted by

$$\{0, 1, 2, 3, 4, 5, 6, 7, 8, 9\}$$

or by $\{x \mid x \text{ is a digit}\}$, which is read "the set of all x such that x is a digit."
□

Example 4.2 $\{1, \frac{1}{2}, \frac{1}{3}, \frac{1}{4}, \ldots\}$ is the same set as $\{\frac{1}{n} \mid n \text{ is a positive integer}\}$.   □

Of course, the description must be good enough to tell us exactly what belongs in the set and what does not. If that is the case, we say the set is **well-defined**. For example, the set {a, b, c} is well-defined because it is clear exactly what is in the set (the first three letters of the alphabet) and what is not (everything else). The set {1, 2, 3, ..., 100} is well-defined, provided we assume that the "..." continues the obvious pattern and thus includes all natural numbers between 3 and 100.

Example 4.3   "The set of all tall American citizens" and "the set of all lovable puppies" are not well-defined sets because their descriptions do not allow us to decide unambiguously what is in each set and what is not. On the other hand, "the set of all American citizens who are more than six feet tall" and "the set of all beagles less than six months old" are well-defined sets.   □

$\{x \mid x \text{ has a certain property}\}$ is a well-defined set if and only if the statement "x has a certain property" always is unambiguously true or false, no matter what x is. If the defining condition for a set is *always* false whenever anything is substituted for the variable, then the set has no elements in it; but it is considered to be a set, just the same. This convention simplifies the language of set theory.

DEFINITION The set containing no elements is called the **empty set** (or **null set**) and is denoted by $\emptyset$.

Example 4.4 The empty set has many possible descriptions, such as:

- The set of all three-headed people,
- $\{x \mid x$ is a planet within 5 miles of the sun$\}$,
- $\{n \mid n$ is a whole number whose square is 3$\}$.

All of these describe the same set, $\emptyset$. □

DEFINITION A set A is a **subset** of a set B if every element of A is also an element of B. We write this as "$A \subseteq B$."

Of course, it may happen that every element of A is in B, but B contains nothing else (that is, every element of B is also in A). In this case, the sets A and B are indistinguishable as sets, because these two collections contain precisely the same things.

DEFINITION Two sets A and B are **equal** if A is a subset of B and B is a subset of A. We write "$A = B$."

DEFINITION A set A is a **proper subset** of a set B if A is a subset of B, but A does not equal B. We write "$A \subset B$."

In Examples 4.5 and 4.6, the given sets consist of letters of the alphabet.

Example 4.5 $\{a, b, c\} \subseteq \{a, b, c, d, e\}$; in fact, $\{a, b, c\} \subset \{a, b, c, d, e\}$ because the two sets are not equal. □

Example 4.6 $\{a, b, c\} = \{c, b, a\}$ because every element of each set is an element of the other. Notice that the order in which the elements are written does not matter; the two sets are collections of exactly the same elements, and hence the sets are equal. □

Example 4.7 Let $A = \{1, 2, 3, 4, 5, \ldots\}$, $B = \{x \mid x$ is a natural number$\}$, and $C = \{2, 4, 6, 8, \ldots\}$. Then the following statements all are true:

$$A \subseteq B, \quad C \subseteq B, \quad B \subseteq A, \quad A = B, \quad C \subset B, \quad A \not\subseteq B$$

$$B \not\subseteq A, \quad B \not\subseteq C, \quad A \neq C, \quad A \subseteq A, \quad A \not\subset A, \quad A = A \quad □$$

Often it is important to know what is *not* in a set. For instance, if we ask for things that are not in the set {1, 2, 3, 4}, many different kinds of answers are possible: 5, 100, $\frac{1}{2}$, you, this book, the lions in the Bronx Zoo, and so forth. Some of these answers are more appropriate in one context, some in another. The appropriate kind of response can be inferred from the general setting of the question. It would be reasonable in this case, for example, to regard the answers "you, this book, the lions in the Bronx Zoo" as probably beyond the intent of the question, but which of the other suggested answers might fit the intent is not clear. Hence, sometimes it is necessary to specify the set of all elements being considered in a particular discussion. This set is called the **universal set** for that discussion; it is denoted by $\mathcal{U}$. With this in mind we can determine unambiguously what is *not* in a set.

DEFINITION   The set of all elements in the universal set that are not in a particular set A is called the **complement** of A, and is denoted by A'. In symbols,

$$A' = \{x \mid x \in \mathcal{U} \text{ and } x \notin A\}$$

Example 4.8   Consider the set $A = \{0, 1, 2, 3, 4, 5\}$. If the universal set consists of the ten digits, then $A' = \{6, 7, 8, 9\}$. If the universal set is the set $\mathbf{W}$ of all whole numbers, then $A' = \{x \mid x \in W \text{ and } x > 5\}$; that is,

$$A' = \{6, 7, 8, 9, 10, 11, \ldots\} \qquad \square$$

It is useful to have a way of picturing relationships among sets. The most common way of doing this is by drawing the sets as regions enclosed by circles inside a rectangular box. The box represents the universal set. If a particular region is to be emphasized, that region is shaded. Figures drawn this way are called **Venn diagrams**.[2] (See, for example, Figure 4.1.)

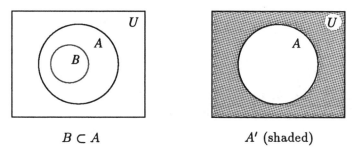

$B \subset A$ $\qquad\qquad\qquad\qquad$ A' (shaded)

Figure 4.1   Venn diagrams for a proper subset and a complement.

---

[2] Named for John Venn, an English mathematician who used them in 1876 in a paper on Boole's algebra of logic.

DEFINITION   The set of all elements that are in both a set A and a set B is called the **intersection** of A and B, and is written $A \cap B$. In symbols,

$$A \cap B = \{x \mid x \in A \text{ and } x \in B\}$$

DEFINITION   If A and B are sets such that $A \cap B = \emptyset$, we say that A and B are **disjoint**. (That is, A and B are disjoint if they have no elements in common.)

Example 4.9   Let $A = \{1, 2, 3, 4\}$, $B = \{3, 4, 5, 6\}$, and $C = \{5, 6, 7, 8\}$. Then $A \cap B = \{3, 4\}$ and $B \cap C = \{5, 6\}$, but $A \cap C = \emptyset$.   □

The intersection of two sets is usually pictured by the Venn diagrams of Figure 4.2.

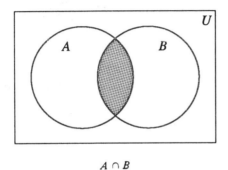

$A \cap B$ $A \cap B = \emptyset$

Figure 4.2   Venn diagrams for intersection.

DEFINITION   The set of all elements that are in either a set A or a set B (or both) is called the **union** of A and B, and is written $A \cup B$. In symbols,

$$A \cup B = \{x \mid x \in A \text{ or } x \in B\}$$

Example 4.10   If $A = \{a, b, c\}$ and $B = \{a, c, e, g\}$, then $A \cup B = \{a, b, c, e, g\}$.   □

Some simple, but useful, consequences of the definitions of union and intersection are listed here for convenience. If A and B are subsets of some universal set $\mathcal{U}$, then:

$$A \cap A = A \qquad\qquad A \cup A = A$$
$$A \cap \emptyset = \emptyset \qquad\qquad A \cup \emptyset = A$$
$$A \cap \mathcal{U} = A \qquad\qquad A \cup \mathcal{U} = \mathcal{U}$$

If $A \subseteq B$, then $A \cap B = A$ and $A \cup B = B$.

The complements of unions and intersections are characterized by a pair of statements known as De Morgan's Laws. They are named after Augustus De Morgan, a 19th-century British mathematician who used a version of these laws in his development of formal logic.

De Morgan's Laws: If A and B are subsets of some universal set, then

$$(A \cup B)' = A' \cap B' \qquad \text{and} \qquad (A \cap B)' = A' \cup B'$$

The Venn diagrams in Figure 4.3 illustrate the first of these laws. Illustration of the second is similar.

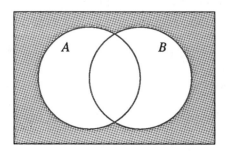

Figure 4.3 *These diagrams illustrate the first of De Morgan's Laws.*

The shaded portion of the left diagram of Figure 4.3 represents $(A \cup B)'$. In the right diagram, A' is the vertically striped region and B' is the horizontally striped region. $A' \cap B'$ is represented by the cross-hatched area, which is precisely the same region as that shaded in the left diagram.

There is another way of combining sets, not quite as obvious as union but just as important. First we must consider the idea of an **ordered pair,** which is simply a pair of elements whose order is specified. We write the ordered pair with first element x and second element y as (x, y). Notice that, because order matters, $(x, y) \neq (y, x)$ unless x and y are the same thing. In general:

DEFINITION   Two ordered pairs (a, b) and (c, d) are **equal** if $a = c$ and $b = d$.

DEFINITION   The **Cartesian product** of two sets A and B is the set of all ordered pairs whose first elements are from A and whose second elements are from B. It is denoted by $A \times B$.[3] In symbols,

$$A \times B = \{(a, b) \mid a \in A \text{ and } b \in B\}$$

---

[3]Named after René Descartes, the 17th-century French mathematician and philosopher who first published a description of analytic (coordinate) geometry.

Example 4.11   If $A = \{a, b, c\}$ and $B = \{\star, \heartsuit, 1, 2\}$, then

$$A \times B = \{(a, \star),\ (a, \heartsuit),\ (a, 1),\ (a, 2),\ (b, \star),\ (b, \heartsuit),$$
$$(b, 1),\ (b, 2),\ (c, \star),\ (c, \heartsuit),\ (c, 1),\ (c, 2)\} \qquad \square$$

Example 4.11 illustrates a general fact about the number of elements in a Cartesian product. In that example, A contains 3 elements and B contains 4 elements, so $A \times B$ contains $12\ (= 3 \cdot 4)$ elements because for each of the 3 elements of A there are 4 ordered pairs containing it in the first position. In general,

If A contains m elements and B contains n elements, then $A \times B$ contains $m \cdot n$ elements.

Example 4.12   If A is any set, then $A \times \emptyset = \emptyset$ because there are no elements to be used as second elements of the ordered pairs. $\qquad \square$

EXERCISES 4.2

In Exercises 1–3, list the elements of each set. What is an obvious universal set in each case?

1. $\{x \mid x$ is an odd digit$\}$

2. $\{x \mid x$ is a month whose name contains the letter "r"$\}$

3. $\{x \mid x$ is a whole number and $x < 20\}$

For Exercises 4–13, let the universal set be all the letters of the alphabet, and let $A = \{a, b, c, d, e\}$, $B = \{c, d, e, f, g\}$, and $C = \{b, c, d\}$. Which of the following statements are true?

4. $A \subseteq B$

5. $A = B$

6. $C \subseteq A$

7. $C \subseteq B$

8. $C \subset A$

9. $C \subset B$

10. $A' \subseteq C'$

11. $C' \subseteq A'$

12. $C' \subset A'$

13. $A' = C'$

For Exercises 14–16, let $A = \{1, 2, 3, 4, 5\}$. Describe A' in each case:

14. $\mathcal{U} = \{0, 1, 2, 3, 4, 5\}$

15. $\mathcal{U} = \{0, 1, 2, 3, 4, 5, 6, 7, 8, 9\}$

16. $\mathcal{U} = \{x \mid x$ is a positive whole number$\}$

For Exercises 17–28, let the universal set be

$$\{0, 1, 2, 3, 4, 5, 6, 7, 8, 9\}$$

and let

$$A = \{1, 2, 3\}, \quad B = \{3, 4, 5, 6\}, \quad C = \{5, 6\}$$

Find each of the following sets.

17. $A \cup B$

18. $A \cap B$

19. $A \cup C$

20. $A \cap C$

21. $B \cup C$

22. $B \cap C$

23. $A \times C$

24. $B \times C$

25. $(A \cup B)'$

26. $A' \cup B'$

27. $(A \cap B)'$

28. $A' \cap B'$

In Exercises 29–31, illustrate the given relationships among three nonempty sets A, B, and C by using a single Venn diagram.

29. $(A \cup B) \subset C$ and $A \cap B = \emptyset$

30. $A \subset B$, $A \cap C = \emptyset$, and $B \cap C \neq \emptyset$

31. A and B are disjoint, but neither A and C nor B and C are disjoint.

For Exercises 32–36, let A and B be subsets of some universal set $\mathcal{U}$. Show by some persuasive argument that each statement is true.

32. $A \cap \emptyset = \emptyset$ **33.** $A \cup \emptyset = A$

34. $A \cap \mathcal{U} = A$ **35.** $A \cup \mathcal{U} = \mathcal{U}$

36. If $A \subseteq B$, then $A \cap B = A$ and $A \cup B = B$.

WRITING EXERCISES

1. The definitions of *complement*, *intersection*, *union*, and *Cartesian product* appear in this section both in words and in symbols. Discuss the advantages and disadvantages of each way of stating definitions, using these as illustrative examples. Which of the two ways do you prefer, and why?

2. What aspect of the philosophy of René Descartes made it natural for him to be interested in pursuing mathematical ideas? Consult whatever sources you wish, but be sure to write most of this answer in your own words and to give appropriate source credit for any ideas that are not your own.

3. (a) Suppose that, for each positive integer n, the set $A_n = \{n - \frac{1}{3}, n + \frac{1}{3}\}$. (That is, we have infinitely many disjoint two-element sets of rational numbers.) Can you define a set consisting of *exactly one* element from each of the original sets $A_1, A_2, A_3, \ldots$? If so, do it. If not, explain why not.

(b) Suppose that, for each positive integer n, there is a set A_n containing two different objects that look exactly the same. (For instance, imagine infinitely many brand-new pairs of socks.) Can you define a set consisting of *exactly one* element from each of the original sets $A_1, A_2, A_3, \ldots$? If so, do it. If not, explain why not.[4]

(c) What is the essential difference between the questions in Parts (a) and (b)? Explain.

4.3   What Is Probability?

In general terms, the field of mathematics called **probability** is the quantitative study of "randomness." Something is **random** if it is not predictable; that is, if it cannot be determined *with certainty* from prior knowledge or experience. The possible outcomes of an unfinished gambling game and the date of death of a living person are random phenomena we discussed in Section 4.1. So are the results of flipping a coin or buying a lottery ticket or predicting next week's stock market prices. The uncertainty of these kinds of things is easier to cope with if we can assign a numerical value to the likelihood that a random situation will turn out one way rather than another. That allows us to measure the various alternatives, using numbers to determine which of several possible outcomes is most likely, least likely, and so forth.

---

[4]This question relates to an important principle in formal set theory called the Axiom of Choice. (See Appendix Section B.9.)

The Bourse, Brussels, is one of the world's major stock exchanges.

One preliminary observation is in order here. Although problems involving games of chance make convenient examples for basic probability theory, they are by no means typical of the important current applications of probability. In fact, it is their *obvious* randomness that makes these examples atypical. Most scientific applications of probability arise in situations caused by or governed by physical, economic, or social principles, and as such these situations should be predictable. For instance, when a coin is tossed, the side it lands on is determined by its weight and center of gravity, the magnitude and direction of the force applied by the thumb that tosses it, the distance it falls, the surface it lands on, and so forth. However, the interaction of these principles is often too complex to analyze easily, so it is convenient to think of the possible results as random occurrences. In other words, the randomness of a situation is often an *assumption* made to allow a complex problem to be handled in a simple way.

Because the theory of probability is based on randomness, its applicability in particular situations depends on the validity of that assumption in each case. Such an assumption together with other simplifying assumptions have the effect of transforming a concrete problem of physical or behavioral science into an analogous abstract problem. The setting for the problem in its abstract form is called a **model** of the original situation. If the assumptions in this abstraction process eliminate only details that are truly irrelevant, then any solution of the model problem will also solve the original "real-world" problem. On the other hand, if the model problem is an oversimplification in some respects, its solution may have little to do with the original problem. For instance, when a die is thrown, the assumption that any one of its six faces is equally likely to turn up is valid only if the die has not been "loaded," or weighted off-center in some way.

The reliability of the abstraction process varies from problem to problem, and is often complex. An investigation of this question is not essential to the development of basic probability theory, so we shall treat the relationship between concrete problems and their models as accurate and shall deal primarily with the models. Thus, for example, we shall generally assume that either side of a tossed coin is equally likely to turn up, that each face of a thrown die is as likely to turn up as any other, and so forth.

Let us begin our formal study of probability by introducing some useful standard terminology. Any situation or problem involving uncertain results is called an **experiment** (regardless of whether or not we have control over the situation). The various possible results are called **outcomes**, and the set of all possible outcomes is called the **sample space** of the experiment. The sample space is usually denoted by S. In the language of sets, S is the *universal set* for its experiment, and the outcomes are all the elements of this universal set.

DEFINITION   Any subset of the sample space is called an **event**.

It is often important to know the total number of events in a sample space — that is, the total number of subsets of the space. In these cases, it is handy to remember that:

If a set contains n elements, then it has 2^n subsets.

(This rule is a special case of a general counting principle presented in the next section, so we leave further discussion of it until then.)

Example 4.13   The tossing of a coin is an experiment with two possible outcomes, *heads* or *tails*. (Our model ignores the possibility that it might land on its edge.) If we denote these two outcomes by h and t, then the sample space is $\{h, t\}$. There are four possible events:

$\{h\}$ — The coin lands heads up.

{*t*} — The coin lands tails up.

{*h, t*} – The coin lands either heads up or tails up.

$\emptyset$ — The coin lands neither heads up nor tails up.   □

Example 4.14   Rolling a single die is an experiment with six possible outcomes. Its sample space can be represented by the set

$$S = \{1, 2, 3, 4, 5, 6\}$$

There are 64 events in this experiment because S has 64 (= 2^6) subsets. Two such events are:

{2, 4, 6} — The die falls with an even number facing up.

{1, 2, 3, 4} — The die falls with a number less than 5 facing up.   □

Example 4.15   A raffle in which a single winning ticket is drawn from a barrel containing 1000 tickets is an experiment with 1000 possible outcomes. Thus, there are 1000 elements in its sample space, and there are 2^{1000} events.   □

Example 4.16   Tossing a coin twice is an experiment with four possible outcomes. Thus, the sample space S can be represented as {*hh, ht, th, tt*}. Here, *ht* represents heads on the first toss and tails on the second, whereas *th* represents tails, then heads. There are 2^4, or 16, events in this experiment. (What are they?)   □

The basic idea that underlies all of probability is this: Somehow, we want to assign numbers to each event (subset) of a sample space to measure the likelihood that the outcome of the experiment (element of the sample space) will lie inside, rather than outside, the event. Don't be put off by the formality of the language here; this is an idea that should be quite familiar from everyday life. For instance, when we say that there's a "50–50" chance of a tossed coin landing heads up, we are assigning the number 50% to the event {*heads*} and 50% to the event {*tails*}. In Example 4.15, the holder of a single raffle ticket would surely say he/she had one chance in a thousand of winning (assuming the raffle is fair), so the number attached to the event consisting of that one ticket would be $\frac{1}{1000}$.

The assumption that a raffle is fair is the same as saying that each ticket is as likely to be drawn as any other. Similarly, we say a coin is fair if either side is as likely to turn up as the other when it is tossed. These are examples of a general situation in which the common-sense approach to probability is simplest and most natural. We say that all the outcomes in a sample space are **equally likely** if each one has the same chance of happening as any other. (This is not a very satisfying formal definition, but it will serve our purposes well enough.) *Whenever possible, we shall try to describe the* *sample space of an experiment in such a way that all outcomes are equally likely* because in this case the assignment of numbers to events is easy.[5]

NOTATION   The number of elements in an event E is denoted by $n(E)$.

DEFINITION   In a sample space S of equally likely outcomes, the **probability** of an event E, denoted by $P(E)$, is the number of outcomes in E divided by the number of outcomes in S. In symbols,

$$P(E) = \frac{n(E)}{n(S)}$$

Example 4.17   If a fair coin is tossed, the probability of its landing "heads up" is the number of elements in the event $E = \{h\}$ divided by the number of elements in the sample space $S = \{h, t\}$. In symbols:

$$P(E) = \frac{n(E)}{n(S)} = \frac{n\big(\{h\}\big)}{n\big(\{h, t\}\big)} = \frac{1}{2}$$

because E contains 1 element and S contains 2.   □

Example 4.18   In Example 4.14, the sample space S contains 6 outcomes. Let us call the two events listed there E and F; that is,

$$E = \{2, 4, 6\} \quad \text{and} \quad F = \{1, 2, 3, 4\}$$

Then

$$P(E) = \tfrac{3}{6} = \tfrac{1}{2} \quad \text{and} \quad P(F) = \tfrac{4}{6} = \tfrac{2}{3}$$

In other words, if a single (fair) die is rolled, there are 3 chances in 6 of rolling an even number, and there are 4 chances in 6 of rolling a number less than 5.   □

Example 4.19   If Ms. Chance buys 25 of the 1000 tickets sold for a single-prize raffle, then the event that she wins the raffle consists of 25 outcomes. If we call that event W, then

$$P(W) = \frac{25}{1000} = \frac{1}{40}$$

That is, she has 25 chances in 1000 — or 1 chance in 40 — of winning the single prize.   □

---

[5]Sometimes, in order to better match a model with its corresponding real-world situation, the "probabilities" (numbers) assigned to events are determined by data drawn from observations. Such an approach is called "empirical" (rather "theoretical"). That approach does not alter the basic rules of probability, but, in our view, the common-sense motivation for those rules is then somewhat obscure. Thus, we make the initial assumption of equally likely outcomes for the sake of simplicity and intuitive comfort.

Example 4.20 As Example 4.16 shows, tossing a coin twice is an experiment with a four-element sample space. If we want the result to be one head and one tail (in either order), then the desired event, which we shall call E, is $\{ht, th\}$, and

$$P(E) = \frac{n(E)}{n(S)} = \frac{2}{4} = \frac{1}{2}$$ □

Example 4.21 There is a tempting, but *wrong*, approach to the experiment described in Examples 4.16 and 4.20 that merits some attention. (See if you can figure out why it's wrong before we tell you.) We might describe the possible outcomes of tossing a coin twice by using the three-element sample space

$$T = \{\text{both } heads, \text{ one } head \text{ and one } tail, \text{ both } tails\}$$

In this case, the event $E = \{\text{one } head \text{ and one } tail\}$ contains just one of the three outcomes in the sample space, so

$$P(E) = \frac{n(E)}{n(T)} = \frac{1}{3}$$

The probability as computed here is different from the probability of the same event in the same experiment as computed in Example 20. Why is that approach correct and this one wrong? As you might already have determined, the difficulty here is that *not all the outcomes in the sample space T are equally likely*. In fact, the "one *head* and one *tail*" outcome is twice as likely to happen as either of the others, so the definition of probability given above will not provide a correct measure of the likelihood of the events in this experiment. □

EXERCISES 4.3

The experiment for Exercises 1–7 consists of tossing a fair coin three times. Let E, F, and G be the following events:

E — the first two tosses are the same (that is, both heads or both tails)
F — the first and third tosses are the same
G — the second and third tosses are tails.

1. List all the equally likely outcomes for this experiment; that is, find a suitable sample space.

2. List the outcomes in each of the events E, F, and G.

3. Find $P(E)$, $P(F)$, and $P(G)$ for the events E, F, and G.

4. List the elements of $E \cap F$; then find $P(E \cap F)$.

5. List the elements of $E \cup F$; then find $P(E \cup F)$.

6. List the elements of $F \cap G$; then find $P(F \cap G)$.

7. List the elements of $F \cup G$; then find $P(F \cup G)$.

The experiment for Exercises 8–12 consists of rolling a single die, then tossing a coin. Let E, F, G, and H be the following events:

E — the number on the die is even
F — the number on the die is 5
G — the coin lands heads up
H — the number on the die is less than 4 and the coin lands tails up

8. List all the equally likely outcomes for this experiment; that is, find a suitable sample space.

9. List all the outcomes in each of the events E, F, G, and H.

10. Find the probability of each of the events E, F, G, and H.

11. List the outcomes in $E \cap G$; then find $P(E \cap G)$.

12. List the elements in $F \cup H$; then find $P(F \cup H)$.

The experiment for Exercises 13–20 consists of rolling a pair of (fair) dice. Let E, F, G, and H be the following events:

E — the sum of the numbers thrown is 2
F — the sum of the numbers thrown is 7
G — the sum of the numbers thrown is even
H — the sum of the numbers thrown is less than 5.

13. List all the equally likely outcomes for this experiment; that is, find a suitable sample space.

14. Does the set $\{2, 3, 4, \ldots, 12\}$ represent a suitable sample space for this experiment? Why or why not?

15. List all the outcomes in each of the events E, F, G, and H.

16. Find the probability of each of the events E, F, G, and H.

17. List the elements in $E \cap F$ and the elements in $E \cup F$.

18. Find $P(E \cap F)$ and $P(E \cup F)$.

19. List the elements in $G \cap H$ and the elements in $G \cup H$.

20. Find $P(G \cap H)$ and $P(G \cup H)$.

The experiment for Exercises 21–27 consists of rolling three fair tetrahedral dice. (Each die has four faces, numbered 1, 2, 3, 4.) Let E, F, G, and H be the following events:

E — the sum of the numbers thrown is 2
F — the sum of the numbers thrown is 7
G — the sum of the numbers thrown is even
H — the sum of the numbers thrown is less than 5.

21. List all the equally likely outcomes for this experiment; that is, find a suitable sample space.

22. List all the outcomes in each of the events E, F, G, and H.

23. Find the probability of each of the events E, F, G, and H.

24. List the elements in $E \cap F$ and the elements in $E \cup F$.

25. Find $P(E \cap F)$ and $P(E \cup F)$.

26. List the elements in $G \cap H$ and the elements in $G \cup H$.

27. Find $P(G \cap H)$ and $P(G \cup H)$.

28. A single card is drawn from a regular deck of playing cards.

  (a) What is the probability of drawing an ace? Why?

  (b) What is the probability of drawing a heart? Why?

  (c) What is the probability of drawing the ace of hearts?

A jar contains seven ping-pong balls, numbered 1 through 7. Two balls are drawn from the jar, and the ball drawn first is *not* replaced before the second draw. Answer Exercises 29–31 for this experiment.

29. What is a suitable sample space for this experiment?

30. Find the probability of each of the following events:

E — both balls have even numbers on them
F — one ball has a 5 on it
G — the sum of the two numbers is 6.

31. What is the probability that the sum of the numbers on the balls is *not* 6? Why?

WRITING EXERCISES

1. Saying that a process or experiment is random is not the same as saying that all possible outcomes are equally likely. Explain this statement, and illustrate it by giving an appropriate example.

2. Is a random process always fair? Explain. (You probably should start by sharpening the question!)

3. (a) Discuss how the mathematical usage of the word "model" as described in this section is analogous to its nonmathematical usage.

   (b) If you have studied Chapter 3, compare and contrast the use of the word "model" in Section 3.4 with its use here.

4.4   Counting Processes

As the examples and exercises of Section 4.3 suggest, analyzing an experiment often depends on being able to count the exact number of outcomes or events in some set. Here we discuss three problems whose solutions typify the most important basic counting techniques. These techniques are then used to illustrate some of the basic laws of probability.

Problem 1
How many three-digit numbers can be made using only the digits 0 and 1?

To form a three-digit number, we must fill three places, *in order*, and in this case only the digits 0 or 1 may be used in any place. Figure 4.4 illustrates all the possible choices. There are two choices for the first digit. For each of these possibilities, there are two choices for the second digit, making a total of four different ways of choosing the first two digits. For each of these four ways there are two choices for the third digit. Hence, there are eight possible three-digit numbers containing only the digits 5 and 6.

Generalizing the reasoning used to solve this problem, we are led naturally to the **Fundamental Principle of Counting**:

(4.1) | If an arrangement of objects requires r choices and if there are n possibilities for each choice, then the number of different arrangements is n^r.

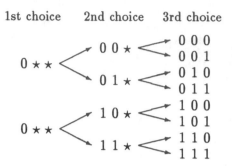

Figure 4.4   *2 choices for the first digit,*
4 choices for the first two,
and 8 choices for all three.

An important special case of this rule gives us the formula for the number of subsets of a set. Suppose we have a set S containing r elements. Any subset of S can be formed by considering each element of S in turn and deciding whether that element should be *in* or *out* of the subset. Thus, the formation of any subset requires r choices from the two-element set

$$\{ \ in, \ out \ \}$$

Applying Rule (4.1), we see that the total number of possible subsets must be 2^r.

Problem 2

Teams A, B, C, and D compete in a playoff tournament. If it is not possible for any two teams to tie for a place, how many different possibilities are there for the final standings of these teams?

This problem is solved much like the first one; again, we have a sequence of choices to make and a set to choose from. In this case, however, once an element has been chosen, it cannot be used again. (For instance, a team cannot be both first and third in the same tournament.) Thus, there are 4 possibilities for first place, and for each of them there are 3 possibilities for second place, making a total of 12 different ways the first two places may be decided. For each of these 12 situations there are 2 possibilities for third place, but once this has been decided, the remaining team must be fourth. Hence, there are 24 possible final standings in this tournament, as shown in Figure 4.5.

In general, if an ordered arrangement of r things is to be formed from a set of n things, and if these things may *not* be used more than once in the same arrangement, there are again r choices to be made. However, although there are n possibilities for the first choice, there are only $n - 1$ possibilities

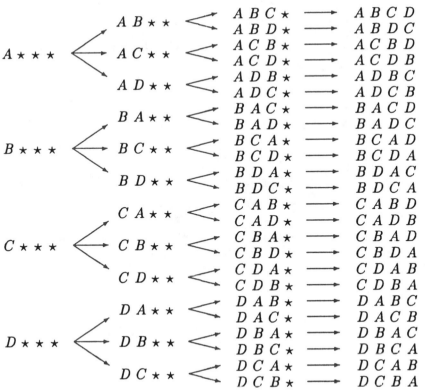

Figure 4.5 *There are 4 possibilities for first place, 12 for the first two places, and 24 for the first three (and for all four) places.*

for the second choice, $n-2$ possibilities for the third, and so on. This pattern continues until the rth choice, for which there are $n-(r-1)$ possibilities. Observing that $n-(r-1)$ can be rewritten as $n-r+1$, we get the following general rule:

(4.2)

> If an arrangement of r objects is to be made from a set of n things and if none of the n things can be used more than once, then the number of possible arrangements is
>
> $$n \cdot (n-1) \cdot (n-2) \cdot \ldots \cdot (n-r+1)$$

DEFINITION   An ordered arrangement of r elements selected from a set of n $(> r)$ elements in such a way that no element may be used more than once is called a permutation of n things taken r at a time. The total number of such permutations is denoted by $_nP_r$.

NOTATION The product of all consecutive positive integers from 1 to n inclusive is denoted by $n!$ and is read "n factorial."

In this notation,

$$
\begin{aligned}
_nP_r &= n \cdot (n-1) \cdot (n-2) \cdot \ldots \cdot (n-r+1) \\
&= \frac{n \cdot (n-1) \cdot (n-2) \cdot \ldots \cdot (n-r+1) \cdot [(n-r) \cdot \ldots \cdot 2 \cdot 1]}{(n-r) \cdot \ldots \cdot 2 \cdot 1} \\
&= \frac{n!}{(n-r)!}
\end{aligned}
$$

To extend this formula to the case of all possible permutations of n elements (that is, n things taken n at a time), we must agree that

$$0! = 1$$

Then

$$_nP_n = \frac{n!}{(n-n)!} = \frac{n!}{0!} = n!$$

Problem 3
If the sample space for an experiment is $\{a, b, c, d, e\}$, how many three-element events are there?

This type of problem differs from the previous ones in that the order of the elements chosen does not matter; that is, any two 3-element events (subsets) that contain the same elements are indistinguishable as sets, regardless of the order of the elements in them. If the order mattered, the solution would be

$$_5P_3 = \frac{5!}{2!} = 5 \cdot 4 \cdot 3 = 60$$

However, in that case the events $\{a, b, c\}$, $\{a, c, b\}$, $\{b, a, c\}$, etc., would be considered different from each other, when, in fact, they are not; that is, they are different representations of the same subset. Now, there are exactly six possible rearrangements of this subset because the number of possible permutations of a three-element set is $3! = 6$. Thus, the 60 ordered subsets may be grouped into collections according to which sets contain the same elements, and each of these collections will contain six sets. Since each collection represents a single unordered 3-element subset, the answer we seek is $\frac{60}{6} = 10$. In general,

(4.3) | If we wish to count all the unordered r-element subsets of an n-element set, we count the number of ordered subsets of that size and then divide by the number of possible orderings of any one of them. Using the formula derived from the preceding problem, we compute

$$\frac{_nP_r}{_rP_r} \quad \text{or} \quad \frac{_nP_r}{r!}$$

DEFINITION   An r-element subset of an n-element set is called a **combination of** n **things taken** r **at a time.** The total number of such combinations is denoted by $_nC_r$.

By the preceding discussion,

$$_nC_r = \frac{n!}{(n-r)! \cdot r!}$$

Example 4.22   To determine how many numerals of at most three digits can be formed in base seven, we use Rule (4.1). There are seven single digits in base seven — 0, 1, 2, 3, 4, 5, 6 — and any one of them can be used anywhere in the three places of such a numeral. Thus, $n = 7$ and $r = 3$ in this case, so there are 7^3 possibilities.   □

Example 4.23   A small club has 7 members. The club decides to elect three officers — President, Secretary, and Treasurer. Assuming all 7 members are candidates for each office, but nobody can hold two offices, how many possible outcomes does this election have?

Combination Lock.

To answer this, we observe that each possible outcome is an ordered triple selected from a set of 7 elements, with no element used more than once. This means we can use Rule (4.2), with $n = 7$ and $r = 3$. Thus, the answer is $7 \cdot 6 \cdot 5$, which equals 210. In other words, there are 210 possible slates of officers for this club. This is the number of permutations of 7 things taken 3 at a time, which is symbolized by $_7P_3$.   □

Example 4.24   A typical poker hand is a random selection of 5 cards from a deck of 52. What is the chance of getting a hand containing a specific royal flush — say, the Ace, King, Queen, Jack, and 10 of hearts?

To answer this we need to know how many different 5-element subsets a 52-element set has. Those subsets (the possible poker hands) form the sample space for this experiment. This means that Rule (4.3) applies, with $n = 52$ and $r = 5$. Thus, we want

$$_{52}C_5 = \frac{52!}{(52-5)! \cdot 5!} = \frac{52 \cdot 51 \cdot 50 \cdot 49 \cdot 48}{5!} = 2{,}598{,}960$$

Because the event we want consists of a single element of this large sample space, its probability is $\frac{1}{2{,}598{,}960}$.   □

Problem. What is the probability of getting a royal flush in hearts?

EXERCISES 4.4

In Exercises 1–8, compute:

1. $_6P_4$ **2.** $_{10}P_6$ **3.** $_{10}P_3$ **4.** $_8P_8$

5. $_6C_4$ **6.** $_{10}C_6$ **7.** $_{10}C_3$ **8.** $_8C_8$

In Exercises 9 and 10, how many numbers of no more than four digits can be made by using:

9. only the odd digits?

10. only 2, 4, 6, and 8?

In Exercises 11–16, how many five-letter license plates can be made by using:

11. all the letters of the alphabet?

12. only the letters A through J ?

13. only the vowels A, E, I, O, U ?

14. all the letters of the alphabet with no repetitions?

15. only the letters A through J with no repetitions?

16. only the vowels A, E, I, O, U with no repetitions?

In Exercises 17–22, how many ways can the letters in each word be rearranged?

17. CAT   **18.** WORDS

19. TICKLE   **20.** ARTICHOKE

21. SCATTER (*Watch out!*)

22. MISSISSIPPI (*Be careful!*)

23. You have wandered by accident into a class in Ancient Greek Literature. A ten-question multiple-choice test is handed out, with each answer to be chosen from four possibilities. Having nothing better to do, you decide to take the test. Because it is written entirely in Greek, a language you know nothing about, you answer each question completely at random. What is the probability that you get all the answers right?

24. The tickets in a single-prize raffle are identified by a five-place "number" in which the first two places are letters of the alphabet and the remaining three places are single digits. All the tickets that can possibly be numbered this way are sold. You buy the tickets AB981 through AB990 (and no others). If the raffle is fair (that is, if all outcomes are equally likely), what is the probability that you will win the prize?

WRITING EXERCISES

1. Compare and contrast the mathematical meanings of the words "permutation" and "combination" with their nonmathematical meanings.

2. Write the Fundamental Principle of Counting entirely in words, without using any letters or other symbols to represent arbitrary quantities. Then, reflecting on how you dealt with this question, comment briefly on the role of symbols in writing mathematical ideas.

3. Construct a clearer and/or simpler symbolic way of representing the concepts denoted in this section by $_nP_r$ and $_nC_r$. Explain why you think your way is better than the one given here. (*Note*: Some mathematical symbolism is used simply because of tradition, not necessarily because it is the clearest or simplest way to represent an idea.)

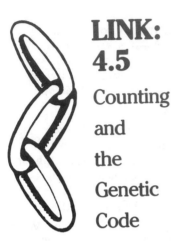

LINK: 4.5

Counting and the Genetic Code

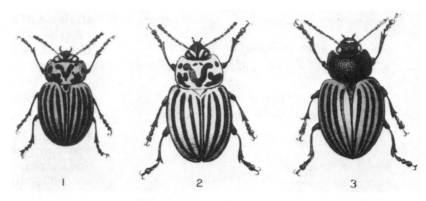

Mutations. Some of the numerous variations in Leptinotarsa multitaeniata.
Figure 2 is the type of the species; figures 1 and 3 are extreme mutants.

Ever since mankind began to speculate about its own existence, one of the most tantalizing problems has been: How do the various life forms — people, animals, plants, even microorganisms — transmit life from generation to generation, and with it transmit the bewildering variety of characteristics that distinguish species from species and family from family? We all have physical characteristics that we say are "inherited" from our parents, but none of us is an exact copy of any ancestor; cats breed cats and dogs breed dogs, and the kittens and pups are similar to, but not exactly like, their parents; rose bushes never blossom forth with lilies, but no two flowers are identical; there are examples almost everywhere we look.

The rise of modern science in Western culture began in the 16th and 17th centuries with the work of men such as Copernicus, Galileo, and Newton. As science sought to turn the observable into the predictable, and technology began to turn the predictable into the constructible, most natural phenomena seemed to conform to unifying principles that govern their behavior. But the propagation of life itself remained a mystery, immune to the increasingly precise mathematical techniques of science. It was almost as if Nature had decreed a fundamental diversity to defy the scientists' thirst for simple logical explanations.

The first breakthrough came in the 1860s when Gregor Mendel, an Augustinian monk in Austria, observed a pattern in the inherited characteristics of peas and set down some basic principles which have since been shown to apply to all living creatures. These "Mendelian laws" described certain factors, subsequently called *genes* (from the Greek word for giving birth), whose behavior could be described precisely enough to make the occurrence of some inherited characteristics fairly predictable. The study of these inheritance patterns soon came to be called *genetics.*

Despite their predictive power, Mendel's laws did not explain exactly *how* these genetic factors are passed on. In particular, *how* do a single sperm cell and a single egg cell carry with them the complex variety of factors that govern the birth and growth of an entire human being? The answer to this question emerged in 1944, after 80 years of biochemical research, with the first direct experimental evidence that an organic acid called DNA contains all that information in its molecular structure. Each DNA molecule has the form of a *double helix*, a barber-pole spiral of two linked chemical chains. Each chain is made up of alternating sugar and phosphate molecules, and the two chains are bonded together at the sugars by linked pairs of chemical bases called *nucleotides.* There are only four nucleotides in DNA — adenine (A), cytosine (C), guanine (G), and thymine (T) — yet the sequence of these bonded pairs in a single DNA molecule carries the entire genetic "blueprint" for its organism! Moreover, the bonding occurs in such a way that the links in one of the chains determine the links in the other, so a single strand of DNA carries all the hereditary information to be passed along. Thus, the entire *genetic code* for all living things is represented by sequences of the four letters A, C, G, and T.

The marvelous simplicity of this device for passing on hereditary characteristics might seem puzzling at first. How (you might ask) is it possible for so few different chemicals to account for the seemingly endless variety of plants and animals? The answer lies in the Fundamental Principle of Counting, explained in Section 4.4, and the fact that DNA chains are quite long. Let us take a simple example. A single virus particle contains only one DNA molecule. Its nucleotide chain may contain 5000 to 200,000 "links," depending on the type of virus. If we take the simplest type, a 5000-link chain of nucleotides, how many different genetic patterns are possible? There are 5000 positions in the chain and each one can be filled in 4 different ways, so the Fundamental Principle of Counting says that there are 4^{5000} different possible chains. Exponential notation can be deceptively brief, so let us do a little estimating to get a better sense of the size of this number:

$$4^{5000} = (4^5)^{1000} = (1024)^{1000} > (1000)^{1000}$$

This last number, if written out, would have a 1 followed by three thousand 0s. If typewritten (in 10-pitch) on a single line, this number, even without commas, would be 25 feet long!

We humans are much more complex than viruses, so our DNA chains are much longer. A DNA chain in a typical human cell might be some five billion nucleotides long. That is, the number of various human genetic patterns is about $4^{5,000,000,000}$, somewhat larger than $1000^{1,000,000,000}$. This number, if typed on a single line, would be a 1 followed by more than 4700 *miles* of 0s; it would stretch from New York City to Moscow. With that many possible genetic variations, it is not surprising that no two people are exactly alike! (By way of comparison, the total number of people who have ever lived has recently been estimated as not more than a mere 100,000,000,000, of whom some 5,000,000,000 are living now.)

A much more modest application of the Fundamental Principle of Counting played a role in "breaking" the genetic code; that is, in seeing how the information contained in the DNA chains determines the way an organism grows. Virtually all chemical reactions that take place in the growth and development of cells are controlled by the presence or absence of specific types of proteins called enzymes. The many, many different kinds of enzymes are formed from long chains of amino acids, but there are only 20 different amino acids. The key question is: How does a chain made up from four different things (nucleotides) determine a specific chain made up from 20 different things (amino acids)? It is clear that some combination of nucleotides is needed to determine a single amino acid, but how many? The Fundamental Principle of Counting provides some important hints:

- If we consider only ordered pairs of nucleotides, there are not enough possibilities because the total number of different ordered pairs is $4 \cdot 4 = 16$;
- there are more than enough ordered triplets of nucleotides because $4 \cdot 4 \cdot 4 = 64$;
- there appear to be far too many ordered quadruples because $4 \cdot 4 \cdot 4 \cdot 4 = 256$.

It was these *numerical* considerations that led researchers to concentrate their attention on nucleotide triplets. In 1961, experimental results began to confirm that the genetic code was indeed based on ordered triplets of nucleotides! The difference between the 64 triplets and the 20 amino acids was explained by the fact that several different triplets correspond to each acid, and within a few years a complete "dictionary" for translating the genetic code into amino acids by way of nucleotide triplets was well on its way to completion.

If you are interested in a more complete, but easily readable, explanation of this topic, you might enjoy Isaac Asimov's book *The Genetic Code*. New York: The Orion Press, 1962.

4.6   Some Basic Rules of Probability

Now that we have some efficient ways of counting the number of outcomes in a sample space, let us return to the concept of probability as described in Section 4.3. Recall that we have been dealing with experiments whose sample spaces contain equally likely outcomes, and in this case the probability of an event is simply the number of outcomes in the event divided by the total number of outcomes in the sample space. Because events are just subsets of the sample space, we can make some easy observations about counting sets of things that will prove to be valuable for generalizing the probability concept to other kinds of experiments. In fact, some of them are the *axioms* of probability; they are statements assumed to be true in every probability situation, from which the other general laws of probability are proved.

First of all, recall that the sample space S of an experiment plays the role of the universal set for that experiment. Because S is a subset of itself, we can ask about its probability. The common-sense answer is that the actual outcome of the experiment is certain to be somewhere in the sample space (by definition of sample space), and thus the number assigned to S should indicate certainty. The definition of probability says that this number must be 1 (or 100%, if you prefer):

$$P(S) = \frac{n(S)}{n(S)} = 1$$

for any sample space S. This becomes our first general axiom.

Axiom 1
The probability of any sample space is 1.

Because any event E is a subset of S, the number of elements in E cannot exceed the number of elements in S, and the fewest elements E can have is none at all. Thus,

$$0 \leq n(E) \leq n(S)$$

Because $P(E) = \frac{n(E)}{n(S)}$, it follows that $P(E)$ must be a number between 0 and 1. This becomes our second general axiom.

Axiom 2
For any event E, $0 \leq P(E) \leq 1$.

The third and final axiom also follows from a common-sense observation about sets: If two sets are disjoint, then the number of elements in their union can be found by counting the elements in each set separately, then adding. That is:

$$\text{If } E \cap F = \emptyset, \text{ then } n(E \cup F) = n(E) + n(F)$$

If these sets are events of a sample space, then saying they are disjoint means that, if the actual outcome of the experiment falls within one of these events, then it cannot also fall within the other. In other words, if either of these events happens, the other cannot. Disjoint events are given a name to reflect this property; they are called **mutually exclusive** events. In a sample space of equally likely outcomes, if we know the separate probabilities of two mutually exclusive events E and F, the probability of their union is easily found by applying the counting observation we just made:

$$P(E \cup F) = \frac{n(E \cup F)}{n(S)} = \frac{n(E)}{n(S)} + \frac{n(F)}{n(S)} = P(E) + P(F)$$

This becomes our third general axiom.

Axiom 3

If E and F are mutually exclusive events, then

$$P(E \cup F) = P(E) + P(F)$$

These three simple axioms allow us to prove some useful rules about how probability works in *any* experiment. Before we develop them, however, it might be useful to review some interconnections among events, sets, and logic.

- An event is a subset of the sample space of an experiment. When we say that an event E happens, we mean that the actual outcome of the experiment lies *within* that particular subset E.
- To say an event E *does not happen*, then, means that the actual outcome lies *outside* of E (but still in the sample space, of course). Consequently, if an event E does *not* happen, its complement E' must happen.
- If we have two events E and F of a sample space, then saying that one or the other of the events happens means that the actual outcome of the experiment lies within $E \cup F$, and saying that both of the events happen means that the actual outcome lies within $E \cap F$.

Thus, knowing how to compute the probabilities of complements, unions, and intersections of events whose individual probabilities are known is an important tool for analyzing experiments and drawing conclusions about them. The rest of this section is devoted to developing and applying the three fundamental probability rules for complement, union, and intersection of events.

The complement of an event E, written E', is the set of all outcomes in the sample space S but not in E. Because E and E' are mutually exclusive, Axiom 3 says that

$$P(E) + P(E') = P(E \cup E')$$

But $E \cup E' = S$ by definition of complement, and Axiom 1 tells us that $P(S) = 1$, so

$$P(E) + P(E') = 1$$

Subtracting $P(E)$ from both sides of this equation, we get the rule for complements:

(4.4)

> If E is any event of a sample space S, then
> $$P(E') = 1 - P(E)$$

Example 4.25 The probability of drawing a heart from a well-shuffled standard deck of cards is $\frac{1}{4}$. (*Hearts* is one of four 13-card suits in the 52-card deck). Thus, the probability of not drawing a heart is $1 - \frac{1}{4}$, or $\frac{3}{4}$. □

Example 4.26 If a pair of fair dice is thrown, there are 6^2 possible outcomes in the sample space. (Why?) Of these 36 possible outcomes, six of them add up to 7. (Which ones are they?) Thus, if we let E be the event of throwing a 7, then $P(E)$ is $\frac{6}{36}$, or $\frac{1}{6}$. This means that $P(E')$, the probability of not throwing a 7, is $1 - \frac{1}{6} = \frac{5}{6}$. □

To compute the probability of the union of two events E and F, we need to generalize Axiom 3 to the case where E and F might have some outcomes in common. Thus, we must consider the case where $E \cap F$ is not empty. Just adding the probabilities of E and F, ignoring the requirement that they be mutually exclusive, will give us incorrect information. A simple example of this is pictured in Figure 4.6. Suppose the sample space S contains the ten outcomes $\{0, 1, 2, ..., 9\}$, and the two events are

$$E = \{1, 2, 3, 4, 5\} \quad \text{and} \quad F = \{4, 5, 6, 7, 8, 9\}$$

so that

$$E \cap F = \{4, 5\}$$

Then $P(E) = \frac{5}{10}$ and $P(F) = \frac{6}{10}$, so $P(E) + P(F) = \frac{11}{10}$, which is bigger than 1 and hence cannot be the probability of *any* event, by Axiom 2. Moreover, counting the outcomes in $E \cup F$ tells us that $P(E \cup F) = \frac{9}{10}$. In this case, the difference between $P(E) + P(F)$ and $P(E \cup F)$ is accounted for by the fact that the two elements in $E \cap F$ were counted in computing $P(E)$ and also in computing $P(F)$. Thus, the sum $P(E) + P(F)$ counts these same elements *twice*, but they only appear once in $E \cup F$.

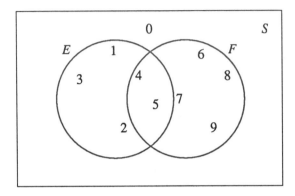

Figure 4.6   P(E ∪ F) < P(E) + P(F)

To resolve the difficulty of counting the "overlapping" elements of E and F twice, observe that there is a way of expressing $E \cup F$ as the union of *disjoint* sets:

$$E \cup F = E \cup (E' \cap F)$$

where $E' \cap F$ is the set of all outcomes in F that are not in E. We can apply Axiom 3 to this equation, and obtain

(∗) $$P(E \cup F) = P(E) + P(E' \cap F)$$

Now, F can be described as the set of all outcomes in F that are not in E together with the set of all outcomes in F that are in E, and it is easy to see that these two parts of F are mutually exclusive. Therefore,

$$F = (E' \cap F) \cup (E \cap F)$$

and, by Axiom 3,

$$P(F) = P(E' \cap F) + P(E \cap F)$$

This last equation can be rewritten as

$$P(F) - P(E \cap F) = P(E' \cap F)$$

Plugging the information from this last equation into Equation (∗), we obtain the general rule for computing the probability of the union of two events:

(4.5) | If E and F are any two events of a sample space, then
$$P(E \cup F) = P(E) + P(F) - P(E \cap F)$$

Despite the apparent complexity of its derivation from the axioms, Rule (4.5) really states a very simple principle: Because the sum of $P(E)$ and $P(F)$ counts the common outcomes in $E \cap F$ twice, one "copy" of $P(E \cap F)$ should be thrown away to get the proper value for $P(E \cup F)$.

Example 4.27   Suppose a single die is thrown. Let E be the event of getting an even number and F of getting a number less than 4. Then

$$E = \{2, 4, 6\}, \ F = \{1, 2, 3\}, \text{ and } E \cap F = \{2\}$$

Moreover, $E \cup F$ represents getting either an even number or a number less than 4; its probability can be computed from Rule (4.5) as follows:

$$P(E \cup F) = P(E) + P(F) - P(E \cap F) = \tfrac{3}{6} + \tfrac{3}{6} - \tfrac{1}{6} = \tfrac{5}{6} \qquad \square$$

Example 4.28   If a single card is drawn at random from a standard deck, the probability it will be

— a heart is $\frac{13}{52}$,
— a face card (king, queen, jack) is $\frac{12}{52}$,
— the king, queen or jack of hearts is $\frac{3}{52}$.

By Rule (4.5), the probability it will be either a heart or a face card is

$$\frac{13}{52} + \frac{12}{52} - \frac{3}{52} = \frac{22}{52} \qquad \square$$

Example 4.29   A survey of results from a recent mathematics quiz showed that

— 30% of the students misspelled "commutative,"
— 20% misspelled "associative,"
— 12% misspelled both words.

The probability that a student chosen at random from that class misspelled at least one of the two words is

$$30\% + 20\% - 12\% = 38\% \qquad \square$$

Working with the intersection of two events is a bit more complicated than handling the union or complement. Clearly, if you know the probabilities of two events E and F *and* the probability of their union, then the probability of their intersection can be found by Rule (4.5). However, if only $P(E)$ and $P(F)$ are known, there is no simple general formula for finding the probability of $E \cap F$. One important special case, however, deserves some attention. That is the case when E and F are "independent." A rigorous discussion of independent events requires considering *conditional probability*, which is the subject of the next section. At this point we shall content ourselves with an informal definition of independence: Two events are said to be **independent** if the occurrence or nonoccurrence of one has no effect on the likelihood of the other. In other words,

if the value of $P(E)$ is the same whether or not we know that the actual outcome of the experiment lies in F, then E and F are independent events.

The rule for computing the probability of the intersection in this case is:

(4.6)

> If E and F are independent events, then
> $$P(E \cap F) = P(E) \cdot P(F)$$

Example 4.30   An experiment consists of flipping a coin three times. If E represents obtaining heads on the first toss and F represents obtaining heads on the third toss, then these are independent events because the occurrence of E has no effect on the occurrence of F. $P(E)$ and $P(F)$ both equal $\frac{1}{2}$. Thus, to find the probability of getting heads on both the first and third tosses, we apply Rule (4.6):

$$P(E \cap F) = P(E) \cdot P(F) = \tfrac{1}{2} \cdot \tfrac{1}{2} = \tfrac{1}{4} \qquad \square$$

Example 4.31   If a single die is thrown, then the event E of obtaining an even number has probability $\frac{1}{2}$. The event F of obtaining an odd number also has probability $\frac{1}{2}$, so

$$P(E) \cdot P(F) = \tfrac{1}{2} \cdot \tfrac{1}{2} = \tfrac{1}{4}$$

However, this is not $P(E \cap F)$ because $E \cap F$ is empty and hence has probability 0. Obtaining an even number and obtaining an odd number are not independent events; if either one happens, the probability of the other drops from $\frac{1}{2}$ to 0. $\qquad \square$

EXERCISES 4.6

In Exercises 1–8, assume that E and F are mutually exclusive events of a sample space S. If $P(E) = .45$ and $P(F) = .30$, compute:

1. $P(E \cap F)$
2. $P(E \cup F)$
3. $P(E')$
4. $P(F')$
5. $P(E \cap E')$
6. $P(E \cup E')$
7. $P(E' \cap F)$
8. $P(E' \cup F)$

In Exercises 9–14, assume that E and F are independent events of a sample space S. If $P(E) = 25\%$ and $P(F) = 60\%$, compute:

9. $P(E')$
10. $P(F')$
11. $P(E \cap F)$
12. $P(E \cup F)$
13. $P(E' \cup F')$
14. $P(E' \cap F')$

In Exercises 15–19, let E and F represent events of a sample space S. Explain why each of the following situations is impossible.

15. $P(E) = .40$ and $P(E') = .75$

16. E and F are mutually exclusive, $P(E) = .5$, and $P(F) = .7$

17. $P(E) = P(F) = .6$ and $P(E \cup F) = .4$

18. $P(E) = .35$, $P(F') = .70$, and E is a subset of F.

19. E and F are independent, $P(E) = .5$, $P(F) = .6$, and E is a subset of F.

In Exercises 20–25, assume that a kindergarten classroom has a box with 40 crayons in it. Five of them are are red, six are blue, and eight are green; the rest are assorted other colors. A girl reaches in without looking and picks a single crayon. What is the probability that she picks

20. a red one?

21. a blue one?

22. one that is not red?

23. one that is not green?

24. one that is neither red nor green?

25. either a red one or a blue one?

In Exercises 26–34, assume that a single card is drawn from a regular deck. Consider the events

D — getting a diamond
F — getting a face card
H — getting a heart.

Find the probability of each of the following events:

26. D
27. F
28. H
29. $D \cap F$
30. $D \cap H$
31. getting either a heart or a diamond
32. getting either a diamond or a face card
33. getting a black card
34. getting a black card that is not a face card

In Exercises 35–40, suppose that E and F are independent events of a sample space S containing 200 equally likely outcomes, and that
$$P(E) = \tfrac{1}{5} \quad \text{and} \quad P(F') = \tfrac{3}{4}$$
Compute:

35. $n(E)$
36. $P(F)$
37. $n(F)$
38. $P(E \cap F)$
39. $P(E \cup F)$
40. $P(E' \cap F)$

WRITING EXERCISES

1. Without referring to coins, dice, or playing cards, describe examples that illustrate each of these concepts:
   (a) mutually exclusive events
   (b) $P(E \cup F) = P(E) + P(F) - P(E \cap F)$
   (c) independent events

2. Write a clear explanation of the difference between mutually exclusive events and independent events. In the course of your explanation, answer these questions, with justification for your answers:

- Are mutually exclusive events always independent?

- Are mutually exclusive events ever independent?

- Are independent events always mutually exclusive?

- Are independent events ever mutually exclusive?

3. Suppose we changed Axiom 1 to: "The probability of any sample space is 5." How would that alter the other material presented in this section? Would it destroy the theory of probability? Discuss.

4.7   Conditional Probability

The intersection of two sets contains only those elements of one set that are also contained in the other. If the sets are the events of an experiment, then the intersection contains only those outcomes of one event that can occur if the other actually happens. This situation is illustrated by the following tale drawn from life in the suburbs:

Unable to locate an eager volunteer to walk the family dog at six o'clock the next morning, a husband-wife couple of mathematicians and their two precocious children were faced with the unpleasant task of choosing an unwilling conscript from among their own ranks. The parents, sensitive to the precepts of modern psychology, were reluctant simply to appoint either child, but at the same time were unwilling to become servants to the whims of their offspring. They decided, therefore, to draw lots and let the matter be decided by chance. As they were finishing dinner, the mother cut four identical slips of paper, marked one with an x, and placed them in a hat. Each member of the family, she explained, would in turn select a slip, and whoever drew the x would have to walk the dog.

At that point the younger child (who was more talented in democracy than in mathematics) protested that this was unfair, claiming that whoever drew first took the least risk. He argued that, while the probability of getting the x on the first draw was $\frac{1}{4}$, the probability of getting it on the second draw rose to $\frac{1}{3}$ because there would only be three slips left to choose from. Similarly, he said, the probability of getting it on the third draw would be $\frac{1}{2}$ and on the fourth draw it would be 1. Because the order of choice affected the amount of risk, he insisted that they find a fair way of determining that order, or else scrap the whole procedure. A heated debate began immediately. Just as the mother was about to restore tranquillity to her strife-torn family by a true analysis of the probabilities involved, it was discovered that the dog had run away and would be gone for at least several days. Since the problem had disappeared with the dog, all discussion ceased and peace returned to the dinner table.

If the mother in this story had been allowed to explain the situation, she undoubtedly would have acknowledged the partial validity of her child's objections. It is true that, if the first slip drawn is not marked x, then there are three slips left as possible outcomes for the second draw, and hence the probability of x appearing is $\frac{1}{3}$. However, it is not certain that x will not appear on the first draw. If it does appear first, then the probability of its appearance on the second draw is 0. (In fact, there is no need for a second draw.)

If "0" is used to represent any of the unmarked slips, then the four equally likely possible outcomes of this experiment may be denoted by

$$(x,0,0,0), \quad (0,x,0,0), \quad (0,0,x,0), \quad (0,0,0,x)$$

Let us call this set of outcomes S. If we denote the event of getting x on the first draw by X_1, then $X_1 = \{(x,0,0,0)\}$ and

$$X_1' = \{(0,x,0,0), \ (0,0,x,0), \ (0,0,0,x)\}$$

X_2, X_3, X_4, and their complements may be defined similarly. Thus,

$$n(X_1) = n(X_2) = n(X_3) = n(X_4) = 1$$

and

$$n(X_1') = n(X_2') = n(X_3') = n(X_4') = 3$$

It follows that

$$P(X_1) = P(X_2) = P(X_3) = P(X_4) = \tfrac{1}{4}$$

and

$$P(X_1') = P(X_2') = P(X_3') = P(X_4') = \tfrac{3}{4}$$

The discrepancy between $P(X_2)$ as given here and the child's evaluation of it can be explained informally by saying that, if this experiment were performed many times, we could expect x to occur second one-third of the times *in which it did not appear first.* But it would not appear first only three-fourths of the total number of times, so it would appear second *one-third of three-fourths of the time* — that is, one-fourth of the time. This line of reasoning may be generalized by the following argument.

Suppose E and F are two events of an experiment and we wish to find the probability that F will occur, assuming E has occurred already. Our assumption restricts the possible outcomes of the experiment to the ones in E, so the only outcomes of F that we may consider are those that are also in E.

DEFINITION   Let E and F be two events of a sample space S (of equally likely outcomes). The **conditional probability of F assuming E** is

$$\frac{n(E \cap F)}{n(E)}$$

It is denoted by $P(F \mid E)$.

Using this definition, we can deal with the probability of the intersection of two events.

(4.7) | If E and F are any two events of a sample space S, then
$$P(E \cap F) = P(E) \cdot P(F \mid E) = P(F) \cdot P(E \mid F)$$

It is easy to check that this is true, directly from the basic definitions:

$$P(E) \cdot P(F \mid E) = \frac{n(E)}{n(S)} \cdot \frac{n(E \cap F)}{n(E)} = \frac{n(E \cap F)}{n(S)} = P(E \cap F)$$

Moreover, since $F \cap E = E \cap F$, we also have $P(F) \cdot P(E \mid F) = P(E \cap F)$.

Example 4.32   In the dog-walking example at the beginning of this section, for the x to be drawn second we must have the occurrence of both X_1' and X_2. Hence, the probability is

$$P(X_1' \cap X_2) = P(X_1') \cdot P(X_2 \mid X_1') = \tfrac{3}{4} \cdot \tfrac{1}{3} = \tfrac{1}{4}$$

The child correctly computed $P(X_2 \mid X_1')$, but did not account for the probability of X_1'.   □

Example 4.33   If two playing cards are drawn from a regular deck and the first is *not* replaced before the second is drawn, the probability that both cards are aces may be computed by using Rule (4.7). The probability that the first card is an ace is $\tfrac{4}{52} = \tfrac{1}{13}$ and, assuming this has happened, the probability that the second card is an ace is $\tfrac{3}{51} = \tfrac{1}{17}$. Hence, the probability that both cards are aces is $\tfrac{1}{13} \cdot \tfrac{1}{17} = \tfrac{1}{221}$.   □

Now independent events can be defined more precisely than they were at the end of Section 4.6.

DEFINITION   If E and F are two events such that $P(F) = P(F \mid E)$, then F is **independent** of E. Two events are **independent** if each is independent of the other. In general, events $E_1, E_2, \ldots, E_n$ are **independent** if every possible pairing of them consists of two independent events.

This definition can be combined with Rule (4.7) to yield a convenient proof and generalization of Rule (4.6), which appeared at the end of Section 4.6. Rule (4.7) states that

$$P(E \cap F) = P(E) \cdot P(F \mid E)$$

for *any* two events E and F. If E and F are *independent*, we can substitute $P(F)$ for $P(F \mid E)$ to obtain

$$P(E \cap F) = P(E) \cdot P(F)$$

In general:

(4.8)

> If $E_1, E_2, \ldots, E_n$ are *independent* events, then
> $$P(E_1 \cap E_2 \cap \ldots \cap E_n) = P(E_1) \cdot P(E_2) \cdot \ldots \cdot P(E_n)$$

Example 4.34 The probability of rolling a 5 in any throw of a single (fair) die is $\frac{1}{6}$, and the fact that a 5 may have turned up before has no effect on the likelihood of its turning up again. Thus, if F_n represents the event that 5 turns up on the nth roll of a single die, then the probability that 5 turns up both times on the first two rolls is

$$P(F_1 \cap F_2) = P(F_1) \cdot P(F_2) = \tfrac{1}{5} \cdot \tfrac{1}{5} = \tfrac{1}{25}$$

If the die is rolled four times, the probability that 5 turns up every time is

$$P(F_1 \cap F_2 \cap F_3 \cap F_4) = P(F_1) \cdot P(F_2) \cdot P(F_3) \cdot P(F_4) = (\tfrac{1}{5})^4 = \tfrac{1}{625} \quad \square$$

Example 4.35 The main assembly line of an automobile factory is fed by two independent auxiliary lines, one for engines and the other for prefabricated body units. Statistical studies have shown that the rate of defective engines is 1 in 400 and the rate of defective body units is 1 in 500. The studies also showed that the appearance of these defects follows no apparent pattern; that is, they appear at random. If we denote the event that a car from this factory has a defective engine by E and the event of a defective body by B, then

$$P(E) = \tfrac{1}{400} \quad \text{and} \quad P(B) = \tfrac{1}{500}$$

The two auxiliary lines are independent of each other, so

$$P(E \mid B) = P(E) \quad \text{and} \quad P(B \mid E) = P(B)$$

Thus, the probability of a car coming off the main assembly line with both engine and body defective is given by

$$P(E \cap B) = P(E) \cdot P(B) = \tfrac{1}{400} \cdot \tfrac{1}{500} = \tfrac{1}{200,000} \quad \square$$

To summarize and unify our discussion of the computational rules of probability, let us consider a fairly complex problem derived from one of the classic games of chance. In one of the common forms of the card game "poker," a *hand* consists of five cards dealt to a player (from a standard deck of 52 cards), with no option for exchanging any of them. The most valuable hand in the game is called a *royal flush*; it consists of the ace, king, queen, jack, and ten, all of one suit. The question we pose is:

> What is the probability that a player will be dealt a royal flush exactly three times in the first ten hands of a game?

To begin with, notice that we could view the dealing of each hand as a separate experiment, or we could consider the whole situation as a single experiment with ten similar parts. To make the necessary distinction when an experimental procedure is repeated two or more times, we call each repetition a **trial**, and reserve the term "experiment" for the entire collection of trials performed. In this case, then, our experiment consists of ten trials. For each trial, there are as many outcomes as there are five-card subsets (hands) of a set (deck) of 52 cards. A single outcome *of the entire experiment* is an ordered set of ten hands.

The next step in solving the problem is to find the probability of getting a royal flush in any one hand. Treating a single trial as an experiment by itself (just for the moment), let us denote the set of all possible hands by H and the event of getting a royal flush by R. If we assume that any card is as likely to be dealt as any other, the definition of probability tells us that

$$P(R) = \frac{n(R)}{n(H)}$$

It is easy to see that $n(R) = 4$, because there is just one royal flush for each suit. Now, $n(H)$ is the number of five-element subsets of a 52-element set, so

$$n(H) = {}_{52}C_5 = \frac{52!}{(52-5)! \cdot 5!} = 2,598,960$$

(See Example 4.24.) Thus, the probability of getting a royal flush in a fair deal is

$$\frac{4}{2,598,960} = \frac{1}{649,740}$$

This says that, on the average, a poker player may expect one royal flush out of every 649,740 hands. (You might like to verify this experimentally if you have a fondness for long poker games.)

The question of getting exactly three royal flushes in ten hands involves ten independent events — getting a royal flush in the first hand, getting a royal flush in the second hand, etc. (The hands are independent of each other because they are dealt to the same person each time, so they must occur on different deals. It is assumed that the cards are shuffled thoroughly before each hand is dealt.) If we denote these events by R_1, R_2, ..., R_{10}, then

$$P(R_1) = P(R_2) = \ldots = P(R_{10}) = \tfrac{1}{649,740}$$

Let us call this number p. Then the probability that any one of these events will *not* happen is $1 - p$, so

$$P(R_1') = P(R_2') = \ldots = P(R_{10}') = 1 - p = \tfrac{649,739}{649,740}$$

If we wanted the three royal flushes in three particular hands, say the first, fourth, and seventh, then Rule (4.8) tells us that the probability of this happening is

$$P(R_1 \cap R_2' \cap R_3' \cap R_4 \cap R_5' \cap R_6' \cap R_7 \cap R_8' \cap R_9' \cap R_{10}')$$
$$= p \cdot (1 - p) \cdot (1 - p) \cdot p \cdot (1 - p) \cdot (1 - p) \cdot p \cdot (1 - p) \cdot (1 - p) \cdot (1 - p)$$
$$= p^3 \cdot (1 - p)^7$$

However, we do not care which three hands are royal flushes, so *any* three-element subset of the ten hands will satisfy our needs just as well as any other. There are $_{10}C_3$ different three-element subsets, each of them representing a desirable event for the experiment as a whole. The probability we seek, then, is the probability of the union of these events. Now, if any particular pattern of three royal flushes turns up, then none of the others can, so these desirable events are mutually exclusive. Therefore, by Axiom 3 on page 140 the probability of their union is simply the sum of the probabilities of the events. But each of these events has probability $p^3(1 - p)^7$, so the probability of getting exactly three royal flushes in ten hands of poker is

$$_{10}C_3 \cdot p^3 \cdot (1 - p)^7 = 120 \cdot \left(\tfrac{1}{649,740}\right)^3 \cdot \left(\tfrac{649,739}{649,740}\right)^7$$

that is, approximately

$$\frac{1}{2,000,000,000,000,000}$$

The method of solution just used typifies the process by which Jacob Bernoulli in the 17th century was able to prove a general theorem about problems of this type:

> **Binomial Distribution Theorem**[6]: Suppose an experiment consists of n independent trials. If the event E being considered is the same for each trial and if its probability for any single trial is p, then the probability that E will occur exactly r times is given by
>
> $$_nC_r \cdot p^r \cdot (1-p)^{n-r}$$

Example 4.36   The probability of obtaining exactly four heads in six tosses of a coin is

$$_6C_4 \cdot \left(\frac{1}{2}\right)^4 \cdot \left(\frac{1}{2}\right)^2 = 15 \cdot \frac{1}{16} \cdot \frac{1}{4} = \frac{15}{64} \qquad \square$$

Example 4.37   The probability of getting exactly three 6s in five throws of a single fair die is

$$_5C_3 \cdot \left(\frac{1}{6}\right)^3 \cdot \left(\frac{5}{6}\right)^2 = 10 \cdot \frac{1}{216} \cdot \frac{25}{36} = \frac{125}{3888} \qquad \square$$

Example 4.38   If a machine that makes Christmas-tree ornaments produces random defects at the rate of 1 in 100, then the probability of finding exactly two defects in a package of six ornaments is

$$_6C_2 \cdot \left(\frac{1}{100}\right)^2 \cdot \left(\frac{99}{100}\right)^4 = 15 \cdot \frac{1}{10,000} \cdot \frac{95,059,601}{100,000,000}$$

or approximately .00144. The package contains *at least* two defects if there are exactly two, exactly three, exactly four, exactly five, or exactly six defects in it; these are five mutually exclusive events. Thus, to find the probability that the package contains at least two defects, we just add the probabilities of these five separate cases:

---

[6]The term *binomial distribution* refers to the patterns of possible results of a repeated-trials experiment for which the outcome of each trial is either a "success" or a "failure"; that is, an experiment whose results can be thought of as *yes–no* lists. The royal-flush question just discussed is such an experiment because each trial (hand) either is a royal flush or it isn't. For more about binomial distributions, see Section 4.10.

$$
{}_6C_2 \cdot \left(\frac{1}{100}\right)^2 \cdot \left(99100\right)^4 + {}_6C_3 \cdot \left(\frac{1}{100}\right)^3 \cdot \left(\frac{99}{100}\right)^3
$$
$$
+ {}_6C_4 \cdot \left(\frac{1}{100}\right)^4 \cdot \left(\frac{99}{100}\right)^2 + {}_6C_5 \cdot \left(\frac{1}{100}\right)^5 \cdot \left(\frac{99}{100}\right)^1
$$
$$
+ {}_6C_6 \cdot \left(\frac{1}{100}\right)^6 \cdot \left(\frac{99}{100}\right)^0
$$

that is, approximately .00146.   □

EXERCISES 4.7

In Exercises 1–6, assume that three playing cards are dealt from a thoroughly shuffled standard deck.

1. What is the probability that the first card is a heart?

2. What is the probability that the second card is a heart?

3. If the first card is accidentally turned over and seen to be a heart, what is the probability that the second card is a heart?

4. If the first card is accidentally turned over and seen to be a club, what is the probability that the second card is a heart?

5. If neither of the first two cards is seen, what is the probability that both are hearts?

6. If none of the cards are seen, what is the probability that all three are hearts?

In Exercises 7–18, assume that one hundred slips of paper, numbered from 1 through 100, are placed in a barrel and mixed thoroughly. The even-numbered slips are green and the odd-numbered slips are red. A blindfolded person draws the slips from the barrel.

7. If a single slip is drawn, what is the probability that its number is even?

8. If a single slip is drawn, what is the proba-

bility that its number is 25?

9. If a single slip is drawn, what is the probability that its number is a multiple of 3?

10. If a single green slip is drawn, what is the probability that its number is a multiple of 3?

11. If a single slip is drawn, what is the probability that it is green and its number is a multiple of 3?

12. If a single slip is drawn, what is the probability that it is green or its number is a multiple of 3?

13. If a single slip is drawn, what is the probability that its number is a multiple of 4?

14. If a single red slip is drawn, what is the probability that its number is a multiple of 4?

15. If a single slip is drawn, what is the probability that it is red or its number is a multiple of 4?

16. If two slips are drawn in succession without being replaced, what is the probability that both numbers are multiples of 3?

17. If three slips are drawn in succession without being replaced, what is the probability that all three are red?

18. A single slip is drawn and then replaced in

the barrel. If this experiment is done seven times, what is the probability that exactly four of the drawings result in red slips?

In Exercises 19–22, assume that the birth rate for boys is approximately 49%. Assume further that each birth is independent of the others and that the sex of a child is a random occurrence.

19. If a couple's first two children are boys, what is the probability that their third child will also be a boy?

20. What is the probability that a couple's first four children will all be girls?

21. If it is known that a couple's first two children are boys, what is the probability that all four of the couple's children are boys?

22. If a couple has six children, what is the probability that at least three are girls?

23. What is the probability of rolling three 7s in three consecutive tries with a pair of fair dice?

24. What is the probability of rolling exactly two 7s in three tries with a pair of fair dice?

25. Let E_1, E_2, ..., E_n be events in a sample space S.
   (a) Prove:
   $$P(E_1 \cap E_2 \cap E_3) =$$

$$P(E_1) \cdot P(E_2 \mid E_1) \cdot P(E_3 \mid E_1 \cap E_2)$$
   (b) Make a conjecture about a formula for $P(E_1 \cap E_2 \cap \ldots \cap E_n)$.

26. (*Recall Exercise 23 of Section 4.4.*) Each question on a ten-question multiple-choice test in Ancient Greek Literature has four possibilities. If you answer each question completely at random, what is the probability that you answer at least seven of the ten questions correctly?

WRITING EXERCISES

1. (a) State in one (complete) sentence the general probabilistic idea that is illustrated by the dog-walking story at the beginning of this section.
   (b) Write a different story that illustrates the same idea.

2. The definition of *independent* in this section appears to define the same word three times. Explain the differences among these three uses of the word.

3. Describe how at least three of the dozen problem-solving tactics of Section 1.2 can be used to help in the solution of Exercise 26. (Even if you can't complete Exercise 26, you should be able to do this writing exercise.)

LINK: 4.8

Probability
and
Marketing

Loteria Nacionale *by John Phillip, 1866.*

A basic problem of marketing a product or service is how to get maximum public response for one's advertising dollar. Large companies have entire departments whose main task is to find innovative ways of solving this problem, and every day we are confronted with the results of their efforts — newspaper ads, television commercials, discount coupons, giveaway games, direct-mail flyers, and the like. Among these modern marketing and advertising techniques there are various forms of a historically popular money-making scheme called a *lottery.*

In its simplest form, a lottery is a game in which people buy tickets and one of these tickets is chosen at random ("by lot") to determine the winner of a prize. Lotteries have been used to raise money for churches and schools, for local and state governments, and for private profit. In the 1800s the use of lotteries to make money from the unwary was raised to a virtual art form by showmen like P. T. Barnum (who once remarked, "There's a sucker born every minute.") Because of the widespread abuses of lottery schemes, private and public lotteries became illegal in this country by 1890. In recent years some states, starting with New Hampshire, have reinstated public lotteries to raise revenue.

One of the most common lottery-type marketing techniques is the giveaway game, the awarding of prizes to a few people chosen by lot from among a much larger audience of potential consumers. Insofar as people do not have to pay (directly) for the chance to win, this game is not a lottery in the strict sense, but it operates on the same principle. In the rest of this section we shall examine a fictional, but typical, lottery-type giveaway situation, focusing on the balance between monetary cost and psychological attractiveness.

The Situation

An automobile manufacturer was bringing out a new type of light truck. The marketing department had targeted a preferred mailing list of 100,000 people as good prospective buyers, provided they could be induced to test-drive the trucks. The company allocated $150,000 of their advertising budget for a direct-mail pitch to these people. $100,000 of that money was needed to cover the direct costs of preparation, printing, and mailing.

The Problem

How could the other $50,000 best be used to induce these people to test-drive the trucks?

After considerable discussion, it was decided that a giveaway scheme should be set up. Each person would be mailed a certificate to be turned in at a local dealership for a test drive and some sort of "reward." However, with only $50,000 available to reward 100,000 people, equal distribution would not provide much of an incentive to visit the dealers because $50,000 ÷ 100,000 is only 50 cents per person. Not many people would spend an hour or so of time and travel for 50 cents.

On the other hand, the entire $50,000 could be offered as a single prize, to be awarded by random drawing from all the certificates turned in. The chance of winning $50,000 should be a much more attractive reward! An objection was raised: Consumer-protection regulations require the mailing to state clearly the chances of winning, which in this case would be only 1 in 100,000 (assuming every certificate is returned). This low probability might well discourage a lot of people, who would feel that the odds of winning were too poor to bother with.

The scheme finally chosen struck a balance between high-payoff possibilities and attractive odds. Different prize values were set:

1st prize:	$10,000	—	1	to be awarded	=	$10,000
2nd prize:	$1000	—	5	to be awarded	=	5,000
3rd prize:	$50	—	200	to be awarded	=	10,000
4th prize:	$5	—	5000	to be awarded	=	25,000
Totals:			5206	prizes		$50,000

The advertising copy could then state truthfully:

> Come in for a test drive. Return this certificate and be eligible to **WIN** one of **MORE THAN 5200 CASH PRIZES**, up to a **GRAND PRIZE OF $10,000!**
>
> Chances of winning better than 1 in 20
> (No prize less than $5)

From a mathematical viewpoint, two aspects of this $50,000 giveaway story are worth a closer look. To gain some precision in our analysis, let us add one more useful, common-sense definition to our store of probability ideas. In situations like this one, where the various outcomes are uncertain and each one results in a monetary payoff if it occurs (even though the payoff sometimes may be $0), the **mathematical expectation of an outcome** is defined to be the payoff for that outcome multiplied by its probability. [Notice that, because the probability is always a number less than or equal to 1, this product *reduces* the amount of money "expected" in accordance with the likelihood that the particular outcome will actually occur.] The **mathematical expectation** of the whole situation (game, lottery, or whatever) is just the sum of the mathematical expectations of all the different outcomes. Let us apply this idea to the auto manufacturer's ad campaign just described.

First of all, although the three proposed ways of distributing the money appear to be quite different, *their mathematical expectations are the same*:

- 50 cents to everyone — The probability of winning is 1 because everyone who turns in a certificate gets the 50 cents. So the mathematical expectation is

$$(50 \text{ cents}) \cdot 1 = 50 \text{ cents}$$

- $50,000 to one winner drawn at random — The probability of winning is $\frac{1}{100,000}$, so the mathematical expectation is

$$\$50,000 \cdot \tfrac{1}{100,000} = 50 \text{ cents}$$

- Multiple-prize scheme — The mathematical expectation is

$$\$10,000 \cdot \tfrac{1}{100,000} + \$1000 \cdot \tfrac{5}{100,000} + \$50 \cdot \tfrac{200}{100,000} + \$5 \cdot \tfrac{5000}{100,000}$$
$$= \ \$.10 + \$.05 + \$.10 + \$.25 \ = \ 50 \text{ cents}$$

Secondly, the probability of winning a prize in the multiple-prize scheme is, of course, the total number of prizes being awarded divided by the total number of certificates, in this case

$$\tfrac{5206}{100,000}, \text{ about } 5.2\%$$

Now, $\frac{1}{20} = 5\%$, so the ad copy's assessment of chances is accurate. However, the probability of winning the Grand Prize is still only $\frac{1}{100,000}$, or .0001%, and the probability of winning more than \$50 is only $\frac{6}{100,000}$, or .0006%. Thus, the mathematical expectation is made to appear better than it actually is.

This persuasive technique is now used in almost every lottery or giveaway promotion. For instance, the Connecticut Weekly Lottery has listed a Grand Prize of \$100,000 and other prizes of \$2000, \$500, \$50 and \$5; every ticket states: "Overall chance of winning: 1 in 25." A recent fast-food company giveaway listed prizes from thousands of dollars to a bag of french fries, and claimed that the overall chance of winning was 1 in 91. The lottery approach to marketing (and taxation) is not dishonest, but a correct understanding of it requires a grasp of the probabilistic principles upon which it is based. The lack of this understanding increases the profit margins of lotteries and enhances the effectiveness of marketing giveaway schemes. There's nothing wrong with trying to "beat the odds," but you really ought to know what odds you're up against before you try.

EXERCISES 4.8

WRITING EXERCISES

1. Find a lottery or marketing giveaway scheme currently active in your locality and determine its mathematical expectation. Discuss how its advertising statements help or hinder prospective participants understand their chances of winning.

2. Invent a giveaway story that is analogous to the truck advertising scheme described in this section. Use a different setting and different numbers, but illustrate all the same mathematical ideas.

4.9   What Is Statistics?

Statistics is a word that is used in a wide variety of senses, and often is involved in attempts to lend credibility to otherwise doubtful opinions. The word itself has two meanings, one plural and one singular. In the plural sense it denotes collections of facts that can be stated numerically, such as mortality tables or census figures. In the singular, it is the science that deals with such facts, collecting and classifying them in such a way that general predictions may be based on them. This second meaning is the one that concerns us here.

Probability and statistics are closely related fields because they investigate the same basic situation from two different viewpoints. Both approaches deal with a large collection of things and a subset of that collection. In terms of the vocabulary we have just developed, probability begins with a *sample space*, a set whose elements are known, and evaluates the likelihood that some outcome of an *event*, a subset of the sample space, will occur in an experiment. Statistics begins with a **sample** — a known subset of a larger set that is mostly unknown — and by an analysis of that sample attempts to infer the composition of the larger set, which is called the *population*.

Because a sample is just a set of numerical facts, there is no reason to expect the elements of the set to exhibit any convenient relationship to each other, except perhaps that they are all derived from a common investigation. Consequently, the data must be organized and analyzed to yield a concise description of the sample as a whole. This is the job of **descriptive statistics**, the subject of the present section. The two key questions in this regard are:

What single number can best be used to characterize the sample?

and

How are the elements of the sample spread about this number?

The first question asks for a way of indicating some sort of "center" for a sample, and any number used in this way is called a measure of the **central tendency** of the sample. A number that describes the way a sample is spread about some central value is said to measure the **dispersion** of the sample.

Of the three common measures of central tendency we shall consider, the first is probably the most familiar. It is the process of finding an average.

DEFINITION   Let A be a sample with elements represented by the numbers $x_1, x_2, \ldots, x_n$. The (**arithmetic**) **mean** of A, denoted by $\bar{x}$, is

$$\bar{x} = \frac{x_1 + x_2 + \ldots + x_n}{n}$$

The numbers representing the elements of a sample may not all be distinct. For example, if the sample deals with exam results for a class of students, it usually happens that several students get the same score. It is sometimes convenient to deal with the distinct numbers of a sample and multiply each one by the number of times it occurs.

DEFINITION   The number of elements of a sample represented by a particular number is called the **frequency** of that number.

Note

The following five examples are continued throughout the section. They are identified by the numbering given here, with letters added to distinguish the various parts.

Example 4.39
(a)

A class of students scored as follows on a short test:

$$50, 60, 60, 65, 70, 70, 70, 75, 80, 80$$

The sum of these ten scores is 680, so the mean score is 68. The frequency of 70 is 3; 60 and 80 have frequency 2; 50, 65, and 75 have frequency 1. □

Example 4.40
(a)

Another class of ten students took the same test and scored

$$30, 40, 50, 65, 70, 75, 75, 80, 95, 100$$

The sum of these scores is also 680, so the mean score is again 68. The frequency of 75 is 2; all the other scores have frequency 1.   □

Example 4.41
(a)

A survey is taken of the net incomes of 25 families. The results, rounded to the nearest thousand, are:

$10,000	—	2 families
$12,000	—	6 families
$14,000	—	2 families
$16,000	—	5 families
$18,000	—	5 families
$20,000	—	3 families
$80,000	—	2 families

The frequency of each income is the number of families that have it, and the mean income may be found by computing as follows:

$$
\begin{array}{rcll}
2 \times \$10,000 & = & \$\ 20,000 \\
6 \times 12,000 & = & 72,000 \\
2 \times 14,000 & = & 28,000 \\
5 \times 16,000 & = & 80,000 \\
5 \times 18,000 & = & 90,000 \\
3 \times 20,000 & = & 60,000 \\
2 \times 80,000 & = & 160,000 \\
\hline
25 & & \$510,000
\end{array}
$$

Mean income = $\$510,000 \div 25 = \$20,400$ ☐

Example 4.42 (a) The same survey as explained in Example 4.41a was made in a small housing development, and it was discovered that:

5 families had net incomes of $10,000;
15 were at the $20,000 level;
the remaining 5 made $30,000.

Clearly, the mean income is $20,000, and the frequencies of the three different income levels are 5, 15, and 5, respectively. ☐

Example 4.43 (a) The improvement of a certain stretch of road was being considered, so the Highway Commission decided to determine the amount of traffic that used the road. It made a daily count of the number of cars using the road during a 30-day test period. The results of the survey, rounded to the nearest hundred, were:

Number of days	Number of cars per day	Frequency times number per day
4	1100	4,400
6	1600	9,600
3	1900	5,700
2	2000	4,000
6	2300	13,800
6	2500	15,000
3	2900	8,700
30		61,200 cars

Mean cars per day = $61,200 \div 30 = 2040$ ☐

Although the mean of a sample is the most common measure of central tendency, and often the most reliable, its use may obscure certain characteristics of the sample that relate to central tendency, as shown by Examples 4.39a and 4.40a. Both classes contain ten students, and the mean test score of both classes is 68. Yet even a cursory glance at the individual scores reveals a vast difference between the two classes. A similar situation exists in Examples 4.41a and 4.42a; although the mean incomes of the groups differ very little, there are many differences between the central tendencies of the two samples.

Other measures of central tendency help to describe these differences. One such measure singles out the "middle number" of a sample arranged in size order. Obviously, this number is unique only if there is an odd number of elements in the sample. A sample containing an even number of elements has two "middle numbers"; the mean of those two numbers is used in that case. This "middle number" is called the **median** of the sample.

Example 4.39(b)
Because the scores are already arranged in size order and there are ten elements in the sample, the median of the sample is the mean of the fifth and sixth scores in the order given. Both of these scores are 70, so the median is 70.   □

Example 4.40(b)
The fifth and sixth scores in this sample are 70 and 75, so the median is 72.5.   □

Example 4.41(b)
If the 25 incomes are arranged in size order, the thirteenth (the "middle") one is $16,000. This is the median.   □

Example 4.42(b)
Again, the median income is the thirteenth one in size order, which is $20,000.   □

Example 4.43(b)
Because there are 30 elements in this sample, the median is the mean of the fifteenth and sixteenth numbers in size order. These are 2000 and 2300, so the median number of cars per day is 2150.   □

It is often helpful to know the most common numerical value in a sample. However, there may be several values that occur at least as many times as any other, implying that this type of measure may not yield a unique result.

DEFINITION   A **mode** of a sample A is an element of A whose frequency is greater than or equal to the frequency of every other element of A.

Example 4.39(c)
The frequency of 70 is greater than the frequency of any other score, so 70 is the (only) mode of this sample.   □

Example 4.40(c)
Similarly, 75 is the mode of this set of scores because it appears twice and all the others appear only once.   □

Example 4.41(c)
The mode of this income sample is $12,000 because more families (6) have this income level.   □

Example 4.42(c)
The mode is $20,000. Notice that, even though the mean incomes in Examples 4.42 and 4.43 differ by only $400, the medians differ by $4000 and the modes differ by $8000.   □

Example 4.43(c)
There are three values whose frequencies are greater than or equal to the frequency of any other element in the sample; thus, this sample has three modes — 1600, 2300, and 2500.   □

The primary value of the median and the mode as measures of central tendency is that neither one is greatly influenced by occasional extreme results. For instance, if we were to delete the two $80,000 incomes from Example 4.41, the mean would then be approximately $15,200, a change of $5200, but the median and the mode would remain the same. This characteristic of the median and the mode is especially useful when the available information indicates that a sample contains a few extreme results, but there is no way of finding their numerical value. In such a case it is impossible to compute the mean, but the median and the mode can still be found.

The information supplied by the measures of central tendency alone does not always describe a sample adequately. For instance, the mean, median, and mode of the income sample in Example 4.42 are all $20,000. If we had a sample consisting of 25 net incomes of $20,000, again the mean, median, and mode all would be $20,000, but these two samples obviously are not the same. Still another sample with the same measures of central tendency is one containing twenty-three $20,000 incomes, one $1000 income, and one $39,000 income.

In order to distinguish among these samples or any others whose central tendencies are similar, we must consider their dispersion. In other words, we must find ways to reflect numerically the way elements of a sample are distributed around some central point. The simplest and most natural measurement of this kind gives the maximum difference between elements.

DEFINITION The **range** of a sample is the difference between the largest and smallest numbers of the sample.

Examples 4.39(d)–4.43(d)

In Example 39, the range of scores is $80 - 50 = 30$.

In Example 40, the range of scores is $100 - 30 = 70$.

In Example 41, the income range is $\$80,000 - \$10,000 = \$70,000$.

In Example 42, the income range is $\$30,000 - \$10,000 = \$20,000$.

In Example 43, the range of cars per day is $2900 - 1100 = 1800$.   □

The range is not a very precise measure of the dispersion of a sample because it is the same for any two samples with the same extreme values, regardless of how many elements take on those values and where the rest of the numbers lie. For instance, the samples

$$1, 1, 1, 5, 5, 5 \quad \text{and} \quad 1, 2, 3, 3, 4, 5$$

have the same range, even though most elements of the latter one are closer to the mean. We might try to describe this situation more exactly by considering the average of the individual differences between the mean and the elements of the sample. Thus, if we define the **deviation** of an element x of a sample whose mean is $\bar{x}$ to be the difference $x - \bar{x}$, then we could compute the average of all the individual deviations. Unfortunately, because the mean of a sample is actually a sort of "balance point" for the elements, the sum of the individual deviations is always zero! Hence, this "average difference" approach must be modified somehow if it is to provide a meaningful description of the distribution.

As you might have guessed already, the average of the deviations of a sample becomes useful if all deviations are considered as distances from the mean, without regard to the directions indicated by their signs. This is done by using absolute values.

DEFINITION Let A be a sample with elements represented by the numbers $x_1, x_2, \ldots, x_n$, and let $\bar{x}$ denote the mean of A. The **mean deviation** of A is

$$\frac{|x_1 - \bar{x}| + |x_2 - \bar{x}| + \ldots + |x_n - \bar{x}|}{n}$$

Examples 4.39(e)–4.40(e)

In both of these cases, $\bar{x}$ (the mean test score) is 68. In Example 4.39, the mean deviation is

$$\frac{|50 - 68| + 2|60 - 68| + |65 - 68| + 3|70 - 68| + |75 - 68| + 2|80 - 68|}{10}$$

$$= \frac{18 + 2 \cdot 8 + 3 + 3 \cdot 2 + 7 + 2 \cdot 12}{10} = \frac{74}{10} = 7.4$$

By a similar computation, the mean deviation for Example 4.40 is 17.4. This larger mean deviation indicates that the scores of second sample are more widely scattered about the mean than those of the first sample are. □

Examples 4.41(e)–4.42(e)

The mean deviations in these two cases are computed in the same way: Multiply the absolute value of each individual deviation by its frequency, add, and divide by the total number of elements (the sum of the frequencies). Thus, for Example 4.41, we have:

freq.		$|x - \bar{x}|$		
2	×	$ 10,400	=	$ 20,800
6	×	8400	=	50,400
2	×	6400	=	12,800
5	×	4400	=	22,000
5	×	2400	=	12,000
3	×	400	=	1,200
2	×	59,600	=	119,200
25				$ 238,400

$$\text{Mean deviation} = \frac{\$238,400}{25} = \$9536$$

The mean deviation of the second income sample is much easier to compute:

$$\frac{5 \cdot \$10,000 + 15 \cdot \$0 + 5 \cdot \$10,000}{25} = \$4000$$

This smaller mean deviation reflects the fact that the second set of incomes is bunched more closely about the mean. □

Example 4.43(e)

You can test your understanding of these ideas by computing the mean deviation for this example. See if you end up with an answer of 460 cars per day. □

The mean deviation describes the dispersion of a sample simply and accurately, but the use of absolute value makes it unwieldy for algebraic manipulation because of the sign changes required. The "either-or" decision required by absolute value is especially annoying if the computation is being done with the aid of a calculator or computer. An alternative way of making the individual deviations positive is by squaring them. This leads to another measure of dispersion that is as descriptive as the mean deviation, but easier to handle mechanically.

DEFINITION   Let A be a sample with elements represented by the numbers $x_1, x_2, \ldots, x_n$, and let $\bar{x}$ denote the mean of A. The **variance** of A is

$$\frac{(x_1 - \bar{x})^2 + (x_2 - \bar{x})^2 + \ldots + (x_n - \bar{x})^2}{n - 1}$$

The **standard deviation** of the sample A is the (nonnegative) square root of its variance.

Note
If the denominator in its definition were n, the variance would be the mean of the squares of the n individual deviations. However, it has been shown that dividing by n leads to a consistent underestimation relative to the population as a whole, which is corrected if we divide by $n - 1$, instead.

NOTATION   The standard deviation of a sample is commonly denoted by s. Thus, the variance of a sample is s^2.

Examples 4.39(f)–4.40(f)
The variance of the first set of scores is

$$\frac{(-18)^2 + 2 \cdot (-8)^2 + (-3)^2 + 3 \cdot 2^2 + 7^2 + 2 \cdot 12^2}{9} = \frac{810}{9} = 90$$

The standard deviation is $\sqrt{90} \approx 9.5$.[7]   For the second set of scores, a similar computation yields a variance of 506.67 and a standard deviation of approximately 22.5. The difference between these two standard deviations provides a clear measure of the difference in dispersion between the two sets of scores.   □

Examples 4.41(f)–4.42(f)
Here, too, the standard deviations provide a clear indication of the difference in the distributions of incomes in these two examples. In the first case, the variance, s^2, is 331,333,333, so the standard deviation, s, is \$18,203. In the second case, where the incomes are bunched more closely around the mean,

---

[7]The symbol $\approx$ means "approximately equal" in this context.

the variance and standard deviation are much smaller: $s^2 = 41,666,667$ and $s = \$6455$. □

Example 4.43(f)
Test your understanding of the computational process here by computing the variance and standard deviation for the number of cars per day in this example. You should get $s^2 = 298,345$ and $s = 546$. □

Besides being more suitable for algebraic work, the standard deviation has a surprising property. To describe it, we need to anticipate a data-distribution concept called "normal distribution" that will be explained more fully in the next section. For now, it will suffice to illustrate the intuitive idea by a typical example.

Suppose you need to measure in centimeters the length of your school's football field. One way to get a reliable figure would be to get each student in the school to measure the field and take the average (the mean) of all those measurements. Now, you would expect that, under "normal" circumstances (no difference in the tape measures being used, no attempt to falsify results, etc.), the individual measurements would be clustered around the mean, with relatively few of them really far off. In fact, if the frequencies of the individual measurements were plotted on a graph, they would look like a so-called "bell-shaped curve," highest at or near the mean and dropping off rapidly and symmetrically as you move away from the mean in either direction. Data clustered around its mean like this is said to be "normally distributed." (See the end of Section 4.10 for a discussion of *normal distribution.*)

Now, if s is the standard deviation of a normally distributed collection of numerical data:

- 68% (approximately $\frac{2}{3}$) of the numbers in the collection have deviations between $+s$ and $-s$;

- 95% of the numbers have deviations between $+2s$ and $-2s$;

- 99.7% of the numbers have deviations between $+3s$ and $-3s$.

Thus, the standard deviation of a set of data provides a precise, efficient way of describing this idea of "clustering" around the mean for all normal distributions. In fact, the term "*standard* deviation" is suggested by these facts, in that all normal distributions can be standardized by using their standard deviations as a common unit of measure. In the next two sections we shall see how this property of the standard deviation makes it a powerful tool for predicting the reliability of statistical results.

EXERCISES 4.9

In Exercises 1–6, find the mean, median, mode(s), range, mean deviation, variance, and standard deviation of each sample. (Round all answers to one decimal place.)

1. Sample: 3, 1, 5, 10, 8, 1, 5, 2, 1

2. Sample: 13, 10, 2, −5, 7, 10, −2

3. Sample: 3.9, 8.05, 5.2, .11, 10.6, 2.0

4. A class of twelve students scored as follows on an exam:

    75, 60, 60, 72, 70, 86, 93, 54, 72, 50, 72, 85

5. The daily totals of shoppers at a market during a certain week are:

    Mon. 700   Tue. 450   Wed. 630
    Thu. 520   Fri. 810   Sat. 1050

6. A college hockey team achieved these scores during a fifteen-game season:

    2, 0, 3, 3, 1, 5, 4, 7, 0, 2, 3, 12, 6, 0, 1

7. (a) Compute the mean and all the individual deviations for the sample:

    1, 3, 7, 4, 5, 4, 9, 1, 2

   (b) Verify that the sum of all the deviations in Part (a) is 0.

8. Prove that the sum of all the individual deviations of *any* sample must be 0.

9. Construct three samples with the same range but with different standard deviations.

10. Construct two seven-element samples with a mean of 5 and a range of 10, but such that the standard deviation of the first sample is at least 1.5 more than the standard deviation of the second.

11. A restaurant makes a two-week survey of the number of customers ordering fried chicken so that it may plan its meat orders. The daily totals are:

    56, 72, 65, 73, 35, 96, 104, 58, 68,
    70, 62, 41, 82, 99

Find the mean, range, and standard deviation of this sample.

12. A school system is recruiting teachers. As part of its publicity it has issued a statement of teacher salaries it pays, each rounded to the nearest thousand, as shown in Table 4.1. Compute all measures of central tendency that are possible for this sample. If you were considering teaching in this school system, which single measure of central tendency would tell you the most about your prospects there, and why?

Number of teachers	Salary
2	$ 19,000
3	20,000
5	21,000
5	22,000
8	23,000
13	24,000
7	25,000
4	27,000
1	30,000
6	above 30,000

Table 4.1   Teachers' salaries rounded to the nearest thousand.

13. A census-taker surveys a (very peculiar) neighborhood and finds only two income levels: 20 families make $100,000 a year and 19 families make $10,000 a year.

    (a) What are the mean, median, and mode of this sample?

    (b) Rechecking his data, the census-taker finds that he made a tallying error. The true figures show that 19 families make $100,000 a year and 20 families make $10,000. Compare the mean, median, and mode of this sample with the figures you found in Part (a). Comment.

WRITING EXERCISES

1. Discuss the relative advantages and disadvantages of using the mean or the median as the primary measure of central tendency of a sample. Describe a situation in which the mean would be better than the median; then describe a situation in which the median would be better than the mean. Explain what "better" means in each case.

2. (a) Write the definition of *variance* of a sample completely in words. Do not use any symbols, not even letters to represent numbers.

(b) Comment in general on the role of notation in mathematics. In particular:

   i. Is mathematical notation a useful "shorthand" for expressing quantitative ideas?

   ii. Is mathematical notation *more than* just a shorthand? Does it play any other role in the subject? If so, what?

   iii. What are the drawbacks to using notation?

4.10   Distributions

As one passes from the purely descriptive part of statistics to the uncertainties of making general predictions from a sample, the mathematical waters suddenly deepen. The background required to discuss predictive statistics rigorously is extensive, and any pretense of completeness in a treatment as brief as this section and the next would be far less than honest. We therefore limit our exploration to a general outline of some fundamentals of **statistical inference**, methods by which the characteristics of a sample are used to obtain a description of the population it represents.

The fundamental idea is very simple: If the size of the population (or some other factor) makes it impossible to examine all of its members, then we select a sample (a subset of those members) and, by analyzing the sample, try to infer something about the entire population. For instance:

- A newspaper that wants to know how all the registered voters in the country are going to vote in an upcoming election will poll a relatively small sample of voters and use the results to predict how the election will turn out.

- A maker of shotgun shells who wants to check that the shells will fire properly can only afford to test relatively few of them, since each one tested is one less that can be sold.

- A meteorologist who wants to know the average snowfall in a region during a storm cannot measure the total amount of snow that falls and divide by the area it covers. Instead, samples taken at various spots around the region are used to infer the average for the region as a whole. (What is the population in this case?)

It would be ideal if we could say that the characteristics of any sample exactly reflect the characteristics of the population from which it was drawn. Unfortunately, this is hardly true. We know from our study of probability that, even in the case of a known population (sample space), the samples drawn from it (the results of an experiment) are not always the same. To get truly reliable information about entire populations, we must interpret the results of samples very carefully in relation to the limitations of the sampling process. Our study of this process begins with *distributions*.

Informally, the idea is this: Suppose we are interested in some characteristic of a large population that can be measured by a single number — the gas mileage of each automobile, the rainfall at each place in the world on a particular day, the height of every living person, the number of copies of the *New York Times* sold at each newsstand in New York City yesterday, etc. Such a numerical characteristic is called a **variable** (because it can vary from member to member in the population). The number associated with a particular member of the population is called the **value** of the variable for that member.[8] Sometimes the values of a variable can be found explicitly; we can measure someone's height, for example. Sometimes, however, the values cannot be determined. For instance, the amount of rainfall in your home town a year from today, which will be a specific number *then*, cannot be stated with certainty *now*. A variable whose numerical values cannot be determined with certainty is called a **random variable**.[9]

Whether or not a variable is random, its values over a large population are hard to comprehend if they are just viewed as a big collection of numbers. Thus, it is natural to seek some pattern that describes in summary form how the numerical values are spread out; such a pattern is called the **distribution** of the variable. A distribution pattern may be represented visually by a graph or algebraically by a formula; the most important kinds of distributions can be represented both ways, as we shall see. While an algebraic formula is essential for drawing numerical conclusions about the underlying population, a graphical picture is often easier to understand intuitively. The most common type of graph used to represent distributions is the **histogram**, a form of bar graph, such as the ones shown in Figures 4.7 – 4.9. A histogram can be described as follows:

---

[8] In other mathematical words, a variable is a *function* from the population to some set of numbers, and its value for a particular element of the population is the *image* of that element.

[9] See the beginning of Section 4.3 for a discussion of randomness.

- The possible values of the variable are grouped into a finite number of adjacent intervals on the horizontal axis. If there are only finitely many different possible values, each is assigned an interval, all of equal length.

- The number of times a member of the population is assigned a value in a particular interval (the *frequency* of that value interval) is represented by the height of a vertical bar over the interval. The frequencies are measured according to the scale on the vertical axis.

If this general description sounds confusing, put it aside until you read the next two examples; it should become much clearer then.

Example 4.44   As you may know, a student's "QPR" (quality-point ratio) is often computed as follows: Assign the numbers 4, 3, 2, 1, 0 to the course grades A, B, C, D, F, respectively; multiply each course-grade number by the number of credit hours for that course; divide by the total number of credit hours the student has taken. Thus, the QPR is a variable whose value for each student is a number between 0.0 and 4.0. The QPR's for the 185 students in the oceanography program at Death Valley College during the decade from 1980 to 1990 were distributed as follows:

$$0.0 \leq QPR < 0.99 \quad 25 \text{ students}$$
$$1.0 \leq QPR < 1.99 \quad 40 \text{ students}$$
$$2.0 \leq QPR < 2.99 \quad 75 \text{ students}$$
$$3.0 \leq QPR \leq 4.0 \quad 45 \text{ students}$$

This distribution is represented by the histogram in Figure 4.7.   □

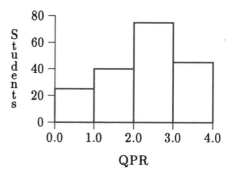

Figure 4.7   A QPR Distribution.

Example 4.45   Ten slips of paper, numbered 1 through 10, are placed in a hat, and then a number is drawn at random. The histograms in Figures 4.8 and 4.9 show the

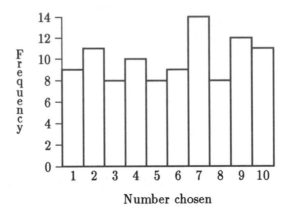

Figure 4.8   100 repetitions.

frequencies of each number drawn in 100 repetitions and in 1000 repetitions of this experiment, respectively.[10]   □

In Example 4.44, the variable (QPR) is not random; The QPR of each student (member of the population) is a specific number that is known at least to the Registrar of the College. In Example 4.45, however, the uncertain outcome of the experiment makes the variable (the number to be drawn) random, at least before the drawings occur. Each repetition of the drawing yields a value of the random variable. Of course, after all the drawings have occurred, each of the 100 (or 1000) repetitions has its assigned

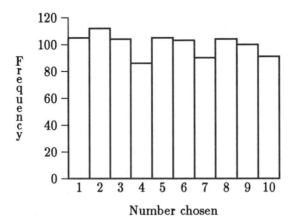

Figure 4.9   1000 repetitions.

[10]These experiments were simulated by using a random-number-generating computer program.

number, so these outcomes are no longer random. Hence, these results can be represented by particular histograms.

However, if we think about these 100 or 1000 repetitions as being *samples* of the larger population of all possible repetitions of this experiment, then we are still dealing with a random variable. Theoretically, this variable assigns each possible drawing a whole number between 1 and 10 (inclusive), but we cannot know with certainty each one of all these possible outcomes. If asked to predict (or bet on) the outcome of the next drawing, what would your common sense say? Think about it for a moment, then see if the next few sentences come close to "reading your mind."

> Since there are ten numbers in the hat, there is only 1 chance in 10 that any particular number will be drawn. Hence, no matter what I say, I have a 90% chance of being wrong. That's a pretty risky bet!

In other words, our common-sense understanding of the "law of averages" would tell us that the expectation of the outcome of this experiment is closely tied to the probability of choosing each number, which (assuming all the slips of paper are the same size, are thoroughly mixed, etc.) is $\frac{1}{10}$. "In the long run," you might say, "each number should turn up about $\frac{1}{10}$ of the time." If you were to guess "7" over and over again for many drawings, for instance, you would expect to be right about 10% of the time.

The mathematical counterpart of this common-sense idea is Bernoulli's **Law of Large Numbers.** Roughly speaking, this law says that the more times an experiment is repeated, the closer the ratio of successes to total trials will approximate the probability of success on a single trial. Applied to statistical sampling, this law essentially says:

> The larger the sample taken, the closer the distribution of the sample will approximate the distribution of the population.

(Of course, the statement of the Law of Large Numbers may be made much more precise than this, but the mathematics required would take us far afield.[11]) In the case of drawing the slips of paper from the hat, this law tells us that larger and larger numbers of repetitions of this experiment should be distributed more and more uniformly, with each of the ten values tending toward a frequency that is $\frac{1}{10}$ of the total number of trials. Figures 4.8 – 4.10 illustrate this idea. (A frequency value expressed as a fraction or percentage of the total number of trials is called a **relative frequency.**)

---

[11]A readable discussion of the Law of Large Numbers appears in Chapter XI of [8] in the list *For Further Reading* at the end of this chapter.

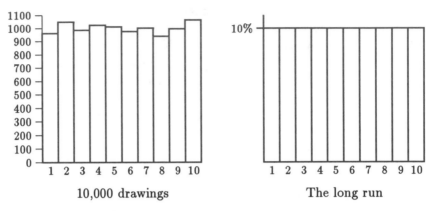

10,000 drawings   The long run

Figure 4.10 *Repeated drawings of numbered slips.*

The validity of the Law of Large Numbers depends to a great extent upon the condition that the sample be chosen in a random manner. In other words, the elements must be selected without regard to any predetermined pattern, in such a way that each element of the population has an equal chance of being chosen. This type of sample is naturally called a **random sample**; these are the samples governed by the laws of probability and statistics. *From now on, any sample discussed is assumed to be random.*

Generalizing from a sample involves two questions:

> What statements can be made about the characteristics of the population as a whole?

and

> How reliable are these statements?

The key to answering both of these questions may be found by taking another look at distributions. In Examples 4.44 and 4.45, we described distributions in terms of the frequencies of the numerical values assigned to the underlying population. Such distributions are called, naturally enough, **frequency distributions**. However, when we extend the number-drawing experiment of Example 4.45 to a consideration of *all possible* outcomes, the distribution can no longer be a frequency distribution because the actual numbers of outcomes of each kind cannot be determined. Instead, with the help of the Law of Large Numbers, we arrived at a uniform distribution based on the *probability* of each possible outcome, as illustrated by the second graph in Figure 4.10. Such a distribution is called (You guessed it!) a **probability distribution**.

Example 4.46   Here is a probability distribution that is not uniform (and hence is more interesting). Suppose we toss a fair coin five times and record the total number of *heads*. There are 32 equally likely possible outcomes of this experiment (Why?), and they may be classified conveniently in terms of six mutually exclusive events:

0 *heads*, 1 *head*, 2 *heads*, 3 *heads*, 4 *heads*, and 5 *heads*

By counting the outcomes in the various events, we can compute their respective probabilities (as in Section 4.3):

$$\frac{1}{32}, \quad \frac{5}{32}, \quad \frac{10}{32}, \quad \frac{10}{32}, \quad \frac{5}{32}, \quad \text{and} \quad \frac{1}{32}$$

These six events and their probabilities constitute a probability distribution for the set of all possible five-toss outcomes (the underlying population in this case). The "$n = 5$" histogram in Figure 4.11 illustrates this distribution.

In general, if the experiment consists of tossing a coin n times, the outcomes can be stated in terms of obtaining exactly r heads, where r can be any integer from 0 to n, inclusive. The corresponding probabilities will be

$$_nC_r \cdot \left(\frac{1}{2}\right)^n$$

for each r from 0 to n, inclusive. (The factor $(\frac{1}{2})^n$ is constant for an experiment of n trials; it accounts for the total number of elements in the sample space.) The $n+1$ mutually exclusive events and their corresponding probabilities form a probability distribution for the set of all possible n-toss outcomes, which is the underlying population. Figure 4.11 also contains the histograms for the $n = 3$ and $n = 10$ distributions.   □

Example 4.46 typifies a **binomial distribution**, a probability distribution with an underlying experiment that can be analyzed in terms of two alternatives — *success* or *failure*, *yes* or *no*, *heads* or *tails*, etc. Binomial distributions are determined by the numbers

$$_nC_r \cdot p^r \cdot (1 - p)^{n-r}$$

which depend on the numbers $_nC_r$ for a particular probability p and some fixed number n of independent trials.[12] If either p or n is altered, the binomial distribution will also change. Distributions of this kind are useful because a wide variety of statistical problems can be structured in terms of two-alternative questions, and also because they have a relatively simple computational form. In Example 4.46, the fact that $p = (1 - p) = \frac{1}{2}$ makes the distributions for each n especially simple to calculate.

---

[12]Compare this with the statement of the Binomial Distribution Theorem, near the end of Section 4.7.

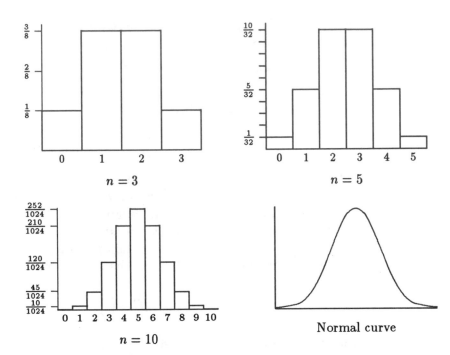

Figure 4.11 *Some binomial distributions and a related normal curve.*

The final picture in Figure 4.11 is a curve that approximates the binomial distribution for $p = \frac{1}{2}$ and large values of n. This curve is known as a **normal curve**; it represents a common and very important type of distribution for infinite populations. Although normal curves may be taller or shorter, wider or narrower than the one pictured here, each one is "bell-shaped" and is symmetric about the vertical line through its highest point, curving downward toward the bottom axis more and more steeply at first, then leveling out as it approaches the axis on either side. The shape of the curve represents the distribution of the population, in much the same way as the shape of a histogram does. For large populations, you might visualize lots of thin relative-frequency rectangles under the curve, generalizing the pictures in Figure 4.11. This description can be made much more precise if we relate each normal curve to the numerical properties of the distribution it represents.

The two keys to describing a **normal distribution** — a distribution described by a normal curve — are the mean and the standard deviation of the values of the random variable on the population being studied. We saw in Section 4.9 how to compute these two numbers for relatively small sets of numbers (samples). The ideas are exactly the same here, except that the population may be very large, so it usually is not possible actually to compute these numbers.[13]   Nevertheless, the numbers exist in theory, at least, and their properties are fundamental to most of statistics. It is traditional to denote this mean and this standard deviation by the lower-case Greek letters μ ("mu") and σ ("sigma"), respectively. (These are the Greek counterparts to our letters m and s.) To distinguish them from the mean and standard deviation of a particular sample, μ and σ are usually called the **population mean** and **population standard deviation**, respectively.

An important fact about a normal curve is that *it is completely determined by the numbers μ and σ*. Pictorially, the highest point of the curve occurs at μ and the curvature changes from becoming steeper to becoming shallower (that is, from *concave downward* to *concave upward*) at the points exactly σ distance away from μ on either side. Figure 4.12 shows two normal curves with different means (μ_1 and μ_2), but with the same standard deviation (σ); Figure 4.13 shows two normal curves with the same mean (μ), but with different standard deviations (σ_1 and σ_2).

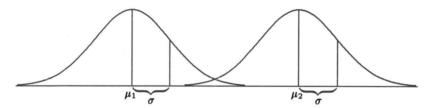

Figure 4.12 Two normal curves with different means, but with the same standard deviation.

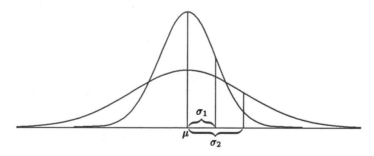

Figure 4.13 Two normal curves with the same mean, but with different standard deviations.

[13]Section 4.11 discusses how the mean and standard deviation of a large population may be deteremined by looking at samples.

Regardless of its mean μ and standard deviation σ, a normal curve has the following remarkable and useful properties:

- Although the curve extends infinitely far in both directions, the entire area between the curve and the horizontal axis is exactly 1;[14]
- about 68% of that area lies between the vertical lines at $\mu - \sigma$ and $\mu + \sigma$;
- about 95% of that area lies between the lines at $\mu - 2\sigma$ and $\mu + 2\sigma$; and
- about 99.7% of that area lies between the lines at $\mu - 3\sigma$ and $\mu + 3\sigma$.[15]

Thus, it is often convenient to "standardize" the graphing of normal distributions so that the mean lies at 0 on the horizontal axis and the horizontal unit of measure is σ. Figure 4.14 is a picture of the standard normal curve.

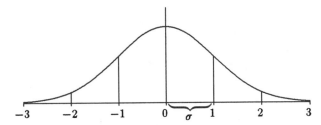

Figure 4.14   The standard normal curve.

Note

If you are curious about the precise formula that defines a normal curve, here it is:

$$f(x) = \frac{1}{\sigma\sqrt{2\pi}} \cdot e^{\frac{-(x-\mu)^2}{2\sigma^2}}$$

where e, which is approximately 2.718, is the base for natural logarithms. (Aagh!) Standardizing to $\mu = 0$ and $\sigma = 1$ gives us the somewhat less formidable expression

$$f(x) = \frac{1}{\sqrt{2\pi}} \cdot e^{\frac{-x^2}{2}}$$

---

[14]Rigorous explanations of how to compute the area under a normal curve and how the area under an infinite curve can be finite require the methods of integral calculus, which are beyond the scope of this book. For our purposes, it is enough to know that such computations can be done and that beyond some finite distance from μ in either horizontal direction, the area under the normal curve is negligible.

[15]Compare this with the end of Section 4.9.

but it, too, takes us beyond the scope of this book. Some calculus is required to derive the properties described above from these formulas. We shall not use either formula in the rest of this chapter; they are included just to satisfy your curiosity.

"So what?" we hear you say. "Why all this fuss about one particular kind of distribution?"

Consider this: The concept underlying all of probability and statistics is *randomness*; that is, the behavior of random phenomena is the cental theme of this area of mathematics. The fact is that

if the outcome of *any* random phenomenon is averaged over many repetitions, its distribution tends to be normal.

This powerful statement is a consequence of the Central Limit Theorem, which we shall discuss at the beginning of the next section. For now, suffice it to say that the normal distribution truly reflects the "normal" behavior of randomness.

EXERCISES 4.10

For Exercises 1 – 8, make a histogram to represent the given data.

1. A single die was thrown 100 times. The frequency that each number appeared is:
   1 – 14 times; 2 – 20 times; 3 – 16 times; 4 – 18 times; 5 – 13 times; 6 – 19 times

2. A single die was thrown 500 times. The frequency that each number appeared is:
   1 – 78 times; 2 – 90 times; 3 – 82 times; 4 – 80 times; 5 – 89 times; 6 – 81 times

3. A single die was thrown 1000 times. The frequency that each number appeared is:
   1 – 159 times; 2 – 165 times; 3 – 173 times; 4 – 164 times; 5 – 161 times; 6 – 178 times

4. A pair of dice was thrown 100 times. The frequency of the sums of numbers that appeared is:
   2 – 4 times; 3 – 6 times; 4 – 11 times; 5 – 8 times; 6 – 15 times; 7 – 13 times; 8 – 9 times; 9 – 14 times; 10 – 12 times; 11 – 8 times; 12 – 1 time.

5. A pair of dice was thrown 500 times. The frequency of the sums of numbers that appeared is:
   2 – 17 times; 3 – 27 times; 4 – 45 times; 5 – 62 times; 6 – 65 times; 7 – 84 times; 8 – 62 times; 9 – 51 times; 10 – 42 times; 11 – 32 times; 12 – 13 times.

6. A pair of dice was thrown 1000 times. The frequency of the sums of numbers that appeared is:
   2 – 31 times; 3 – 49 times; 4 – 84 times; 5 – 114 times; 6 – 141 times; 7 – 166 times; 8 – 138 times; 9 – 114 times; 10 – 87 times; 11 – 53 times; 12 – 23 times.

7. The grades of 30 students on a 10-point quiz are:

   8, 6, 9, 10, 0, 6, 7, 9, 8, 4, 7, 10, 8, 6, 2, 9, 9, 8, 5, 7, 8, 8, 2, 9, 6, 10, 4, 6, 8, 10

8. The grades of 30 students on a 100-point exam are:

   79, 81, 94, 76, 75, 87, 57, 55, 93, 62, 87, 86, 86, 91, 83, 78, 53, 91, 84, 98, 65, 82, 39, 67, 87, 81, 64, 80, 74, 66

   (*Hint*: You might want to group the grades in some way so that the visual representation is informative.)

9. Relate the data sets in Exercises 1 – 3 to the Law of Large Numbers.

10. Relate the data sets in Exercises 4 – 6 to the Law of Large Numbers. (*Hint*: What is the probability of the occurrence of each number? Why? See Example 4.26 in Section 4.6.)

11. Consider the experiment of tossing a fair coin four times, and suppose that we are interested in the number of *heads* that turn up.

    (a) Compute the probabilities of getting each possible number of heads.

    (b) Draw a histogram to represent the probabilities you computed in Part (a).

    (c) Take a coin and actually do this experiment 10 times; then draw a histogram to represent your results.

    (d) (For people with patience.) Repeat this experiment 20 more times, and draw a histogram to represent the results of these twenty trials.

    (e) Combining your data for Parts (c) and (d), draw a histogram representing the results of the thirty trials.

12. A computerized advertising display contains a row of 8 lights which are programmed to switch on and off as follows: At 2-second intervals, the computer assigns 0 or 1 to each light at random with a probability of $\frac{1}{2}$ for either choice. The lights assigned "1" go on (or stay on) and the lights assigned "0" go off (or stay off).

    (a) What is the probability that exactly 4 lights will be on at some particular time?

    (b) What is the probability that exactly 6 lights will be on at some particular time?

    (c) What is the probability that exactly 0 lights will be on at some particular time?

    (d) Describe the binomial distribution that represents the probabilities that exactly n lights will be on at any given time, for each possible n from 0 to 8.

    (e) Represent the distribution of Part (d) by a histogram.

    Suppose that the computer is reprogrammed to be twice as likely to assign a "1" as to assign a "0" to each of the lights.

    (f) What is the probability that exactly 4 lights will be on at some particular time?

    (g) What is the probability that exactly 6 lights will be on at some particular time?

    (h) What is the probability that exactly 0 lights will be on at some particular time?

    (i) Describe the binomial distribution that represents the probabilities that exactly n lights will be on at any given time, for each possible n from 0 to 8.

    (j) Represent the distribution of Part (i) by a histogram.

13. Open your local phone book to a page in the residential listings (usually the white pages), and choose two full columns of phone numbers on that page.

    (a) Make a histogram showing the distribution of the *last* digits of the phone numbers in the two columns. (This histogram should have ten bars.)

    (b) Make a histogram showing the distribution of the *first* digits of the phone numbers in the two columns.

    (c) Do the last digits appear to be randomly distributed? Do the first digits appear to be randomly distributed? Relate your answers to your histograms.

(d) If you were to make histograms for the first-digit and last-digit distributions of *all* the numbers in your phone book, what do you think their shapes would look like? Why?

14. (For students who have access to a computer with BASIC and can use it.) The BASIC program given at the end of this exercise will choose a random sample of five single-digit numbers, compute the mean of the sample, and print it. (*Note*: Each time the program is run, your computer will ask:

Random number seed (-32768 to 32767)?

Just choose a number in the given range and type it in.[16] Each time you run the program, choose a different number.)

(a) Run the program 30 times, and represent your results by a histogram.
(*Note*: If you are comfortable with programming in BASIC, feel free to modify the program so that it automatically repeats the process 30 times and records all 30 means.)

(b) Run the program 30 more times, and represent your results by a histogram.

(c) Combine your results from Parts (a) and (b), and represent them by a single histogram.

(d) If you think of these samples as being drawn from a very large population of single-digit numbers, with equally many of each digit, what do you think the population mean would be? Why? How well does your conjecture fit with your results in Parts (a)–(c)?

(e) Modify the program to handle ten-number samples, then repeat Parts (a) through (d).

```
10   RANDOMIZE
20   LET A = 0
30   FOR I = 1 TO 5
40   LET B = INT(RND*10)
50   A = A + B
60   NEXT I
70   PRINT A/5
80   END
```

WRITING EXERCISES

1. (*Note*: Situations used as examples in this section are "off limits" as answers to these questions.)

(a) Describe a situation in everyday life that is an example of a normal distribution of a random variable. Explain.

(b) Describe a situation in everyday life that is an example of a random-variable distribution which is *not* normal. Explain.

2. Is it appropriate to grade a class by using a normal distribution curve? What characteristics of a class might make this grading procedure more or less meaningful? Discuss.

3. (For students who are comfortable with programming in BASIC.)

(a) Write a computer program that will simulate tossing a coin 5 times and record the total number of heads.

(b) Modify your program from Part (a) so that it will simulate tossing a coin n times, where n is a number chosen and entered by the user, and will record the total number of heads.

(c) Write instructions for your program of Part (b) so that it can be used to generate data by someone who is a beginner in using computers.

---

[16]Some advanced forms of BASIC (such as True BASIC) will randomize without your help.

4.11   Generalization and Prediction

Because the idea of distribution applies both to sample spaces and to populations (known and unknown collections) it forms a natural link between probability and statistics. "Wait a minute," you say. "If we don't know the elements of a population, how do we know anything about its distribution?" That is a perceptive natural question at this point (We're glad you thought of it!) and its answer actually determines how statistics is applied in the real world. The key to the answer is **sampling** — selecting at random small samples (subsets) of the population and using the characteristics of these samples to infer the makeup of the population as a whole. Two major results, mentioned in Section 4.10, provide the theoretical foundation for sampling: The Law of Large Numbers and the Central Limit Theorem. It is time to take a closer look at how they work.

Roughly speaking, the Law of Large Numbers (as applied to sampling) says that, the larger the sample taken, the more likely it is that the characteristics of the sample will approximate the characteristics of the population from which it was drawn. For instance, suppose a big-city newspaper wants to predict the outcome of an upcoming election between two mayoral candidates, Abbott and Baker (whom we shall sometimes call A and B, for short) by polling registered voters in its circulation area. Common sense and the Law of Large Numbers tell us that asking 10 voters how they will vote is better than asking only one, and asking 100 voters is better than asking 10. In fact, the best sample would be the *entire* set of voters, but clearly the constraints of time and money make this option impossible; there are, say, 500,000 eligible voters! For the management of the newspaper, then, knowing that "bigger is better" is not sufficient; they also need to know "How large is large enough?" That is, when will the results of the poll they take approximate the actual voting preferences of the entire population within some reasonable margin of error, and how large is that margin of error? Answering these questions depends on the Central Limit Theorem.

To understand the main idea of the Central Limit Theorem, we must distinguish between two populations — the population $\mathcal{P}$ we are sampling and the population $\mathcal{S}$ of all possible random samples of a particular size.[17] Here is some notation to help keep the necessary distinctions straight:

- We shall use x as the random variable representing the property of $\mathcal{P}$ that interests us. In our election example, for instance, we can let x take on the value 1 when the voter chooses Abbott, say, and 0 when the voter chooses Baker. (To keep the example simple — at the expense

---

[17]Since there is a population $\mathcal{S}$ for *each* sample size n, it would be more precise to denote these populations by $\mathcal{S}_n$. However, this notational refinement is not essential to understanding the main idea.

of some realism — we shall assume that every eligible voter votes, and that every vote cast is either for A or for B.)

- Recall from Section 4.10 that the mean and standard deviation of x in the population $\mathcal{P}$ are denoted by μ and σ, respectively. In our election example, μ is the proportion of the votes to be cast for Candidate A. This number is what the newspaper wants to determine; it specifies both the winner of the election and the relative margin of victory.

- In Section 4.9, we denoted the mean of a sample by $\bar{x}$. In the language of Section 4.10, $\bar{x}$ is a random variable that takes on values from the population $\mathcal{S}$. That is, each possible random sample (member) of the population $\mathcal{S}$ has a mean $\bar{x}$, a numerical value that depends on the actual random choice of the elements of the sample. (The means of different random samples would, presumably, be different.)

The Central Limit Theorem relates the distribution of the random variable $\bar{x}$ over $\mathcal{S}$ to the distribution of the random variable x over $\mathcal{P}$. Loosely stated, it says that, for "large enough" sample sizes, $\bar{x}$ is approximately normally distributed around the mean μ of the original population $\mathcal{P}$ — regardless of how $\mathcal{P}$ itself is distributed — and the standard deviation of $\bar{x}$ can be approximated in terms of the standard deviation σ for $\mathcal{P}$. More precisely:

> **The Central Limit Theorem**[18]: Let $\mathcal{P}$ be a population with mean μ and standard deviation σ, and let $\mathcal{S}$ be the population of all n-element samples of $\mathcal{P}$. Then the distribution of the random variable $\bar{x}$ of means of these n-element samples is approximated by a normal distribution that has mean μ and standard deviation $\frac{\sigma}{\sqrt{n}}$.

Thus, the random variable $\bar{x}$ over the population $\mathcal{S}$ and the random variable x over the population $\mathcal{P}$ have the same mean — μ — and the standard deviation of $\bar{x}$, which we shall call $\bar{\sigma}$, is found by dividing the standard deviation of x by $\sqrt{n}$:

$$\bar{\sigma} = \frac{\sigma}{\sqrt{n}}$$

Notice that the effect of the denominator $\sqrt{n}$ is to make $\bar{\sigma}$ smaller as the sample size n gets bigger. This provides a numerical measure of the natural expectation that, the larger the sample, the less its mean, $\bar{x}$, should deviate from the the true population mean, μ.

---

[18]The first general version of this theorem was proved in 1810 by Pierre Simon Laplace, an outstanding mathematician and astronomer who was also a prominent political figure in France during the time of Napoleon.

The Central Limit Theorem holds for "reasonably large" samples in a "large enough" population. The detail involved in making these quantifying phrases exact is unnecessary for our purposes. The statisticians' working "rule of thumb" will serve us quite nicely; they consider the theorem to be reliable in situations where the sample size is at least 30 and the population is at least 20 times as large as the sample size. In such a situation, the Central Limit Theorem tells us:

1. μ is in some sense the "most likely" value for the mean $\bar{x}$ of any sample.

2. $\bar{x}$ is symmetrically distributed about μ; that is, any particular value of $\bar{x}$ is as likely to be smaller than μ as it is to be greater than μ.

3. About 68% of the sample means will fall between $\mu - \frac{\sigma}{\sqrt{n}}$ and $\mu + \frac{\sigma}{\sqrt{n}}$; about 95% will fall between $\mu - 2\frac{\sigma}{\sqrt{n}}$ and $\mu + 2\frac{\sigma}{\sqrt{n}}$; about 99.7% will fall between $\mu - 3\frac{\sigma}{\sqrt{n}}$ and $\mu + 3\frac{\sigma}{\sqrt{n}}$.

Let us apply these ideas to our election example. The newspaper plans to ask a randomly chosen sample of voters whether they will vote for Abbott or for Baker. Of course, it would be foolish for the editors to predict that the proportions of all the votes cast in the election will be *exactly* the same as the proportion they find in their sample. For instance, if they were to question 30 voters and find that 18 (= 60%) choose Abbott and 12 (= 40%) choose Baker, it would be very unwise to predict that Abbott will get *exactly* 300,000 votes (60% of the total). However, the principles listed in the preceding paragraph can be used to provide some idea of a likely range of results, along with some numerical estimate of the probability that the election results will fall within the predicted range.

A numerical mean $\bar{x}$ for any voter sample can be found by assigning 1 to each of the a votes for Abbott and 0 to each of the b votes for Baker; that is,

$$\bar{x} = \frac{a \cdot 1 + b \cdot 0}{a + b}$$

where $a + b$ is the total number of voters surveyed.[19] Now, Statement **3** implies that the mean μ of the voting population — the number the editors want — has

$(\star)$

- about a 68% chance of being within $\frac{\sigma}{\sqrt{n}}$ of $\bar{x}$;
- about a 95% chance of being within $2\frac{\sigma}{\sqrt{n}}$ of $\bar{x}$;
- about a 99.7% chance of being within $3\frac{\sigma}{\sqrt{n}}$ of $\bar{x}$.

---

[19] Compare this with the definition of *mean* in Section 4.9.

Thus, if they knew σ, the standard deviation of the voting population, then they could predict a likely range for the outcome of the election and have some measure of the confidence to place in their prediction.

Now, recall from Section 4.9 that the standard deviation is the square root of the variance, and that the variance, v, is the "average" of the squares of the differences between the individual values of x and the mean, μ. In this case, there is an easy way to approximate the variance.[20] Since each value of x is either 1 or 0, the only two differences between them and the mean are $1 - \mu$ and μ, respectively. Now, μ also represents the proportion of the 500,000 voters who will choose Abbott, and consequently $1 - \mu$ represents the proportion of voters who will choose Baker; that is, $500,000\mu$ votes will be cast for Abbott and $500,000(1 - \mu)$ votes will be cast for Baker. Thus, the variance is

$$v = \frac{(1 - \mu)^2 \cdot 500,000\mu + \mu^2 \cdot 500,000(1 - \mu)}{500,000}$$

Cancelling the factor 500,000 from both the numerator and the denominator, we get

$$\begin{aligned} v &= (1 - \mu)^2 \cdot \mu + \mu^2 \cdot (1 - \mu) \\ &= \mu(1 - \mu)(1 - \mu + \mu) \\ &= \mu(1 - \mu) \end{aligned}$$

The standard deviation, σ, is the square root of this value:

$(\star\star)$ $$\sigma = \sqrt{\mu(1 - \mu)}$$

But this means that the editors need to know μ in order to find σ, so they seem to be back where they started!

Fortunately, they don't need to know σ exactly in order to estimate how good their prediction is; an approximation will do. Moreover, the form of Equation $(\star\star)$ is particularly nice in this situation because it provides a pretty close approximation for σ even when the estimate of μ is relatively crude. For instance, suppose their survey of 30 voters came out as described before, with a sample mean of .6. Then, using .6 as an estimate for μ, they would get

$$\sigma \approx \sqrt{.6(1 - .6)} = \sqrt{(.6)(.4)} = \sqrt{.24} \approx .49$$

Now, suppose that this sample mean is pretty far off from the actual value of μ; let's say the actual value is .5. In that case,

$$\sigma \approx \sqrt{.5(1 - .5)} = \sqrt{(.5)(.5)} = \sqrt{.25} = .50$$

---

[20]This is where the simplifying assumption that there are only two possible choices pays off. We have a binomial distribution, so the computations are much simpler.

a difference of only .01. Thus, if they use the sample mean as an approximation for μ in Equation $(\star\star)$, the resulting value for σ will be close enough to be used for determining just how good an approximation it really is.

To see what the value $\sigma = .49$ would tell them about the reliability of this poll, refer to the three statements of $(\star)$. They say that

- about 68% of such sample means are within $\frac{\sigma}{\sqrt{n}}$ of μ;
- about 95% of such sample means are within $2\frac{\sigma}{\sqrt{n}}$ of μ;
- about 99.7% of such sample means are within $3\frac{\sigma}{\sqrt{n}}$ of μ;

Now, because the distance between two points is the same, regardless of the point from which you measure, an interval of a particular size centered at μ will contain $\bar{x}$ if and only if an interval of the same size centered at $\bar{x}$ contains μ. Thus, we can use the sample mean, .6, as the midpoint of each interval and, recalling that the sample size n in this case is 30, we can compute these value ranges:

$$\frac{\sigma}{\sqrt{n}} \approx \frac{.49}{\sqrt{30}} \approx .09$$

Therefore,

- $\approx$ 68% confidence: μ is between .51 and .69;
- $\approx$ 95% confidence: μ is between .42 and .78;
- $\approx$ 99.7% confidence: μ is between .33 and .87.

This means that newspaper editors have a problem. Remember that the winner of this election will get more than 50% of the votes cast, so μ must be larger than .5 in order for their prediction that Abbott will win to be correct. But only in the first of these three cases is the entire interval of possible values for μ greater than .5, and the confidence for such a prediction is only about 68%. A prediction that has about 1 chance in 3 of being wrong isn't very reliable. A prediction that has a 95% chance of being correct would be good enough for the editors, but the interval of possible values for μ is too big (because it includes values below .5). The only way to shrink the size of the interval is to gather a bigger sample.

How big is big enough? Well, we want to be sure that $\bar{x} - 2\frac{\sigma}{\sqrt{n}}$ is greater than .5, so a precise answer to that question depends on knowing in advance what the mean of this new sample will turn out to be. In this case, if the new sample mean is not very far from .6, then a sample of 100 or so will do, because

$$.6 - 2 \cdot \frac{.49}{\sqrt{100}} \approx .6 - 2 \cdot \frac{.49}{10} = .502$$

Here is a summary of the ideas typified by this polling example.

▷ The general problem it represents is that of trying to estimate the mean of a large population by examining a relatively small random sample.

▷ As we saw, it is impossible to determine the exact population mean with certainty. The best we can hope for is a range of values, along with a numerical measure of the "confidence" we can have in our prediction that the actual mean lies within that range. The range of values is called a **confidence interval**; the measure of confidence is called a **confidence coefficient**.

▷ Common confidence coefficients are (approximately) 95% and 99.7%. These choices are natural consequences of the characteristics of normal distributions and the fact that the distribution of sample means tends to be approximately normal for "large enough" samples. (This is the main point of the Central Limit Theorem).

▷ The size of a confidence interval depends on the desired confidence coefficient. For the two most common coefficients, the corresponding confidence intervals are $(\bar{x} - 2\bar{\sigma}, \bar{x} + 2\bar{\sigma})$ and $(\bar{x} - 3\bar{\sigma}, \bar{x} + 3\bar{\sigma})$.

▷ Since $\bar{\sigma} = \frac{\sigma}{\sqrt{n}}$, the standard deviation $\bar{\sigma}$ gets smaller as the sample size n gets larger. That is, the distribution of the sample means tends to be more tightly grouped around the population mean. Therefore, for a fixed confidence coefficient, the size of the confidence interval can be made smaller by making the size of the sample larger.

▷ Using this description of the standard deviation $\bar{\sigma}$ of the sample means in terms of the standard deviation σ of the population $\mathcal{P}$ depends on being able to approximate σ in some way.

- If $\mathcal{P}$ is *binomially distributed* — that is, if there are only two possible outcomes, say c and d, for the random variable x as it ranges over $\mathcal{P}$ — then a working approximation of σ is easy to describe. Assigning 1 and 0 to the outcomes c and d, respectively, we observe that μ and $(1 - \mu)$ are the respective proportions of c and d outcomes for the entire population and hence that $\sigma = \sqrt{\mu(1 - \mu)}$. (See the discussion leading up to Equation (★★) in the election example.) As we have seen, this formulation shows that changes in the value of μ result in much smaller changes in σ, so we can approximate σ fairly well by using the mean $\bar{x}$ of a reasonably large sample in place of μ. That is,

$$\sigma \approx \sqrt{\bar{x}(1 - \bar{x})}$$

- If $\mathcal{P}$ is not binomially distributed, then σ must be determined in some other way. Further discussion of finding σ for other distributions would quickly take us well beyond the intended scope of this book.[21] In some of the examples and exercises that follow, we shall simply presume that a given value of σ has been found or approximated appropriately.

Example 4.47   **Problem:** A large supply of old bean seeds is found in the back storeroom of an agricultural supply store. Before selling them (at a reduced price), the store manager wants to be able to tell his customers what percent of the seeds are likely to germinate. He chooses 200 seeds at random and plants them in his greenhouse. Within two weeks, 167 of the seeds have sprouted. What is the likely germination rate of the supply of bean seeds?

Solution: The information sought is a confidence interval for the likely proportion of viable seeds (i.e., seeds that will germinate). If we assign 1 and 0 to viable and nonviable seeds, respectively, then this proportion is the mean μ of a binomial distribution with $n = 1$ from which a 200-element sample was drawn and tested. (Note that, as stated in the Central Limit Theorem, the mean proportion of viable seeds for all 200-element samples is also μ.) The first step in the process is to decide on an acceptable confidence coefficient; let us assume that the manager decides that a 95% confidence coefficient is good enough. Then, using $\bar{x}$ — the proportion of sample seeds that germinated — as an estimate for μ, the endpoints of the interval $(\bar{x} - 2\bar{\sigma}, \bar{x} + 2\bar{\sigma})$ are computed as follows:

$$\bar{x} = \frac{167}{200} = .835$$

$$\sigma \approx \sqrt{\bar{x}(1-\bar{x})} = \sqrt{(.835)(.165)} \approx .371$$

$$\bar{\sigma} = \frac{\sigma}{\sqrt{n}} \approx \frac{.371}{14} = .0265$$

$$(\bar{x} - 2\bar{\sigma}, \bar{x} + 2\bar{\sigma}) \approx (.835 - 2(.0265), .835 + 2(.0265)) = (.782, .888)$$

Of course, this interval either contains μ or it doesn't. However, the fact that this is a 95% confidence interval means, roughly speaking, that the likelihood of getting such a sample when the actual germination rate of the seeds falls *outside* the interval (.782, .888) is only about 5%. But a normal distribution is symmetric about its mean, so any outcome outside this interval would be equally likely to be above it as below it. Thus, there is only about a 2.5% chance that such a sample would occur if the actual germination rate is *below* .782. This means that the manager can predict with at least 97.5% confidence that the germination rate for these bean seeds is at least 78%.

□

---

[21] Detailed examinations of various types of distributions may be found in most standard introductory texts on statistics, such as [1] and [7] in the list *For Further Reading* at the end of this chapter.

Example 4.48   A breakfast-food company wants to know if the mean net weight of the boxes of corn flakes filled by its machinery is within .1 oz. of the stated net weight of 12 oz. — that is, between 11.9 oz. and 12.1 oz. It carefully weighs a random sample of the contents of 10 boxes and finds a mean net weight $\bar{x} = 11.97$ oz. and a standard deviation $s = .11$ oz. Wanting to be quite sure, the company decides to use 99.7% as the confidence coefficient. Using s as an estimate of the population standard deviation σ, it computes the appropriate confidence interval:

$$(\bar{x} - 3\bar{\sigma}, \bar{x} + 3\bar{\sigma}) \approx (11.97 - 3\tfrac{.11}{\sqrt{10}}, 11.97 + 3\tfrac{.11}{\sqrt{10}}) \approx (11.87, 12.07)$$

Since the lower end of this confidence interval is more than .1 oz. from the target weight of 12 oz., the company cannot declare with 99.7% confidence that its machinery is working properly. Rather than stopping production to make adjustments, however, it decides to check a larger sample. Weighing a random sample of 50 boxes, it gets a mean net weight of 11.96 oz. and a standard deviation of .13 oz. This is initially worrisome, since the new sample mean is farther off target than the old sample mean and the new standard deviation suggests a slightly broader dispersion than the old one. However, computing the 99.7% confidence interval, the company observed

$$(\bar{x} - 3\bar{\sigma}, \bar{x} + 3\bar{\sigma}) \approx (11.96 - 3\tfrac{.13}{\sqrt{50}}, 11.96 + 3\tfrac{.13}{\sqrt{50}}) \approx (11.905, 12.015)$$

Since this entire interval lies between 11.9 and 12.1, the company can conclude with 99.7% confidence that the machinery is functioning properly. □

The corn-flakes problem just discussed illustrates the idea of **quality control**, an application of statistical theory to industrial production that uses regular sampling to check acceptable product characteristics. Other examples of applied statistics abound. The life-insurance industry depends heavily on statistical predictions, as do advertisers and commercial broadcasting companies. Since most sciences are experimental by nature and proceed from a limited number of observations to the enunciation of general laws, statistics plays a major role in deriving these laws. For instance, the normal curve is an essential tool in genetics, and the application of the normal distribution to the movement of particles explains the physical phenomenon of Brownian motion. The social sciences and education depend on statistics for the proper formulation and interpretation of surveys and tests. The list goes on and on.

The peculiar strength of statistics is its ability to deal quantitatively with uncertainty, thereby bridging the gap between theoretical results and practical applications. But that strength is also its weakness. Many applications of statistical theory depend on the *assumption* that a normal distribution is right for a situation, as well as the *choice* of an appropriate confidence coefficient and several other choices. These assumptions and choices, if not made properly, render the statistical conclusions based on them invalid and thus useless or misleading. The LINK section that ends this chapter describes a particular application of statistics. Whether this application represents a strength or a weakness is left for you to decide.

EXERCISES 4.11

For Exercises 1–12, you are given a sample mean $\bar{x}$, a standard deviation $\bar{\sigma}$, and a confidence coefficient. Specify the confidence interval determined by these values.

1. $\bar{x} = 0$, $\bar{\sigma} = 1$; 68% confidence

2. $\bar{x} = 0$, $\bar{\sigma} = 1$; 95% confidence

3. $\bar{x} = 0$, $\bar{\sigma} = 1$; 99.7% confidence

4. $\bar{x} = 100$, $\bar{\sigma} = 10$; 68% confidence

5. $\bar{x} = 100$, $\bar{\sigma} = 10$; 95% confidence

6. $\bar{x} = 100$, $\bar{\sigma} = 10$; 99.7% confidence

7. $\bar{x} = 8.1$, $\bar{\sigma} = 1.25$; 68% confidence

8. $\bar{x} = 8.1$, $\bar{\sigma} = 1.25$; 95% confidence

9. $\bar{x} = 8.1$, $\bar{\sigma} = 1.25$; 99.7% confidence

10. $\bar{x} = .32$, $\bar{\sigma} = .2$; 68% confidence

11. $\bar{x} = .32$, $\bar{\sigma} = .2$; 95% confidence

12. $\bar{x} = .32$, $\bar{\sigma} = .2$; 99.7% confidence

For Exercises 13–18, you are given a sample mean $\bar{x}$, a standard deviation $\bar{\sigma}$, and a confidence interval. Specify the confidence coefficient for the given interval.

13. $\bar{x} = 5$, $\bar{\sigma} = 1$; interval $(3, 7)$

14. $\bar{x} = 34$, $\bar{\sigma} = 3$; interval $(31, 37)$

15. $\bar{x} = 7.33$, $\bar{\sigma} = .5$; interval $(5.83, 8.83)$

16. $\bar{x} = 1.71$, $\bar{\sigma} = .01$; interval $(1.68, 1.74)$

17. $\bar{x} = .4$, $\bar{\sigma} = .25$; interval $(-.1, .9)$

18. $\bar{x} = 74$, $\bar{\sigma} = 8$; interval $(66, 82)$

For Exercises 19–22, you are given the standard deviation σ for a large population from which an n-element random sample is to be drawn. Determine a minimum size for n so that there is 95% confidence that the mean of the sample is less than .5 unit away from the actual population mean μ.

19. $\sigma = 3.5$ **20.** $\sigma = 4.98$

21. $\sigma = 2.2$ **22.** $\sigma = 7.1$

For Exercises 23–28, $\sigma = 5.2$ is the standard deviation of a large population from which an n-element random sample is to be drawn. In each case, approximate with 95% confidence the maximum "margin of error" between the sample mean and the actual population mean. (Round your answer to 2 decimal places.)

23. $n = 1$ **24.** $n = 2$

25. $n = 5$ **26.** $n = 10$

27. $n = 30$ **28.** $n = 50$

For Exercises 29–32, assume that σ is the standard deviation of a large binomially distributed population (with $n = 1$) from which a 50-element sample has been drawn. In each case, compute an approximate value for σ using the given sample mean, $\bar{x}$. (Round your answer to 2 decimal places.)

29. $\bar{x} = .7$ **30.** $\bar{x} = .62$

31. $\bar{x} = .45$ **32.** $\bar{x} = .18$

33. The label on a package of 200 corn seeds states that the germination rate is 94% ±

3%. Assuming that the stated variation represents a 95% confidence interval,

(a) What is the standard deviation for the germination rate for the seed population?

(b) Predict with 99.7% confidence the maximum number of seeds from this package that will germinate.

34. In an attempt to get an early indication of the outcome of a citywide bonding referendum vote on election day, a Kansas City television station conducted an exit poll of 300 voters. They found that 168 voted *yes* and 132 voted *no*. With what level of confidence can the station predict that the referendum will pass? Justify your answer.

35. A candidate in a primary needs to get more than 50% of the votes cast to win without a runoff election. On the evening before the election, a random telephone survey of 50 voters indicated that 29 of them would vote for her.

(a) The candidate's campaign manager, seeking at least 95% assurance of victory, was not content with this survey. Why not?

(b) The campaign staff phoned 100 more voters. Again, the same proportion of voters contacted — 58 of the 100 — said they would vote for the candidate. This time, the campaign manager was satisfied. Why?

36. A standardized examination given nationwide is known to have a mean score of 72 and a standard deviation of 8.

(a) A group of 30 students took the exam and got a mean score of 75. How likely is it that these students' scores represent a random sample of all scores on the exam? Justify your answer.

(b) A group of 100 students took the exam and got a mean score of 74.5. Compared with the group in Part (a), is it more or less likely that these students'

scores represent a random sample of all scores on the exam? Justify your answer.

37. The machinery of a toy manufacturer cuts small brass axles to a mean length of 15 cm with an allowable variation of $\pm$ .1 cm. Every two hours, the quality-control department takes a random sample of 9 axles, measures them, and records the mean and standard deviation.

(a) One such sample has a mean of 15.04 cm and a standard deviation of .06 cm. How confident can the company be that the machinery is working properly?

(b) Another such sample has a mean of 14.95 cm and a standard deviation of .12 cm. How confident can the company be that the machinery is working properly?

WRITING EXERCISES

1. In your own words, explain the meaning of the word "confidence" as it is used in this section.

2. (a) Which idea of this section interests you the most? Why?

(b) Which idea of this section interests you the least? Why?

(c) Which idea of this section confuses you the most? In what way is it confusing to you?

3. Do some library research on the subject of *quality control*; then write a nontechnical description (about 1 page long) of how sampling is used to monitor quality in industrial production. Be sure to provide a bibliography that lists any sources you used, and be careful that you do not simply copy or paraphrase the material from your sources.

LINK: 4.12

Statistics
in the
Psychology
of
Learning

Testing.

The IQ test is one of the most widely known measurement instruments in educational psychology. Almost all of us have had our IQs measured at least once, and some of us may even recall the score(s). Yet relatively few people know what an IQ score really means (Do you?), except for some vague sense that it measures intelligence level in some way. In fact, the IQ test is a psychological testing tool whose meaning depends almost entirely on the concepts you studied in this chapter. Here we shall explain briefly the statistical basis for IQ's, then tie this to current public policy in a particular area of education.

"IQ" is an abbreviation for *Intelligence Quotient*. A quotient is, of course, a number formed from the division of one number by another. In this case the quotient is formed by dividing a person's mental age by his or her chronological age. It is standard practice to multiply the resulting fraction by 100 and then round after the decimal point, so that IQ scores appear as whole numbers. Thus,

$$IQ = \frac{\text{mental age}}{\text{chronological age}} \times 100$$

rounded to the nearest whole number. For example, a nine-year-old child with a mental age of 10 would have an IQ of 111 because

$$\frac{10}{9} \times 100 = 111.11\ldots$$

The obvious question is: What is *mental age*? This is where statistics enters the scene. In order to measure a child's mental age, educational researchers devise a test that can be administered to children over a broad span of ages.[22] To be standardized, these tests are administered to a large random sample of children of every age group, and the mean score in each age group is used to represent the mental age. Thus, a child whose individual test score matches the mean score for twelve-year-old children has a mental age of 12, regardless of that child's actual (chronological) age.

For instance, suppose the mean score for ten-year-old children on a particular intelligence test is 92. Then

- an eight-year-old child who scores 92 on that test has a mental age of 10 and, therefore, has an IQ of 125 [that is, $\frac{10}{8} \times 100$];
- a twelve-year-old child who scores 92 also has a mental age of 10 and thus has an IQ of 83 [obtained from $\frac{10}{12} \times 100$ by rounding];
- a ten-year-old child who scores 92 has a mental age of 10 and an IQ of 100 [$= \frac{10}{10} \times 100$].

In general, a person with an IQ of 100 is someone whose intelligence, or intellectual ability, tests out at the mean for his/her age group.

Notice that the computation process just described has actually *defined* mental age in a completely statistical way. That's fine, *if* mental age is understood solely in that way. Unfortunately, even the term itself suggests something more, as if mental age were some sort of fixed characteristic of individual minds that is discoverable by testing. It is quite clear, however, that an eight-year-old child with a mental age of 10 and a fifteen-year-old child with a mental age of 10 have very different kinds of minds.

Moreover, the IQ formula implies that an eight-year-old girl with an IQ of 125 is two years ahead in mental age [because $\frac{125}{100} \times 8 = 10$], but when that same child reaches age twelve (presumably with the same IQ), she must be three years ahead in mental age [because $\frac{125}{100} \times 12 = 15$]. According to the

---

[22]The most common IQ tests in use today are the Wechsler Intelligence Scale for Children (WISC), spanning ages 6 through 15, and the Stanford-Binet Intelligence Scale, spanning all ages from preschool through high school.

formula, on her 24th birthday this adult with an IQ of 125 would be fully 6 years ahead in mental age! The obvious extension of this idea leads to absurd conclusions, so the interpretation of variance from the norm is adjusted to mean different things at different ages. Such difficulties with the conceptual validity of "mental age" have led in recent years to IQ testing based on a more formal statistical foundation. Some of the major intelligence tests have now abandoned "mental age" in favor of scoring tables designed so that the (standardized) mean for each age group is 100 and the standard deviation is 15 or 16.

This little exercise in applied statistics can affect people's lives in many ways. Let us look at one important example. Since about 1970, many school systems around the country have begun to provide supportive educational services to a special class of handicapped children known as "learning disabled." These special services are now required (and partially funded) by the federal government. But children qualifying for these services are vaguely defined by federal regulations, which say only that such a child must have "a severe discrepancy between achievement and intellectual ability" in oral and/or written language or mathematics [*Federal Register*, 42 (240): 65083, Dec. 29, 1977]. The question is: What is a "severe discrepancy"? The answer to this question has profound implications for parents and children, school boards and teachers, towns and taxpayers. Who qualifies for this help? How much service must be provided? How much will it cost? All these answers flow from the definition of "severe discrepancy," a definition the federal government does not provide.

Among the states attempting to provide legal definitions for themselves is Connecticut, whose answer is a striking illustration of the effect of statistics on public policy:

> [A] severe discrepancy is said to exist whenever the difference between academic achievement (AA) and intellectual functioning (IQ) (each expressed as a standard score) is equal to or greater than one-and-one-half standard deviations—that is,

$$IQ - AA = 1\tfrac{1}{2} \text{ SD}$$

> [*Guidelines for Identification and Programming of Learning Disabilities.* Hartford, CT: Connecticut Department of Education, 11/27/81 draft, p. 30.]

In other words, to qualify for special services in this area, children must satisfy a statistical criterion that rests, in turn, on statistically based test measurements. Their intellectual potential is measured by an IQ test and then their achievement is measured by other tests whose scores are standardized for purposes of comparison. (The Connecticut guidelines even provide a formula for converting other scoring scales to a mean of 100 and a standard deviation of 15.) Only if the achievement scores fall $1\frac{1}{2}$ standard deviations (22 points) below the IQ scores do the Connecticut guidelines declare these children to be learning disabled. Thus, the concept "learning disabled" in Connecticut is defined in a fundamentally statistical way.

The foregoing is but one illustration of how statistical concepts may be so thoroughly intermixed with ideas from another discipline that there is no way to separate them. As you encounter the behavioral and social sciences in your education, you might find it helpful to look for concepts that are defined fundamentally by statistics. Recognizing them is a critical first step in truly understanding the theories that use them and the implications of those theories in our lives and in our world.

Topics for Papers – Chapter 4

1. This is an independent research project to illustrate how the statistical methods of this chapter are used. It is suitable for either an individual or a small group.

    (a) Formulate a *yes-no* question that can be asked of all the students at your institution. (Before going any further, check with your instructor to be sure that your question is suitable as a basis for this paper.)

    (b) Choose a random sample of students and ask them the question.

    (c) Determine with 95% confidence the proportion of the entire student body that would answer *yes* to your question.

    (d) Write a detailed description of the methods and conclusions of your survey, beginning with a clear statement of the survey question. Be sure to cover *at least* the following points in your paper:

    - How did you decide on the size of your random sample?

    - How did you choose the people in your sample? From what population are they drawn? Why is the choice process you used random?

    - Did any unexpected complications arise when you were conducting your survey? If so, what were they and how did you handle them?

- What is the proportion of the population that you predict would say *yes* to your question? What is the confidence interval for your result? (Include the computations that justify your answers.)

- Now that you have completed the project, what stands out as the most interesting thing you learned from it?

2. Write a research paper elaborating on or extending the ideas in one of the three LINK sections of this chapter. Consider your reader to be an anonymous student in the class, a person who knows the material of the chapter. Be sure to provide a bibliography that lists any sources you used in writing the paper, and be careful that you do not simply copy or paraphrase the material from your sources. Any material not explicitly quoted should have been mentally digested by you, so that your own words describe what really are your own ideas when you write them down.

For Further Reading

1. Alder, Henry L., and Edward B. Roessler. *Introduction to Probability and Statistics*, 3rd ed. San Francisco: W. H. Freeman and Co., 1964.

2. COMAP (Solomon Garfunkel, Project Director). *For All Practical Purposes*, Part II. New York: W. H. Freeman and Co., 1988.

3. Diamond, Solomon. *The World of Probability.* New York: Basic Books, Inc., 1964.

4. Freund, John E. *Modern Elementary Statistics*, 4th Ed. Englewood Cliffs, NJ: Prentice-Hall, Inc., 1973.

5. Johnson, Robert. *Elementary Statistics*, 4th Ed. Boston: Duxbury Press, 1984.

6. Kac, Mark. "Probability," *Scientific American* 211 (3): 92-108, 1964.

7. Moore, David S., and George P. McCabe. *Introduction to the Practice of Statistics.* New York: W. H. Freeman and Co., 1989.

8. Robbins, Herbert. "Theory of Probability," *NCTM 23rd Yearbook, Insights into Modern Mathematics.* Washington, DC: National Council of Teachers of Mathematics, 1957.

9. Weaver, Warren. *Lady Luck.* Garden City, NY: Doubleday & Co., 1963.

10. Willerding, Margaret. *A Probability Primer.* Boston: Prindle, Weber & Schmidt, Inc., 1968.

Optical Character Reader at a Supermarket.

MATHEMATICS OF MACHINES: MICROCOMPUTERS AND PROGRAMMING

5.1 What Is a Computer?

A computer is a machine that can accept, store, and manipulate data, perform arithmetic and logical operations without human intervention, and report the results of these operations. In today's world we are literally met at every turn by computers. When we flip on a light switch, we are tapping into a power system that is constantly monitored and controlled by computers. The utility company maintains records of its thousands of customers by computer and sends out computer-generated bills. Every phone call we make engages a computer-controlled switching system. For many of us, the car we drive has computer components that report to us on fuel, battery, headlights, and other functions. We buy our groceries at supermarkets with electronic sensors that "read" a code on each package, consult a computer for the correct price and for inventory control, and print and total our bill. We use hand calculators; we play video games; we use credit cards; we receive letters that have been "personalized" by word processors. In these and dozens of other instances, we are interacting with computer technology, directly or indirectly.

We tend to think of computers as a very recent phenomenon, and certainly ours is the age of computers. A more accurate perspective, though, is that the technology has mushroomed in the last twenty to thirty years, but the history of computing machines is extremely old. Man's oldest aid in counting and computing are his fingers, almost certainly the motivation for our current decimal system. And even if we discount this example as not being an external or mechanical aid to computation, we must allow the *abacus* as a bona fide instance of a computing machine, one that dates from approximately 3000 B.C. and is still in use today.

Abacus. This ancient computation device is still used in parts of the world today.

There are hints throughout early recorded history of attempts to devise machines that would calculate, though there is little evidence of success. Contact between European and Oriental cultures in the Middle Ages not only introduced the Hindu-Arabic numeration system to the Western world, but also the abacus. In the 13th century, Ramon Lull borrowed ideas from the *zairja*, an Arab "thinking machine," to design the *Ars Magna*, a partly mechanical, partly logical scheme that attempted to arrive at truth in a "machine-like" way.

In 1617 Lord Napier, the originator of logarithms, devised a calculating device known now as *Napier's Bones*, a set of marked sticks made from bone, which was the precursor of the *slide rule*, developed by William Oughtred in 1630. The first mechanical adding machine was built in 1642 by mathematician and philosopher Blaise Pascal, and was improved and extended to handle multiplication and division later in that century by Gottfried Leibniz, one of the discoverers of the calculus. (See Appendix Section B.6.)

Certainly one of the most illustrious niches in the computer hall of fame must go to Charles Babbage (1792–1871). An outstanding mathematician, Babbage was fascinated by machines. In 1822 he built a small working

Napier's Bones. *At left: Napier's Reckoning Board.*
At right: Napier's Reckoning Rod.

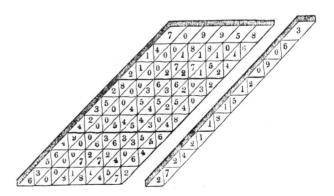

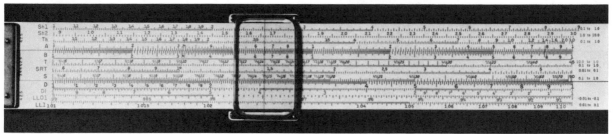

Slide Rule.

model of his *Difference Engine*, a mechanical device to compute and print tables of logarithms. The British government was very interested in this machine, due to the importance of precise tables for accurate navigation, and it supported Babbage's attempts to produce a larger machine. But Babbage became interested in an even grander scheme, his *Analytical Engine*. This machine was a general-purpose computing machine, remarkably like today's computers in its basic design. It had a section devoted to control, another section for logical and computational processes, and one for memory (of 50 digit numbers). The design called for punched cards to be used in the computing section to control which operation was performed, and other cards were to control the transfer of data to and from the memory. Babbage's idea of using punched cards was borrowed from Joseph Jacquard, who had

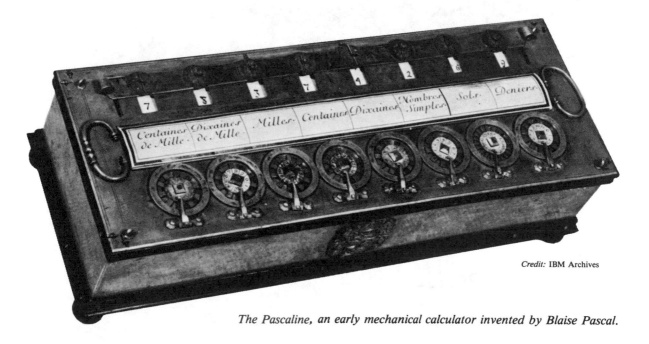

The Pascaline, an early mechanical calculator invented by Blaise Pascal.

Leibniz Calculator

Key-punch. This is now an obsolete means of input.

Credit: IBM Archives

Jacquard Loom.

used them successfully to control weaving machines at the turn of the century (1801).[1]

For about ten years, Babbage was assisted in his work by Augusta Ada Lovelace (1815–1852), the only child of the marriage between the English poet George Gordon, the sixth Lord Byron, and Anna Isabella Milbanke. While translating a French paper about Babbage's Analytical Engine, Lovelace expanded the work with copious notes explaining and extending the ideas of the original paper. In her notes, she developed in detail the

---

[1] Jacquard's punched-card idea was also the inspiration for Herman Hollerith, who revised the idea in 1887 and adapted it for use in counting machines for the 1890 U. S. census. Hollerith's machine was the prototype for others in his Tabulating Machine Company, which later became International Business Machines. Hollerith cards were used by many IBM machines until the early 1980s.

Credit: IBM Archives

Charles Babbage.

idea of "programming" the machine. Because of this work, Ada Lovelace is considered to be the "inventor" of computer programming.[2]

The difference between the Analytical Engine and 20th century computers is not in any essential component, but in their forms. The Babbage machines were mechanical rather than electronic, and relied on a monstrous collection of gears, cams, etc., all driven by steam power. This proved an insurmountable stumbling block, as the state of the art in the production of standardized-precision machine parts was insufficient to match Babbage's vision. Consequently, the Analytical Engine was never built.

Many advances in the precision and sophistication of machines that ultimately made computing machines a reality were made in the 19th and 20th centuries. The first practical typewriter was patented by Samuel Soule in 1868. Ten years later W. T. Odhner devised the "Odhner wheel," which played a role in the development of the adding machine and the cash register of the following decade. Important, too, were the theoretical work of Sir William Hamilton, Augustus DeMorgan, and George Boole. Boole's "An Investigation of the Laws of Thought on Which are Founded the Mathematical Theories of Logic and Probabilities," published in 1854, laid the mathemat-

---

[2]As a tribute to her pioneering work, the programming language Ada, developed in the 1980s, is named after her. (See page 137 of [8] in the list *For Further Reading* at the end of this chapter.)

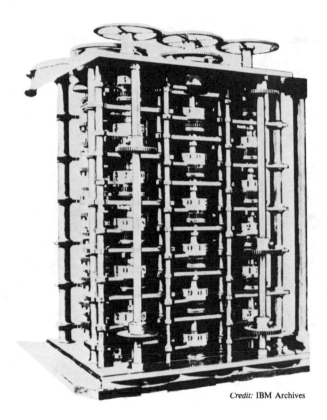

Credit: IBM Archives

Babbage's Difference Engine.

ical groundwork for computer logic. Claude Shannon's 1937 master's thesis at M. I. T. applied Boole's ideas, now known as Boolean algebra, to describe the behavior of relay and switching circuits, and showed that electronic devices could mimic mathematical logic.

The first general-purpose computer, the IBM Mark I, developed by IBM and Harvard scientists in 1944, was more mechanical than electronic. Two years later, John Mauchly and J. Presper Eckert of the University of Pennsylvania completed the first electronic computer, the ENIAC (Electronic Numerical Integrator And Calculator). And in 1949 John von Neumann's proposal that a computer could be controlled by a "stored program" was implemented at Cambridge University on the EDSAC (Electronic Delayed Storage Automatic Computer); thus, the century-old plan of Babbage was finally fulfilled. The first commercial computer was the Univac I, designed by Mauchly and Eckert and sold to the U. S. Census Bureau in 1951.

These "first generation" machines depended on vacuum tubes and relays to store and process data. They were large, they consumed sizable amounts

The ENIAC.

From Vacuum Tubes to Transistors to Integrated Circuits.

Credit: IBM Archives

The Mark I.

of energy, and they were subject to frequent component failures. The transistor, invented at Bell Laboratories in the early 1950s, was successfully used in computer design by the end of that decade, thus marking the "second generation" of computers. The "third generation" machines, introduced in the mid-sixties and in use today, are those that employ integrated circuitry. The past twenty-five years have seen extensive refinements in solid-state circuitry, especially microminiaturization of integrated circuits. This has led from the minicomputers of the sixties and the microcomputers of the seventies to the workstations, desktops, laptops, and palmtops of the eighties and nineties, which are thousands of times faster and cheaper than their forty-year-old grandparents.

Exercises 5.1

WRITING EXERCISES

1. When you hear the word "computer," what are the first ideas or images that come into your mind? Write a paragraph exploring this theme.

2. What is the literal meaning of the word *com-*

puter? (*Hint*: Look it up in a dictionary that was published before 1940 or so.) How does that meaning relate to the way the word is used in this electronic age?

3. Are *compute* and *calculate* synonyms? Are *computer* and *calculator* synonyms? Explain your answers; then comment on any aspect of this question that you find interesting.

5.2   Computer Overview

Our overall aim is to try to get some appreciation of what a computer is, and what it can and cannot do. Specifically, we consider a typical microcomputer (home computer) along with one of the most common programming languages, BASIC, to get some idea of what is entailed in the design and execution of a computer program. We shall develop enough familiarity with the **hardware** (the physical machine) and **software** (language and programs) to begin to understand how people can use computers to store and use information, solve problems, and even play games.

Of course, this will not be a thorough treatment of programming, or of the BASIC language or of any particular microcomputer. We shall get a taste of these things, though — hopefully enough of a taste to dispel some of the mystery and misinformation, such as the notion that a computer is merely a very fast adding machine, or the equally erroneous notion that a computer is an "intelligent" machine.

There are three basic structural units to any computer: the input units, the processor, and the output units.

- The **input units** are the devices by which data is entered into the computer. Some computers are designed to accept data from paper tape, others from magnetic tape or discs. The most common input device for a microcomputer is a keyboard, very much like a typewriter keyboard, although frequently with additional keys for special functions.

- The **processing unit** consists of two parts. The **central processing unit (CPU)** contains the electronic hardware to carry out the actual processing of data, including interpreting directions to the computer, performing arithmetic and logical operations, and controlling access to the internal memory and to the input and output devices. The **internal memory** is the other part of the processing unit; it is a temporary storehouse for both data and programmed instructions, which are accessed by the CPU in carrying out the required processing. In most microprocessors, the CPU and internal memory are miniaturized "chips" or "cards" containing electronic circuitry.

- The **output units**, as the name identifies, are the devices to which the results of the computing or processing are sent. Cards, tapes, or discs may all serve as output devices. The most common output device for a microcomputer is a **cathode ray tube (CRT)**, very similar to a television screen. Another common output device for a microcomputer is a **printer**, which produces a permanent copy, or "hard copy," of the output on paper.

To extend its capabilities or the quantity of data that can be processed, a computer is often provided with **auxiliary memory**, in the form of tape or disk. Auxiliary memory can be accessed by the CPU during the running of a program and thus can be used as an extension of the internal memory. Auxiliary memory is also used to store data and programs for later use. Microprocessors usually employ a "floppy disk" for this function, and many utilize a "hard disk" with considerably more storage capability. In addition, a computer may be connected to special **peripherals**, such as special input controls (a "mouse," videogame controls, heat-sensing devices, etc.) or output devices (an alarm, a lock, some servomechanism, etc.)

In terms of function, a computer performs three basic processes. In the input/output process it translates data (numeric or linguistic) into machine-recognizable form in memory, and vice versa. In arithmetic processes the computer performs arithmetic operations on numbers or it performs "string operations" on verbal data (such as combining two words or phrases to form a longer one, or deleting a word, phrase or character). In logical processes the computer compares two numbers or words and follows different subprocedures, depending on the result of the comparison.

Internally, a computer stores data and instructions for manipulating that data as a series of electronic pulses or states, essentially corresponding to a series of *on-off* switches. This can be represented mathematically as a string of 1s and 0s, that is, as a *binary number*.

Binary numbers are so called because they are based on the number two, rather than the base ten of our familiar decimal system. In the decimal system we represent a number with one or more of the *digits* 0, 1, 2, 3, 4, 5, 6, 7, 8, 9, with the understanding that the *position* of each digit indicates the power of ten it is to be multiplied by. Thus, the decimal number 2058 is understood to represent

$$2 \times 10^3 + 0 \times 10^2 + 5 \times 10^1 + 8 \times 10^0$$

That is, two thousands plus zero hundreds plus five tens plus eight. Because each decimal position in a decimal number corresponds to a power of 10, the system requires ten distinguishable digits.

On-line Computerized Library.

The binary system follows exactly the same structure and logic, with the sole distinction that the positions of digits correspond to powers of 2, and thus requires only two distinguishable digits, 0 and 1. Thus, the binary number 1011001 is understood to represent

$$1 \times 2^6 + 0 \times 2^5 + 1 \times 2^4 + 1 \times 2^3 + 0 \times 2^2 + 0 \times 2^1 + 1 \times 2^0$$

That is, one sixty-four plus zero thirty-twos plus one sixteen plus one eight plus zero fours plus zero twos plus one. In decimal notation, this number would be represented as 89.

Relative to the decimal system, the binary system "trades off" longer strings of digits for fewer digits. For purposes of electronic computers, the tradeoff is more than justified, as the two digits can be represented by the two states of a relay, a diode, or a transistor. Sequences of such electronic devices, then, can store the value of a number, which may represent itself or be a "code" value for a letter or a symbol. And properly designed circuits can effectively add, multiply, or compare two such numbers.

Binary numbers, then, are the form in which data is stored in the computer. And binary numbers are also the symbols (or alphabet) of the "lowest-level" language, or **machine language**, by means of which the computer controls its operations. Long strings of such numbers, corresponding to frequently repeated processes (such as adding two numbers, obtaining a number from memory, etc.) are collected and accessed in a "middle-level" language called **assembler language.**[3] This language, in turn, is used by a "high-level" **programming language**, which uses a word or phrase close to English to call on the computer to execute hundreds or even thousands of machine-language steps to accomplish a process. We shall only be concerned with one such high-level language, and will make no attempt to trace the manipulations going on within the computer.

A large number of programming languages exist; these languages have been designed and modified as the state of hardware has improved and knowledge of effective programming has progressed. Among the most widely used languages are

ALGOL   APL   BASIC   C   COBOL
FORTRAN   LISP   LOGO   PASCAL   PLATO

Despite some significant differences, each of these languages provides the user with structural ways to communicate with the computer, directing it in relatively simple terms to carry out complex procedures. Here we will use part of the language BASIC, an acronym for Beginner's All-purpose Symbolic Instruction Code. BASIC was designed in the mid-1960s by the mathematicians John Kemeny and Thomas Kurtz at Dartmouth College. It was intended to be a simple, first computer language for students. It has since been modified and improved, and has been adopted as the primary language of most microcomputers.

Some microcomputers, when first turned on, are immediately ready to accept directions in BASIC. Part of their internal memory (the ROM, or "read-only memory") has been permanently programmed to **interpret** BASIC statements in machine language. Other microcomputers require a BASIC interpreter program to be run first, to "teach" the computer how to understand BASIC. We shall assume that if you are using such a machine, then the BASIC interpreter program has been loaded and run. With many larger computers and some microcomputers, a program written in BASIC must be **compiled** (translated) into machine language by a separate compiler program prior to the execution of the BASIC program. If you are working with such a machine, you will need to consult its manual to learn how to compile your programs.

There are different **dialects** (minor variations) of BASIC among computer manufacturers, and the language constantly is being expanded and improved, but the essentials of BASIC that we will see are quite standard.

---

[3]The logic and terms used in assembler language are very similar to those used in some programmable hand calculators.

In particular, the programs of Sections 5.3 through 5.5 are compatible with virtually all dialects of interpreted BASIC. Some modification would be necessary to use these programs with S-BASIC (structured BASIC), but this is rarely the only version implemented on the most popular microcomputers.

EXERCISES 5.2

1. What is a computer?

2. Name three situations in addition to those mentioned in the chapter in which a computer is used.

3. Name three basic units of a computer. Describe the function of each.

4. What is the most common input unit for a microprocessor?

5. What is the most common output unit for a microprocessor?

6. What is the CPU? What does it do?

7. What is the difference between internal memory and auxiliary memory?

8. What are the three basic processes that a computer can perform?

9. What are the three levels of computer language?

10. What does BASIC stand for? What is it, and where did it come from?

11. Which binary numbers correspond to the decimal numbers from one to sixteen?

In Exercises 12–15, which binary numbers correspond to the following decimal numbers?

12. 22   **13.** 31   **14.** 45   **15.** 60

In Exercises 16–23, which decimal numbers correspond to the following binary numbers?

16. 1100   **17.** 10010
18. 11011   **19.** 111111
20. 1100100   **21.** 101100101
22. 1001001001   **23.** 11111000000

WRITING EXERCISES

1. Using complete sentences, write a half-page summary that identifies the main theme of this section and outlines how that theme is carried out.

2. Describe your personal experience with and attitude toward computers: Are they useful? interesting? mysterious? familiar? Add other adjectives to this list, if you wish.

3. Describe your impression of how our society views the use of computers. Are they beneficial? problematic? constructive? destructive? indispensable? inescapable? good? bad? How do these societal views differ depending upon a person's age? occupation? education? other factors?

5.3   Some Basic BASIC

Now begins your introduction to BASIC and to programming. You should be aware of the fact that much will be left unsaid; we do not intend this to be a comprehensive treatment. The goal is to give you some sense of what computers are, not to teach you programming. In order to keep the presentation brief, we will not introduce *all* the ways some procedure might be accomplished, and not always the most efficient way. For that knowledge, you should consult a text on BASIC programming. We hope that in following this material you have access to and use of a microcomputer. Although this is not essential, and the understanding of the text and exercises do not require a computer, it is desirable to have a computer and to learn with it. We assume that your (real or imaginary) microcomputer has at least a standard keyboard and CRT, and we will refer to these for the remainder of the chapter.

In the real world of a particular computer you may have to consult a manual to determine how to transfer what you have typed to the CPU. (This is normally done by touching a key labelled ENTER or RETURN.) You will also want to learn how to backspace to correct an inadvertent wrong entry (usually a backspace key or left-arrow key). A flawed line in a program that has been stored in memory can be corrected by reentering the line (see Section 5.4) or by using the computer's editing capabilities (see the computer manual).

When the microcomputer is turned on (or programmed, if necessary, to interpret BASIC) it will signal by displaying on the CRT a **cursor** (a line or small block, which possibly blinks) or some message (such as "OK" or "HELLO" or "READY"). The computer is then said to be in **BASIC mode**, in which state it will accept any BASIC command and execute it. For example, if you type in

PRINT 5 + 8

and ENTER it, the computer will print the answer, 13, on the CRT. In this mode the microcomputer can be used exactly like a hand calculator. Although this use falls far short of the computer's capabilities, it is a good way to introduce some of the symbols and words of BASIC. Table 5.1 gives the BASIC equivalents of the standard arithmetic symbols of operation.

If several symbols of operation appear in a given expression, the computer will first simplify within parentheses, if any. Then the operations are executed in the following order:[4]

- first, exponentiation is carried out;

---

[4]In this regard, computers follow the customary procedures of arithmetic and algebra.

	Symbol		Example	
Arithmetic	BASIC	Arithmetic	BASIC	
$+$	+	$5 + 8$	5 + 8	
$-$	-	$8 - 5$	8 - 5	
$\times, \cdot$	*	2.3×4.5	2.3 * 4.5	
$\div$	/	$-17 \div 3$	-17 / 3	
exponentiate	^ or **	2^5	2\^5 or 2**5	

Table 5.1   BASIC equivalents of standard arithmetic symbols.

- second, multiplication and division are carried out, from left to right;
- last, addition and subtraction are carried out, from left to right.

For example, the BASIC expression 2 + 3 * 6 / - 3 ^ 2 would be computed, successively, as

$$2 + 3 * 6 / - 9$$
$$2 + 18 / - 9$$
$$2 + - 2$$
$$0$$

Of course, the computer does not show these intermediate steps. As a matter of fact, it would perform none of these steps, if we merely ENTERed 2 + 3 * 6 / - 3 ^ 2, since we have not told the computer to *do* anything. However, if we ENTER

$$\text{PRINT } 2 + 3 * 6 / - 3 \verb|^| 2$$

we will see 0 appear almost instantaneously.

To take a first step toward programming, we might wish to store one or more values in memory, before or after some computations, for later use or output. To do this, we **assign** the value (from the keyboard or as the result of a computation) to a **variable**. This is accomplished with a LET command. For example, the statement

(5.1)   LET A = 3 (or simply A = 3)

assigns to the variable A the value 3. Practically, you may think of the computer's internal memory as a vast array of unlabeled pigeonholes. The assignment statement (5.1) places a "3" in some pigeonhole and labels it "A," so that it can be retrieved at any later time by using the label.

We can now PRINT the value of A or use this variable in some computation. For example, we might now ENTER

Microcomputer.

Credit: Radio Shack

$$\text{LET B = A * A}$$

and the variable B will be assigned the value A × A. (In this case, if A has not been reassigned, then A = 3, and thus B = 9.)

An acceptable name for a BASIC variable that is assigned a numeric value is any letter, or any pair of letters, or any letter followed by a number. (Some dialects allow variable names to be longer than two characters.) This provides an enormous set of possible numeric variables. There are also specialized forms for special categories of numeric variables. A BASIC text or computer manual will show their form and use. We will not need these special forms for our modest introduction to BASIC.

It should be noted that a variable can only have one value at a time. If we want to store several numbers, it will be necessary to assign them to variables with distinct names. On the other hand, the value of a variable can be changed as often as we like — a most useful feature in programming. It is acceptable, even desirable in many instances, to use the current value of a variable in computing a new value. For example, the statement

(5.2)   LET A = 2 * A + 3

has the effect of assigning to the variable A the number that is 3 more than twice the "old" value of A. Thus, if Statement (5.1) were first ENTERed, assigning to A the value 3, then the result of ENTERing Statement (5.2) would be to compute 2 × 3 + 3, or 9, and to assign this as the "new" value of A. This use of equality in Statement (5.2) differs from the usual one in mathematics. In algebra, the statement $A = 2A + 3$ would be read as an equation, whose solution is -3. In BASIC, the statement directs the CPU to obtain the current value in pigeonhole "A," perform a computation, and place the newly computed value in that pigeonhole, simultaneously pushing out the old value.

BASIC can also assign, manipulate, and print **alphanumeric** data (such as names, text, dates, etc.). Such data may be entered by typing it between a pair of quotation marks and assigning it to a **string variable**, which is identified as such by adding a dollar sign after a variable name. For example, if you **ENTER** the statements,

$$\text{LET A\$ = ``GREETINGS!''}$$
$$\text{PRINT A\$}$$

the computer will print **GREETINGS!** on the CRT. ("GREETINGS!" is stored in the computer as a 'string' of symbols, filed under the label "A\$.") A **string** may consist of any sequence of symbols (though certain punctuation marks pose problems), provided the number of characters is not too big. The maximum length of a string is a property of the microcomputer used, but is usually 255 at least characters.

Note, in the previous examples, that the computer distinguishes the numeric variable A from the string variable A\$. It should also be noted that an assignment statement cannot mix strings and numbers, nor can a computation be carried out if it mixes string variables and numeric variables. Thus, each of the following statements would be interpreted by the computer as nonsense and would produce an **error message** on the CRT:

$$\text{LET X = "WORD"} \qquad \text{LET M\$ = 8.5} \qquad \text{PRINT A + A\$}$$

In addition to the arithmetic operations, BASIC provides all the standard functions of algebra and trigonometry, plus a large collection of functions that are specially tailored to computing (functions that manipulate strings or convert from string to numeric form, and many others). For example, the BASIC function **ABS** will yield the absolute value of a number, or a numeric variable, or any expression composed of them.

The form of a BASIC function is its *name*, and it usually consists of three or more letters, followed by a pair of parentheses that enclose the argument of the function. The argument may be a number, a variable or some expression involving two or more of these. We will have occasion to use only a few; they are identified in Table 5.2. For other functions available, consult a BASIC text or the manual for the microcomputer you are using.

	FUNCTION		EXAMPLES	
		BASIC	Algebra	Value
absolute value		ABS (-3)	$|-3|$	3
		ABS (7.5)	$|7.5|$	7.5
		ABS (X*Y)	$|X \cdot Y|$	
greatest integer		INT (-3)	$[-3]$	-3
		INT (7.5)	$[7.5]$	7
		INT (X + Y)	$[X + Y]$	
square root		SQR (49)	$\sqrt{49}$	7
		SQR (7.5)	$\sqrt{7.5}$	2.7386
		SQR ((A-B)/C)	$\sqrt{\dfrac{A-B}{C}}$	

Table 5.2 *Some algebraic functions in BASIC.*

EXERCISES 5.3

In Exercises 1–12, write the corresponding BASIC expression for the given arithmetic expression.

1. $7 + 4$
2. $-3 + 4.6$
3. $2 - 7$
4. $-12.35 - (-8.1)$
5. 5×8
6. $11.3 \div (-4.2)$
7. $(-9) \div 2$
8. $(-7)(4.2)$
9. 3^4
10. $8^{\frac{1}{4}}$
11. $13 - 51$
12. $B^2 - 4AC$

In Exercises 13–22 write the corresponding arithmetic expression for the given BASIC expression.

13. 2.4 - 3.5
14. 10 / -4
15. 17 * -4
16. -6/4
17. 17 ** -4
18. -5 ^ 3
19. ABS(6 * -3)
20. SQR(2 ^ 3)
21. ABS(X - Y)
22. INT(3.7)

In Exercises 23–32, write the corresponding arithmetic expression for the given BASIC expression, and calculate the number by hand or with a hand calculator. If you have access to a computer, check your answer by having the computer PRINT the expression.

23. 18 + 3 * 2
24. 15/3 - 2
25. 2 ^ 3 * -2
26. 2 ** -3 + 5/8
27. 2 ^ (3 * -2)
28. 2 * -3 + 5/8
29. 2 * (-3 + 5)/8
30. 2 * -(3 + 5/8)
31. ABS(2 ^ 3 * -2)
32. INT(2 * -(3 + 5/8))

In Exercises 33–52, describe what, if anything, would happen in the computer's memory, and what, if anything, would appear on the CRT if the given statement or statements were ENTERed. If you have access to a computer, ENTER the statement(s) and check the results against your predictions.

33. LET S = 8   34. Q = 2 ^ 5

35. LET F$ = 19   36. LET F$ = NUMBER

37. F$ = "FINAL EXAM"

38. PRINT 2 X 3

39. PRINT 2 * 3   40. PRINT "2 * 3"

41. 2 * 3 =   42. PRINT 2 * 3 =

43. PRINT ABS(-1.21)

44. PRINT INT(-1.21)

45. PRINT SQR(-1.21)

46. PRINT ABS(INT(-1.21))

47. PRINT INT(ABS(-1.21))

48. PRINT INT(SQR(ABS(-1.21)))

49. LET M = 3   50. LET N = 2
    LET N = 2   LET N$ = "TWO"
    PRINT M * N   PRINT N * N$

51. B1 = 8   52. ED = 1.1
    B2 = 3   ED = ED * ED
    B2 = 1/B2   ED = ED + 2
    PRINT B1 ^ B2   PRINT "ED"

WRITING EXERCISES

1. Compare and contrast the way the word "let" is used in ordinary English with the way it is used in mathematics and computer programming.

2. In what ways do *string variables* differ from *numeric variables*? In what ways are these two concepts similar?

5.4   Programming

With the few BASIC statements we have seen so far, we can begin to write a BASIC program. A **program** is a sequence of statements that the computer executes, one after another, when directed to do so. The BASIC interpreter identifies a statement as a program statement if it is preceded by a number.[5] The numbering has the additional function of ordering the steps of the program and labelling the program lines for possible reference elsewhere in the program. For example, if we ENTER

100 PRINT "HELLO"

the BASIC interpreter does not execute the instruction to print the word "HELLO." Instead, it stores this instruction in its memory and will execute it (alone or as part of a longer program) when it receives the command "RUN."

In a BASIC program, then, each BASIC statement will have a labelling number, and when the computer receives the RUN command, it will exe-

---

[5] Versions of BASIC that allow unnumbered program statements are becoming more prevalent, although the most common versions (GW-BASIC and BASICA) do not. We have opted for the numbered-statement format in this chapter because it can be understood (interpreted or compiled) by virtually every version of BASIC.

cute the statements consecutively starting with the lowest numbered statement. Each numbered program statement must be typed in separately and ENTERed.[6]

Example 5.1   The following program is a cumbersome way to compute $3^2 + 4^2$.

```
10   X = 3
20   Y = 4
30   Z = X * X + Y * Y
40   END
```

When RUN, this program will assign X the value 3 in line 10, assign Y the value 4 in line 20, compute $X \times X + Y \times Y$, and assign that value to Z in line 30, then stop.   □

Note that the numbering of the program statements increases but the integers used need not be consecutive; in fact, it is common practice to number in multiples of ten or twenty, or larger. This allows for easy alteration of a program by insertion of one or more new steps between a pair of existing steps. For example, if we RUN the program of Example 5.1, no result is output because we failed to include a PRINT statement. This is easily amended by ENTERing:

```
35   PRINT Z
```

The computer automatically amends the existing program by including this statement in its proper place. To see the expanded, reorganized program, we can ENTER the command "LIST." This command lists the program currently in the memory on the CRT. In our example the computer would print the following on the CRT:

(5.3)
```
10   X = 3
20   Y = 4
30   Z = X * X + Y * Y
35   PRINT Z
40   END
```

It is not necessary to LIST the program; this is a convenience to us as we program, allowing us at any time to display the current program. If we now RUN our program (5.3), it will compute the value 25, assign it to Z, and print that number on the CRT.

---

[6]Most forms of BASIC allow several statements, separated by appropriate punctuation, per program line; we shall not use this option. Some versions allow "full-screen" editing; staying with the most commonly acceptable processes, we do not assume this flexibility.

The IBM PS/2 has 3-dimensional graphics capabilities.

The "END" statement is a common way to terminate execution of a program. The command "STOP" has the same effect. (There are some differences, but as program terminators, either is effective.) In many versions of BASIC, this statement is optional; in others, an error message will appear if the "END" or "STOP" is omitted.

An alternate way of achieving the same result as in Program (5.3), which exemplifies an alternate method of entering data, is to use the following program.

Example 5.2

```
10   READ X
20   READ Y
30   Z = X * X + Y * Y
35   PRINT Z
40   END
50   DATA 3, 4
```

(We might ENTER this program in its entirety, line by line. However, if Program (5.3) were stored in the computer memory, we could simplify our task by ENTERing only the new lines — 10, 20, and 50. When we ENTER lines 10 and 20, the old lines are erased from memory and replaced by the new ones, in the same way that an assignment of a new value to a variable erases the old value.)   □

The "READ" statement in line 10 causes the computer to find the "DATA" statement (or the first such statement, if there were more than one), to read the first item of data there and to assign it to the variable named X. The next "READ" statement will cause the next data item to be read and assigned. Note that several data items, separated by commas, can be listed within one DATA statement. Also, the location of the DATA statement within the program is immaterial; it may even come after the END statement because the DATA statement is not an executed statement.

Let us now change the program of Example 5.2 by replacing lines 10 and 20 with the following two statements:

```
10   INPUT X
20   INPUT Y
```

The "INPUT" statement is a third, and more versatile, way to enter data. When the computer encounters an INPUT statement in a program, it pauses, displays a question mark on the CRT and awaits a keyboard entry of data. The ENTER key is used to signal the end of the data entry, at which point the entered data is assigned to the variable named in the INPUT statement.

In our example, as currently amended, we have a superfluous statement in line 50; we have no need for a DATA statement now that we have replaced the READ statements. This is easily corrected by the command "DELETE 50," which, when ENTERed, deletes line 50. If we now LIST our amended program, we find:

Example 5.3

```
10   INPUT X
20   INPUT Y
30   Z = X * X + Y * Y
35   PRINT Z
40   END
```

If we RUN this program, the CRT will display a question mark, awaiting data in line 10. Let's say we ENTER 5. The computer will now complete execution of line 10 by assigning X the value of 5, and will go on to line 20. Again, the CRT will display a question mark and await input. Let's say we ENTER 8. The computer will assign Y the value of 8 (line 20), compute $5 \times 5 + 8 \times 8 = 89$, then assign Z the value 89 (line 30), print that value (line 35), and stop (line 40). □

We could now RUN the program again as often as we like, and input new values of X and Y each time. This is not a complex computation, but we see in its form one value of the computer. It easily allows us to repeat a calculation many times with different data. Actually, if we wanted to use this program many times, we might amend it as follows:

$$40 \quad \text{GO TO 10}$$

As you might expect, the "GO TO" statement (or "GOTO", without a space) directs the computer to go to, or branch to, another location in the program (the line corresponding to the number specified). In this example, after PRINTing Z, the computer would GO TO step 10, and start the program over again. This program is now unending, and can only be terminated by the appropriate keyboard override, which depends on the particular hardware. (This might be a BREAK key, an ESCAPE key, or a combination of a CONTROL key with some letter key.) As another example of the use of the GOTO statement, consider the following program.

Example 5.4

```
100   P = 0
200   INPUT S
300   P = P + S
400   PRINT P
500   GOTO 300
```

When RUN, this program will await an input at line 200. Suppose we ENTER 2. The computer will then assign S this value, add it to 0 (the initial value of P), obtaining 2, assign this as the new value of P in line 300, and then PRINT this value in line 400. In line 500 the computer branches back to line 300, where it adds P (now equal to 2) to S (which still equals 2), obtaining 4, which is printed. Following this process over and over, the computer will print 2, 4, 6, 8, 10, etc. until it reaches its capacity for the magnitude of a number, or until it is interrupted by a manual override from the keyboard. □

At this point it is appropriate to introduce one of the conveniences BASIC provides, which economizes on the amount of program language we must use. If we wish to PRINT, READ, or INPUT several variables, we need not have a separate program statement for each, but may simply list all the variables, separated by commas. For instance, the statement

PRINT P, Q, R

will cause the values of these three variables to be printed on the CRT, on one line, separated by a few spaces.[7] If we want the three variables to appear on separate lines, three separate PRINT statements should be used. As an example of a simple program that uses this command, consider the following.

Example 5.5

```
 20  INPUT N
 40  LET NA = N * N
 60  LET NB = N * NA
 80  LET NC = N * NB
100  PRINT N, NA, NB, NC
120  END
```

This program will accept a number, compute its square, cube, and fourth power, and print these four numbers on one line of the CRT.   □

In a similar vein, the statement

READ V1, V2, V3, V4

has the same effect as four consecutive READ statements, one for each of the four variables listed; and a statement such as

INPUT N, M, L

allows input of three variables. When the computer encounters this last statement, it prints the usual prompting question mark. You may ENTER values for all three variables in order, separated by commas. If you enter fewer values than the INPUT statement calls for, the computer will accept these and prompt with another question mark (or a double question mark) for additional values. If you ENTER more values than called for, the computer will ignore the excess ones, or, in some forms of BASIC, it will reject the entry and prompt you for a corrected entry.

In all three of these multiple-variable statements, it is allowable to mix the variable types; for example,

READ B, C$, D

---

[7]There is a wide variety of PRINT statements, allowing carefully designed formatting of output to the CRT. The use of a comma in the PRINT statement has the effect of a tab key on a typewriter. (These tab settings are predetermined by the hardware.) This can be used to line up output in columns. The use of semicolons between variables in a PRINT statement causes the values to be output consecutively with no space between. For the many other options on output, consult a text on BASIC (such as [7] or [8] in the list *For Further Reading*) or the manual for the machine you are using.

is acceptable syntax in BASIC. It directs the computer to read a number from a DATA statement and assign that number to B, then read a string and assign it to C\$, and then read another number and assign it to D. Of course, the DATA statement must match the variable types, in the proper order. A similar caution applies to INPUT statements. In a PRINT statement it is permissible to mix variables and constants, or even to use implied operations. Consider the following program.

Example 5.6

```
 5   INPUT X, Y
10   PRINT X, Y, "SUM IS"; X + Y
15   END
```

If this program were RUN, and the values of X and Y input were 5 and 8, respectively, execution of line 10 would produce the following display on the CRT:

```
5              8              SUM IS 13                  □
```

Another nonessential, but very useful, variation on the INPUT statement is the option of a prompting message. For example, the statement

```
INPUT "WHAT TWO NUMBERS"; X, Y
```

when executed will cause the message within quotes to appear in front of the question-mark prompt.[8] Such prompting messages can be very useful in directing a user as to how many values are to be entered or what type of data is to be entered; for example:

```
INPUT "ENTER YOUR NAME AND AGE, SEPARATED BY A COMMA"; N$, A
```

As a final point, we address how to store a program for future use. Storage varies widely with the hardware and operating system, but it is commonly accomplished with the command "SAVE" or some variation on that word, which transfers a copy of the program to external memory. Conversely, a comparable variation on the command "LOAD" will erase the internal program memory and load into it a program from external memory. With or without such transfer of program, the internal memory can be erased and readied for a new program with the command "NEW."

---

[8]Some versions of BASIC require a variant command, INPUT PROMPT, for this feature.

EXERCISES 5.4

For Exercises 1–4, describe what the given program would accomplish if RUN, assuming, where necessary, that appropriate data is ENTERed when run. (For example, the program of Exercise 1 will accept two numbers as input and print their sum on the CRT.)

1.
```
10   INPUT N1
20   INPUT N2
30   PRINT N1 + N2
40   END
```

2.
```
100  READ A, B, C, D
200  S = A + B + C + D
300  S = S/4
400  PRINT S
500  END
600  DATA 87, 73.4, 89.5, 78.1
```

3.
```
20   PI = 3.1416
30   INPUT "WHAT RADIUS"; R
40   C = 2 * PI * R
50   PRINT "CIRCUMFERENCE IS"; C
60   END
```

4.
```
300  INPUT "WHAT IS YOUR NAME"; N$
310  PRINT
320  PRINT
330  PRINT "HOW DO YOU DO,"; N$;"."
340  PRINT "I'M PLEASED TO MEET YOU."
350  PRINT
360  END
```

5. What changes should be made in the program of Exercise 1 so that it will accept three numbers and PRINT their sum?

6. Change the program of Exercise 1 so that it automatically runs again after printing.

7. Change the program of Exercise 2 so that the values for the four variables can be input from the keyboard.

8. Change the program of Exercise 3 so that it will compute and PRINT the area of a circle whose radius is entered.

9. Change the program of Exercise 3 so that it will compute and report both the circumference and the area of a circle whose radius is entered.

10. If the program of Example 5.4 is RUN and the number input is −3, what will be PRINTed on the CRT?

11. Write a program similar to that of Example 5.4 that will accept a numerical input and then PRINT successive powers of that number.

12. Write a program that will accept the entry of two numbers and then compute and PRINT both their sum and their product on the same line of the CRT.

13. Write a program that uses the same input and calculation as does Exercise 12, but that PRINTs the sum and the product on separate lines, with appropriate identifying messages.

14. Write a program that will calculate and PRINT the volume of a rectangular solid of length 11.3, width 6.22, and height 4.7.

15. Write a program that will allow the entry values for the length, width, and height of a rectangular solid, and will then calculate and PRINT the volume of that solid.

16. Write a program that will prompt the user to enter the number of miles traveled and the number of gallons of gasoline used in an automobile trip, and then will calculate and PRINT the correct miles per gallon.

17. Write a program that will accept the entry of values for the angle measure (in degrees) of two angles of a triangle and then will calculate and PRINT the angle measure of the third angle.

18. Write a program that will accept the entry of values for the lengths of the two legs of a right triangle and then will calculate and PRINT the length of the hypotenuse.

WRITING EXERCISES

1. Which of the examples in this section is most closely analogous to Exercise 15? Explain.

2. From Exercises 16–18, choose the one that appears to be the most difficult for you. Write three simpler questions that you think might provide helpful steppingstones for solving the exercise you chose. Be as specific as possible.

3. In what ways are Exercises 15–18 all analogous? In what ways are they different?

5.5   Logic and Loops

Note

In studying this section it might be useful to consult Appendix A on logic, especially Section A.3.

We have seen the capacity of the computer to perform computations very quickly or to provide a format for repeating similar computations many times. But we have not really tapped the essential power of the computer because we have not used its ability to repeat parts of a program or perform different operations selectively, depending on the values of one or more variables.

Such "choices" of the computer are made on the basis of a numerical comparison (less than, equal to, greater than) of two variables or of a variable and a constant. Depending on the result of the comparison, which is a *true* or *false* logical response, the computer performs some operation (PRINTs something, assigns a value to a variable, GOes TO some other location in the program). Consider the following program, for example.

Example 5.7

```
10   INPUT A, B
20   IF A < B THEN PRINT A
30   IF A > B THEN PRINT B
40   END
```

In executing line 20, the computer compares the values of A and B. If the result of this comparison is *true* (A is less than B), then the computer executes the command following "THEN." If the result is *false*, it proceeds to the next line of the program. In this example, the net result will be the printing of the smaller of A and B. If the same value were ENTERed for A and B, the result of both tests would be *false* and nothing would be printed.   □

An alternate form of this conditional statement is

IF... THEN... ELSE...

In this form a command is executed if the test result is *true* (the command after "THEN") *or* if it is *false* (the command after "ELSE"). If the command so executed is not a "GOTO," then the computer continues with the next program step. As an example, consider the following program.

Example 5.8

```
10  INPUT A, B
20  IF A < B THEN PRINT A ELSE PRINT B
30  END
```

This program has almost the same result as that of Example 5.7. If A has a value less than that of B, then A will be printed. Otherwise, B will be printed, as called for by the ELSE command in line 20. The difference between the two programs occurs if A and B have the same value. In this program the failure of the test in line 20 causes the ELSE clause to be executed, so that B is printed.   □

Admittedly, neither of these programs is particularly useful in terms of the information generated. We generally don't need a computer to tell which of two numbers is larger. However, the programs do begin to exemplify the form of conditional statements. As another example of the use of conditional statements, consider the following, which also demonstrates several other useful features of programming.

Example 5.9

```
10  INPUT "ENTER AN INTEGER"; I
20  IF I = INT(I) THEN GOTO 50
30  PRINT I; " IS NOT AN INTEGER."
40  GOTO 10
50  IF I/2 = INT(I/2) THEN A$ = "EVEN." ELSE A$ = "ODD."
60  PRINT I; " IS "; A$
70  END
```

When RUN, this program prompts us to "ENTER AN INTEGER." Suppose, for example, that we ENTER 1.5. Then the test in line 20 will yield *false* because I = 1.5 but INT(I) = 1, the greatest integer in 1.5. The computer thus executes line 30 and line 40, which sends it back to line 10. On the CRT we would see

```
1.5 IS NOT AN INTEGER.
ENTER AN INTEGER                                    □
```

Lines 20 through 40 of the program of Example 5.9 form a **trap** in this program that tests the input to see if it is of the type desired. If not, the computer informs us of this and cycles back to the start. It will not allow an entry other than the type desired.

Continuing to use the program of Example 5.9, we suppose now that we **ENTER** 15 when prompted. In this case the result of the test at line 20 is *true* and the program branches to line 50, where another test is encountered:

$$I/2 = 15/2 = 7.5 \quad \text{but} \quad INT(I/2) = INT(7.5) = 7$$

Consequently, the test fails (yields *false*), and the "ELSE" clause is executed, thereby assigning the string variable **A$** the value "ODD." Thus, when line 60 is executed, the CRT shows:

```
15 IS ODD.
```

You should note the use of lines 50 and 60 to allow the printing of two slightly different messages, depending on the results of the test in line 50. Also note the role of the spaces within the quotes in line 60. Without them, the CRT display would show:

```
15ISODD.
```

BASIC provides great versatility in its conditional statements by allowing pairs of comparison symbols to be used together (with the word *or* understood between the pair. For example, the computer interprets

```
IF X <= 3   THEN...
```

as a test for X less than or equal to 3. Similarly, the computer interprets

```
IF K <> 0   THEN...
```

as a test for K less than or greater than 0; that is, for K not equal to 0.

Simultaneous tests of two or more conditions connected by "AND" or "OR" are also acceptable. For example,

```
IF X <= 3 AND K <> 0   THEN...
```

will yield *true* and will execute the command following "THEN" provided that both conditions are met. The *true* or *false* results of such combinations follow the usual logical conventions (which are explicitly defined and explained in Appendix A).

As another example of logical branching, consider:

Example 5.10

```
10   INPUT "ENTER AN INTEGER''; K
20   N = 1
30   PRINT N; " TIMES ''; K; " EQUALS ''; N * K
40   IF N = 10 THEN END ELSE N = N + 1
50   GOTO 30
```

Suppose, for example, that we ENTER the value 7 for K when prompted in line 10. When line 30 is executed, the CRT will show:

1 TIMES 7 EQUALS 7

Now, in line 40, because the result of the comparison is *false*, the ELSE command is executed and N is assigned the value 2. Then line 50 sends the computer back to line 30, which now produces the output to the CRT:

2 TIMES 7 EQUALS 14

The computer will continue to execute lines 30, 40, and 50, producing a table of consecutive multiples of 7, until $N = 10$. At that point, the result of the test in line 40 will be *true* and the END statement will be executed.

□

This process of repeating several steps of a program for a predetermined number of times is called a **loop**; it is one of the most important procedures in programming. In fact, it is so frequently used that high-level computer languages provide one or more ways to program it easily. In BASIC, several types of loops are available; we shall only work with the one most universally available, the so-called "FOR-loop." A sequence of steps which are to be "looped" through several times is set off by the pair of statements

"FOR..."   and   "NEXT..."

In particular, the result of the program in Example 5.10 could have been accomplished with the FOR... NEXT... format as follows:

Example 5.11

```
10   INPUT "ENTER AN INTEGER "; K
20   FOR N = 1 TO 10
30   PRINT N; " TIMES "; K; " EQUALS "; N * K
40   NEXT N
50   END
```

This program accomplishes the same thing as that of Example 5.10. The only difference is that the computer automatically increases the value of N by 1, compares it to the maximum value to be used (the "10" specified in line 20) and iterates the loop until finished. At that point, the program continues with the line immediately following the NEXT statement.   □

Note carefully the format of the statements in Example 5.11. The FOR statement identifies the name of the variable to be incremented, its initial value, and its limiting value. The NEXT statement identifies the end of the loop and sends the computer back to increment the loop variable. BASIC allows additional versatility in the FOR... NEXT... loop in that the size of the increment (the step change in the loop variable) can be declared. It may be positive, negative, integral, or fractional. For instance, the command

```
FOR T = 5 TO -3 STEP -2
```

starts a loop with variable T, initially assigned the value 5. T decreases in value by 2 each time through the loop until it has value −3. The command

```
FOR J = 0 TO 5 STEP .3
```

starts a loop with variable J, which has initial value 0 and which increases by .3 each time through. Note in this example that J will not take on the value 5. On the seventeenth time through the loop, J will have value 4.8. When J is incremented by .3 to 5.1, that value is greater than 5. The loop will then be bypassed and the program will resume until the first line after the NEXT J statement. If no STEP is specified in the FOR statement (as in Example 5.11), the computer uses the "default value" of 1.

As an example of the use of a loop with a specified STEP, consider:

Example 5.12

```
100   INPUT "ENTER AN INTEGER BETWEEN 1 AND 99"; S
200   K = 0
300   K = S + K
400   IF K < 10 THEN GOTO 300
500   PRINT "THE TWO-DIGIT MULTIPLES OF "; S; " ARE:"
600   FOR P = K TO 99 STEP S
700   PRINT P,
800   NEXT P
900   END
```

Suppose, for example, that 6 is ENTERed and assigned to S in line 100. Then, at line 300, K is assigned the value $6 + 0$, or 6, and the test in line 400 results in a "true." This sends the computer back to line 300. Now, K is assigned the value $6 + 6$, or 12, which is not less than 10, and so the computer goes on to line 500, which prints the indicated message.

In line 600, a loop is entered with initial value $K = 12$ and step $S = 6$, which will effect the printing of 12, 18, 24, ... , 96. (The comma after P in line 700 will engage the tab routine, causing the printing to occur across a screen line. When the end of the line is reached, printing resumes at the beginning of the next line.) When P is now incremented by 6 to 102, this value is greater than 99, so that the loop terminates and line 900 is executed, thereby ending the program.   □

In most forms of BASIC the steps between "FOR..." and "NEXT..." are always executed with the initial value of the loop variable before any incrementing and comparison is done. This is the case even when the initial value is beyond the specified limiting value of the loop variable.

Example 5.13

```
10   FOR H = 25 TO 1 STEP 3
20   PRINT H
30   NEXT H
40   END
```

If this program is RUN, line 20 will be executed and the initial value of H, 25, will be printed. Only when H is incremented to 28 will a comparison with the loop limit of 1 be made and the loop ended.   □

Of course, one would not intentionally write a program such as the one in Example 5.13, but the principle is the same for loops with initial or terminal values that are variables. Thus, care must be taken in using such loops in order that undesired results do not follow. Some of the exercises examine this latter possibility.

EXERCISES 5.5

1. Amend the program of Example 5.12 by adding one or more lines designed to act as a trap between lines 100 and 200 to prevent the use of a nonintegral entry.

2. What will happen in the program of Example 5.12 if −6 is ENTERed? Write a trap to prevent this.

3. What will happen in the program of Example 5.12 if an integer greater than 99 is ENTERed? Write a trap to prevent this.

For Exercises 4–11, consider the following program.

```
10   INPUT "ENTER A POSITIVE INTEGER"; S
20   T = SQR(S)
30   IF T = INT(T) THEN GOTO 60
40   PRINT S; " IS NOT A SQUARE."
50   END
60   PRINT S; " IS THE SQUARE OF "; T
70   END
```

4. What is the result if you ENTER 38809?

5. What is the result if you ENTER 3880?

6. What is the result if you ENTER 0?

7. Describe, in general, what the program accomplishes if RUN, assuming that appropriate data is ENTERed.

8. What is the result if you ENTER 1.44?

9. How can the program be changed to provide a more accurate output for Exercise 8?

10. What is the result if you ENTER −16?

11. How can the program be changed to prevent the problem encountered in Exercise 10?

For Exercises 12–17, consider the following program.

```
10   INPUT "ENTER A POSITIVE INTEGER"; N
20   C = 0
30   FOR I = 1 TO N
40   C = C + I
50   NEXT I
60   PRINT C
70   END
```

12. What is the result if you ENTER 1?

13. What is the result if you ENTER 4?

14. Describe, in general, what the program accomplishes if RUN, assuming that appropriate data is ENTERed.

15. What is the result if you ENTER 100?

16. What is the result if you ENTER −16?

17. How can the program be changed to prevent the error encountered in Exercise 16?

For Exercises 18–20, consider the following program.

```
10   FOR I = 1 TO 8
20   FOR J = 1 TO 5
30   PRINT I * J,
40   NEXT J
50   PRINT
60   NEXT I
70   END
```

18. Describe what the program accomplishes if RUN.

19. Amend the program so that the limits of the loops are not fixed at 8 and 5, but can be input from the keyboard.

20. Write appropriate traps for the program of Exercise 19 so that only positive integers can be used in the program.

For Exercises 21–27, consider the following program.

```
10   INPUT "ENTER A POSITIVE INTEGER"; P
20   N = 0
30   T = P
40   T = T/2
50   IF INT(T) < T THEN GOTO 80
60   N = N + 1
70   GOTO 40
80   PRINT P;" HAS ";N;" FACTORS OF 2."
90   END
```

21. What is the result if 12 is ENTERed?

22. What is the result if 15 is ENTERed?

23. Describe, in general, what this program accomplishes when RUN, assuming appropriate data is entered.

24. What is the result if −6 is ENTERed?

25. What is the result if 0 is ENTERed?

26. What is the result if 0.5 is ENTERed?

27. Amend the program so that it will compute and report the number of times 3 occurs as a factor of the ENTERed number, instead of the number of times 2 occurs as a factor.

28. Write a program that will accept the input of two positive integers and then will report whether or not the first is a multiple of the second.

29. Write a program that will accept the input of a positive integer and then will calculate and report the sum of that number of odd integers, starting with 1. (For instance, if

6 is entered, the program should compute $1 + 3 + 5 + 7 + 9 + 11$. See Exercise 14.)

30. Write a program that will accept the input of two calendar dates (month, day, and year) and then will compute and report the difference of the two in years, months, and days.

31. The Fibonacci numbers are a sequence of integers defined as follows: the first is 1, the second is 1, and every successive one is the sum of the previous two. (Thus, the first few integers in the sequence are 1, 1, 2, 3, 5, 8, ...) Write a program that will compute and PRINT the first thirty Fibonacci numbers.

32. A Pythagorean triple is a set of three positive integers, A, B, and C, such that
$$A^2 + B^2 = C^2$$
For example, 3, 4, 5 is a Pythagorean triple because

$$3^2 + 4^2 = 5^2$$

Write a program to find all Pythagorean triples in which all three integers are less than 100.

WRITING EXERCISES

1. Describe the role of "traps" in a computer program. Why are they included? Are they always necessary? Are they always desirable?

2. Identify a nonmathematical situation in which something analogous to a "trap" is designed into a system or process. Briefly critique the accuracy of the analogy.

3. Defend the statement:
   "Computers can make choices."

4. Defend the statement:
   "Computers cannot make choices."

5.6   CRT Graphics

So far, we have used only a small portion of the PRINT capabilities of the computer. Most microprocessors can specify the format of output data in great detail. Thus, output can be left or right justified, tabs can be set, numbers can be printed in columnar form, with or without commas, plus or minus signs, or dollar signs, and much more. Such features are valuable for reports and tables, and in general, can be useful for displaying output clearly. For our purposes, the ability to format output is superfluous, and we will not pursue it.

In addition to the formatting of alphanumeric data, many microcomputers have the capability to produce nonalphanumeric symbols (such as dots, lines, and boxes) and to PRINT them anywhere on the screen. This is a topic of interest, but it is an area highly dependent on the particular hardware. Some microcomputers require that each such symbol be called up by a numeric code, although the numeric coding may vary from one machine to another. Other microcomputers provide no special graphics. Machines may also differ in the number of lines that can appear on the CRT and the number of characters per line.

For all the above reasons, the compromise treatment of computer graphics given here may have to be altered significantly to fit a particular machine, and may not give a true picture of the versatility of graphics available. On the other hand, it will provide some idea of how a line or curve might be formed and how a video game can give the illusion of motion and react to keyboard (or other) control.

In our imaginary microcomputer, we shall assume that the size of the screen is 20 lines, from top to bottom, each line being 80 characters in length. We shall further assume that any keyboard symbol can be printed at any of these 1600 locations by the command PRINTAT. We shall number the lines 1 through 20, from top to bottom, and the character positions 1 through 80, from left to right. The syntax of our PRINTAT statement is, for example,

PRINTAT (3,20) "HERE"

which would cause the word "HERE" to be PRINTed on the third line from the top of the CRT with its initial letter at the 20th position in from the left.

("PRINTAT" is *not* a standard BASIC command, but is similar to commands available on many microcomputers. See the **Programming Note** at the end of this section.)

Instead of a quoted string, we might use a string variable whose value would be PRINTed starting at the specified location. We shall assume that only one alphanumeric character can be PRINTed AT a given location or, more precisely, that the PRINTAT command would delete and replace any character already on the screen in a location newly referred to. Thus the program

```
10   PRINTAT (2,10) "A = B"
20   PRINTAT (2,12) "/"
30   END
```

would produce a final CRT display of:

A / B

(As with typewriters, most printers can overstrike one character with another, but most versions of BASIC do not provide a way to do this on the CRT. In more elaborate screen graphics almost anything that can be visualized can be simulated.)

Suppose we set ourselves the task of drawing some simple figure on the screen, say a rectangle. Given our restricted set of characters (alphanumeric only), we will get only a crude result. Let's use the underline bar for the top and bottom of our rectangle and a capital "I" for the sides. The following program, then, would produce the desired result.

Example 5.14

```
10   CLS
20   FOR I = 10 TO 70
30   PRINTAT (3,I) "_"
40   NEXT I
50   FOR J = 4 TO 17
60   PRINTAT (J,10) "I"
70   PRINTAT (J,70) "I"
80   NEXT J
90   FOR K = 11 TO 69
100  PRINTAT (17,K) "_"
110  NEXT K
120  END                                              □
```

The command "CLS" in line 10 is available in some dialects of BASIC and is an abbreviation for "CLear Screen." As its name suggests, this command clears the CRT, a useful command in many situations other than graphic displays. The rest of the program should be self-explanatory, and is left as an exercise for you to analyze.

Example 5.15

```
10   CLS
20   LET A$ = "--->"
30   LET B$ = "    "
40   FOR I = 1 TO 20
50   FOR J = 1 TO 77
60   PRINTAT (I,J) A$
70   PRINTAT (I,J) B$
80   NEXT J
90   NEXT I
100  END
```

Line 20 assigns to the string variable A$ a 4-character string composed of three dashes and a ">" symbol which together resemble an arrow. Line 30 assigns a string of 4 blanks to B$. The purpose of this becomes clear in lines 60 and 70, where first the "arrow" is PRINTed AT position (I, J), followed immediately by the blanks in the same location. The visual effect on the CRT is that an arrow appears, then disappears. (Actually, it is just overprinted with blanks). The two loops have the effect of having the arrow appear and disappear at locations successively further to the right on the screen, producing the illusion of an arrow moving across the screen in the first row, then the second, etc.   □

In practice, the illusion of the moving arrow described in Example 5.15 may be flawed — because the computer is too fast! Depending on the hard-

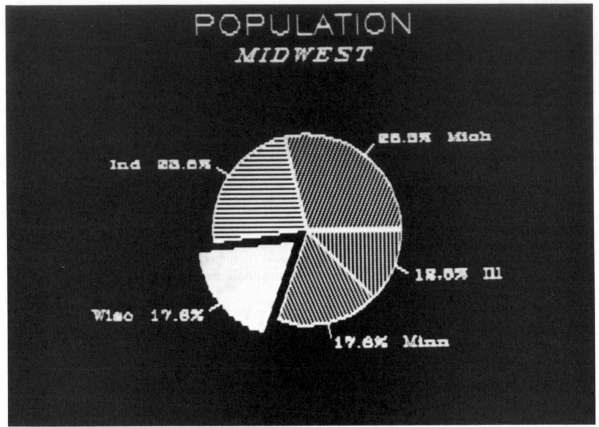

POPULATION
MIDWEST

25.0% Mich

Ind 23.6%

12.0% Ill

Wisc 17.6%

17.6% Minn

Credit: Radio Shack

Computer Graphics.

ware, it may well be that the arrow is printed and erased so quickly that the human eye doesn't have time to form the desired retinal image. This can be remedied by putting in a delaying procedure between lines 60 and 70. For example, we might insert:

```
63   FOR K = 1 TO 100
67   NEXT K
```

This loop contains no statements between the FOR statement and the NEXT statement, but it accomplishes the purpose of occupying the CPU in the task of setting K equal to 1, incrementing it 100 times, and making 100 comparisons. This all takes only a fraction of a second on most microprocessors, but during that brief time the arrow is on the screen long enough to produce an image on the human retina. We can set the value of the upper limit of the loop to produce the desired visual effect by a little trial-and-error experimentation.

Next, we introduce the "INKEY$" command of BASIC. This command provides a way of entering information from the keyboard without halting the program (as with INPUT). When INKEY$ is encountered in a program, the computer scans the keyboard (for a few milliseconds), returns the (string) value of whatever key is depressed and assigns it to the string variable named, as in an INPUT statement. If no key is pressed at that instant, a **null value** is returned; that is, the computer records a value indicating that no key was pressed.

Example 5.16

```
10   LET T$ = INKEY$
20   IF T$ = "" THEN GOTO 10
30   PRINT T$
40   IF T$ = "9" THEN END ELSE GOTO 10
```

Line 10 assigns the value of INKEY$ to the string variable T$. If no key is depressed while this line is being executed, T$ acquires the null value. This is tested for in line 20, which sends the computer back to line 10. Thus, the computer cycles between lines 10 and 20 until some key is pressed, at which point it goes on to line 30 and prints the value of T$. The value "9" was arbitrarily chosen as a way to end the program.   □

Note that 9 is within quotes in line 40 of Example 5.16 because T$ is a string variable. INKEY$ always returns an alphanumeric character, even if it is a numeral, and must always be assigned to a string variable. INKEY$ may be used in any program in which a single keystroke input is wanted. It requires one keystroke rather than two (keystroke, ENTER) and is a faster way to return control to the computer than is the INPUT statement. In a graphics program INKEY$ provides a dynamic way to interact with the graphics on the screen. Consider the next program as an example of this capability.

Example 5.17

```
10    CLS
20    I = 10
30    J = 40
40    PRINTAT (I,J) "X"
50    M$ = INKEY$
60    IF M$ = "" THEN GOTO 50
70    IF M$ = "U" THEN I = I - 1
80    IF M$ = "D" THEN I = I + 1
90    IF M$ = "L" THEN J = J - 1
100   IF M$ = "R" THEN J = J + 1
110   IF M$ = "E" THEN END
120   GOTO 40
```

In this program an "X" is printed near midscreen as a result of lines 10 through 40, and an INKEY\$ is encountered in line 50. If no key is depressed, M\$ will equal the nullstring and line 60 will recycle the computer back to line 50 to check the keyboard again. If any key other than one of those tested for is pressed, the results of all the tests in lines 60 through 110 will be *false* and the program will recycle to line 40, where "X" will be reprinted in the same place. In either case we would see no change, except for a possible flickering on the screen.

Suppose, though, that the "L" key is pressed. The variable M\$ takes on this value, which is positively matched in line 90, where J is diminished by 1, from value 40 to 39. When it reaches line 120, the computer GOes TO line 40 and an "X" is printed at $(10, 39)$, one space to the left of the original location. You should trace out what happens for the other options: U, D, R, and E.   □

As a final example, let us simulate a very simple video game by combining some of the ideas in Example 5.15 with the interactive capabilities of the INKEY\$ command.

Example 5.18

```
 20   A$ = "--->"
 40   B$ = "    "
 60   CLS
 80   PRINTAT (20,40) "!"
100   S$ = ""
120   H = 20
140   FOR I = 1 TO 77
160   PRINTAT (1,I) A$
180   FOR K = 1 TO 100
200   IF S$ = "F" THEN GOTO 240
220   S$ = INKEY$
240   NEXT K
260   IF S$ <> "F" THEN GOTO 480
280   H = H - 1
300   PRINTAT (H,40) "!"
320   IF H > 1 THEN GOTO 500
340   IF I < 37 OR I > 40 THEN GOTO 420
360   PRINTAT (1,I) "BOOM"
380   PRINTAT (9,37) "YOU WIN!"
400   END
420   FOR H = 1 TO 19
440   PRINTAT (H,40) " "
460   NEXT H
480   S$ = ""
500   PRINTAT (1,I) B$
```

```
520   NEXT I
540   GOTO 60
```

Suppose we RUN this program and do nothing else. Lines 20 and 40 assign A$ and B$ as an "arrow" and its "eraser," as in Example 5.15. Line 60 clears the screen and line 80 prints an exclamation point at the middle of the bottom line. Lines 100 and 120 initialize two variables, S$ and H, whose purpose will become clear later. S$ is set equal to the **nullstring** — the string with no symbols — and H is set equal to 20.

Line 140 starts a loop with variable I and line 160 prints the arrow at the upper left of the screen. Lines 180 through 240 form a loop with variable K. The test in line 200 fails each time through the loop because S$ is continually set to the nullstring by line 220; thus, nothing changes in the memory or on the CRT during this loop. Now, since the test in line 260 is positive, the computer branches to line 480, where S$ is again set equal to the nullstring. In line 500 the eraser is printed over the arrow, line 520 closes the I-loop, and the computer returns to line 140. Thus, with no interaction the program will effect the "moving" arrow across the top line of the CRT and then branch back at line 540 to line 60, and do it all over again.

Suppose, though, that we interact with the computer as the arrow moves across the CRT. If we press a key while the computer is executing line 220, S$ is assigned the value corresponding to that key. (It would be virtually impossible to "miss connections" because the computer is executing this line several hundred times per second, far faster than our speediest button-pushing.) If that value is anything other than "F," the test at line 260, when next encountered, sends the computer into the same pattern as if no key had been pressed. But if the "F" key is pressed, the K-loop finishes with S$ = F and the test in line 260 fails (S$ is *not* unequal to F). Now, in line 280, the value of H is decreased by one and a second exclamation point is PRINTed AT $(19,40)$, just above the initial one. The test in line 320 will cause the program to branch to 500, the eraser. Then I is incremented, the arrow "moves" to the right, and the K-loop is encountered again. But the value of S$ is still F and line 220 is skipped over each time; in other words, the K-loop now becomes just a time-delay loop. Again, H is decreased, another exclamation point is PRINTed, and the arrow moves another space.

The effect of pressing the "F" key, then, is to produce an ascending series of exclamation points (lines 280, 300) until one is printed at location $(1,40)$, the top line. The "F" key is the "firing" key to "shoot" at the passing arrow. We will leave to you the remaining analysis of how the program tests for a "hit" or a "miss," and how it changes the video display, depending on the results of the test. Some variations on this very simple "shooting gallery" game are considered in the exercises.   □

The program of Example 5.18 is considerably more complex than those we have encountered before, but if RUN, would produce a very trivial video game compared to those you have encountered in home or commercial games. This game not only has crude graphic images, but has limited control and slow apparent motion. Use of special graphics and higher resolution (the ability to access smaller portions of the screen) would create a better image, but would require more in terms of computer memory and programming. The use of color images increases the memory requirement by a factor of four or more. (A modestly high-resolution color-graphics image may require over two million "bits" of memory!) Clearly, a keyboard is not the best design for video-game input, and the `INKEY$` function allows the one signal to be read only at specific times in the program. The slow speed must be credited to the fact that the computer is spending most of its time interpreting BASIC commands and translating them into machine language.

A commercial video game uses a microprocessor chip dedicated to the processes of that game. The chip need not have the ability to deal with a high-level computer language. The hardware and machine-language software dedicates virtually all of its capabilities to handling screen graphics and specialized interactive input from buttons, dials, and joysticks. The programming involved is complex and lengthy; the handling of screen graphics is probably the most demanding computer work for hardware and programmer alike.

Programming Note

Here is how to implement, on three of the most common types of microcomputers, the fictional BASIC command,

`PRINTAT (I, J) A$`

used in this section to cause the string `A$` to be printed I lines from the top of the screen with its initial letter J spaces in from the left. In each case, two command lines are used for this task.

On IBM-PC compatible machines:
```
LOCATE I,J
PRINT A$
```

On the Apple:
```
VTAB I:HTAB J
PRINT A$
```

On the (Radio Shack) TRS-80:
```
PL = I*64+J
PRINT @ PL, A$
```

(The TRS-80 prints a 16-by-64-space screen, each space indexed by its own number, starting with 0 in the upper left and ending with 1023 in the lower right. Hence, a formula is needed to translate I and J into a single location number, PL.)

If you are using a computer that is not one of these types, consult the BASIC manual for that machine to see how it positions the cursor at a particular screen location.

EXERCISES 5.6

1. In the program of Example 5.14, which ordered pairs identify the four corners of the rectangle?

2. Modify the program of Example 5.14 so that the corners of the rectangle occur at $(5, 20)$, $(5, 60)$, $(15, 20)$, and $(15, 60)$.

3. In the program of Example 5.14, why are the limits of the loop in line 20 from 10 to 70, but from 11 to 69 in line 90?

4. Write a program that will print on the CRT a line of Os running diagonally from

   (a) $(1, 1)$ to $(20, 20)$
   (b) $(1, 20)$ to $(20, 58)$
   (c) $(20, 20)$ to $(1, 58)$

5. Write a program that will print on the CRT a border of Xs around the edge of the screen.

6. In the program of Example 5.15, why (in line 50) is the upper limit of the variable J set at 77?

7. What would be the effect on the program of Example 5.15 if line 30 were DELETEd? How would the output to the CRT change?

8. Write a program that will accept any single-key input and then will fill the screen with that symbol in every position, filling each line from left to right and the lines from top to bottom. (See Example 5.15 for some ideas on an efficient way to do this.)

9. Modify the program of Exercise 8 so that the order of filling the screen is column by column, rather than line by line; that is, the first column, from line 1 to 20, should be PRINTed first, then the second column, and so forth.

10. Modify the program of Exercise 8 so that the program ENDs if the character entered and printed is "E"; otherwise, the program should recycle for a new input without clearing the screen.

11. Write a program similar to that of Example 5.15, but in which an arrow "moves" from left to right on line 1, then from right to left on line 2, from left to right on line 3, and so forth. Make sure your arrow points in the direction of motion.

12. Although our PRINTAT statement is only hypothetical, we may assume that, like other BASIC functions, an error would occur if the parameters were outside the meaningful range. For instance, a command to PRINTAT $(45, 20)$ or to PRINTAT $(12, -3)$ would produce an error. Amend the program of Example 5.17 to make sure that this error cannot occur.

For Exercises 13–17, suppose that the program of Example 5.17 were amended by adding a new line 35 and by changing line 40 as follows:

```
35   M$ = "X"
40   PRINTAT (I,J) M$
```

What happens on the CRT in this amended version—

13. when the command RUN is given?

14. when the key "D" is pressed?

15. when the key "A" is pressed?

16. when the space bar is pressed?

17. when the key "E" is pressed?

For Exercises 18–21, suppose that the program of Example 5.17 were amended as described in the directions for Exercises 13–17 and then RUN. What would appear on the screen as a result of the following sequence of entries? ("*sp*" indicates the space bar.)

18. A, R, T

19. R, –, R, R, A, R, Y

20. M, R, D, *sp*, U, R, U, *sp*, D

21. L, L, R, A

For Exercises 22–25, refer to the directions of Exercises 18–21 and determine which sequence of entries would cause the following to appear on the CRT.

22. CAT   23. OX
24. DULL   25. RIDDLE

Exercises 26–33 refer to the videogame program of Example 5.18. It will be helpful to make a table or chart to keep track of the values of H, I, and S$ as you work your way through the program.

26. What happens if the "F" key is pressed while line 220 is being executed and the value of I at that moment is 37?

27. What happens if the "F" key is pressed while line 220 is being executed and the value of I at that moment is 21?

28. What would be the effect if line 140 were amended by adding "STEP 2"?

29. The effect of lines 280 and 300 is the printing of an ascending string of exclamation points. How would you change the program to give the illusion of a single "moving missile"?

30. In this program the "firing location" is marked by the arrow printed in line 80, which remains fixed throughout the game. How would you change the program so that this arrow could be "moved" left by entering "L" or right by entering "R" prior to firing? Make sure that the vertical line of firing (in program line 300) corresponds to the firing location.

For Exercises 31–33, recall that in the program of Example 5.18 the game continues until a "hit" is made. Describe how you would change the program so that:

31. The game ends when the moving arrow has crossed the screen 10 times or has been hit, whichever comes first.

32. The game ends if the moving arrow is missed three times.

33. The games ends if the moving arrow crosses

the screen without a shot being fired, or if the arrow has been missed a total of 3 times. A "hit" does not end the game, but is counted. At the end of the game, the player's number of hits is reported on the CRT.

WRITING EXERCISES

1. Study the "Arrow Paradox" of Zeno, described in Appendix Section B.3. Does our discussion (starting with Example 5.15) about creating an illusion of motion on a CRT have any bearing on that paradox? Explain your answer. (Think about this with a little care before deciding on your answer.)

2. Analyze the following program and describe what it would do. What would appear on the CRT, assuming appropriate input? (*Hint*: Make a table or chart to keep careful track of the values of each of the variables.)

```
10   S = 1
20   E1 = 1
30   E2 = 80
40   E3 = 2
50   E4 = 20
60   X = 1
70   Y = 80
80   PRINT "PRESS ANY KEY."
90   A$ = INKEY$
100  IF A$ = "" THEN GOTO 90
110  CLS
120  FOR I = E1 TO E2 STEP S
130  PRINTAT (X,I) A$
140  NEXT I
150  IF X = 11 THEN END
160  FOR I = E3 TO E4 STEP S
170  PRINTAT (I,Y) A$
180  NEXT I
190  S = -1 * S
200  E2 = E1
210  E1 = Y + S
220  Y = E2
230  X = E4
240  E4 = E3
250  E3 = X + S
260  GOTO 120
```

LINK: 5.7

Computer Simulation

Courtesy of AT&T

Computer Simulation. Smaller than a telephone push button, this microchip is used in voice recognition and in synthetic-speech systems.

In this chapter, we have seen some very limited uses of the computer. We know that computers are used extensively in fields such as business, communications, education, government, and the military. We know that all of these entities process large amounts of data and must perform computations on this data in activities such as bookkeeping and statistical analysis. Perhaps the most significant application of computers, though, is their use in simulating physical and sociological situations. According to the dictionary, to simulate is to give the same appearance as or have the same effect as that which is copied. In many everyday experiences we are met by simulation, rather than by reality. The voice we hear on the telephone, the music we play on the stereo, the lifelike images we view on television or movie screens — all, of course, are clever simulations. But we become so accustomed to the simulacrum that we may forget it is not "the real thing."

The use of computers in video games is something of this sort. A cleverly designed game can give us the illusion of being in a maze or actually being at the controls of a spaceship. And these examples are not far from a comparable, if more serious, simulation: that of an automobile simulator or a flight simulator to train drivers or pilots by testing their reactions in mock emergencies. The scope of simulation possible with modern computers has opened new vistas in research, especially in the social sciences. Computers have been designed and programmed to simulate such diverse matters as the traffic in New York City, the complete life of a small business, the weather patterns over the Northern Hemisphere, the growth and decline of supplies in a warehouse, mock versions of World War III, and hundreds of other multivariable situations.

A computer simulation is a program that represents in the computer memory all the significant features of the phenomena being simulated. For example, in the New York City traffic simulation, a memory location might be assigned to represent a car or perhaps a group of cars. These "cars" could have certain values assigned to them that would identify their location, direction of travel, and speed. Other variables would represent traffic signals, and still others would be limiting parameters corresponding to speed limits or one-way streets. By varying these parameters, the programmer can try out hundreds or thousands of patterns of one-way streets, timing of traffic signals, and speed limits, with a variety of traffic demands, to determine an optimal way to move traffic.

The traffic model is one that has minimal human intervention, other than the programmer's changes of conditions, and even these may be planned into the program so that the simulation automatically tries a sequence of different situations. Each car and traffic signal is essentially "on automatic." Other forms of simulation may be highly interactive, especially where instruction is part of the goal. This is clearly the case in the cockpit simulator mentioned before, whose purpose is not merely to simulate all the things an airplane can do, but to present specific simulations to pilots and to measure their reactions. Interactive simulations are common, too, for managers or military leaders. Rather than trying every possible business venture (a costly way to learn), a corporation can simulate its operations on a computer and allow its managers to learn the best choices and avoid the worst disasters within the circuitry of the computer. Similarly, generals can try out different battle strategies at the computer console and get an accurate simulation of the effects of their choices.

One of the most interesting stories in the brief history of computer simulation has been in the area of meteorology — though the most profound results have been strikingly and ironically at odds with the earliest hopes in this field. As early as the 1950s, John von Neumann saw the problem of weather simulation as an ideal one for the newly born electronic computer, and he suggested the possibility of long-range weather forcasting and even weather control as a reality that would shortly be within our grasp. The physical laws were known; the equations could be programmed into the computer; all that was needed was accurate data and sufficient computing power and speed.

By 1960, M.I.T. mathematician and meteorologist Edward Lorenz had successfully programmed his computer with a simple system of twelve equations that did a surprisingly good job of simulating the gross features of weather. Of course, it was hypothetical weather, initiated by entering a set of numbers to represent temperature, wind speed and direction, air pressure, humidity, etc. The output was page after page of more numbers, representing the changing values for those quantities; but those numbers communicated to the meteorologist an imaginary weather system that was remarkably like the real one he knew: large-scale cycles of temperature and wind corresponding to seasonal changes, smaller cycles corresponding to daily fluctuations, all with recognizable patterns, but not with machine-like repetition. Von Neumann's dream was taking shape in Lorenz's computer — or was it?

By 1961, Lorenz had altered his computer output slightly by translating some of the numerical values into a rough graph to help him visualize the simulated weather — an undulating curve tracing across the paper. One day he wanted to extend an interesting curve beyond the previous run; to save time, he restarted the program with data corresponding to the situation near the end of the preceding run. (He had the numerical output data as well as the graphical output.) The resulting graph should have matched the end of the preceding one and then gone on to extend it, but, to his amazement, it did not! The new graph started out virtually identical to the old one, but then diverged from it, at first slightly, then more and more, until the two graphs were describing dramatically different weather patterns in what would correspond to only a few days in the simulation. After some reflection, Lorenz tracked down the source of the discrepency: he had programmed the computer to perform its calculations with numbers accurate to six decimal places; but for brevity, the output printed with only three-decimal-place accuracy. When he restarted the program, he had copied in the three-decimal-place numbers, which were slightly different from those the computer had been working with on its prior run. Identifying the source of the discrepency in no way diminished the shock, though. The initial difference in numerical values could be no greater than one part in a thousand, corresponding to a minute, virtually immeasurable, difference in the physical quantity it represented; the effect on the simulated weather within a very short time was totally out of proportion. Mathematician Lorenz might have concluded that his simulation was seriously flawed and that he needed to modify his program. Meteorologist Lorenz, however, recognized that the problem lay not with the simulation, crude as it was; rather, this chaotic behavior of his model was an accurate representation of the chaotic behavior of the weather itself. Von Neumann's dream had turned into a nightmare.

Science had long known that there was an inherent degree of error in applying its theories to reality. No prediction would ever be perfect, even though the underlying theory might be, because there was always some miniscule error in making physical measurements. Physical law could foretell the time of the next solar eclipse, but no one was surprised at an error of a few seconds in a prediction years in the future. Reliable theory could determine the proper trajectory to place a man on the moon, but no one was surprised if the actual flight did not perfectly match the computed orbit. The actual initial values — for time and direction of launch, for mass of the rocket, for engine thrust, etc. — no matter how carefully measured, would differ slightly from the theoretical values, and there were small influences in the real world, such as the gravitational effect of Pluto, that were so trivial that the theory ignored them; such facts would explain the slight discrepency in the orbit, and would be offset by a midcourse correction. But such experiences lent support to one of science's most fundamental tenets, that small variations in initial conditions would have only a small effect on the outcome. Lorenz was one of the first to suspect that this tenet does not always hold. Some phenomena, such as global weather, exhibit what is now known formally as "sensitive dependence on initial conditions" and whimsically as "the butterfly effect" — the notion that something as apparently trivial as the fluttering of a butterfly slightly altering the air motion in one place could gradually escalate until it dramatically changed the weather halfway around the world days later.

But the story should not be viewed as ending in defeat for either meteorology or mathematics. Von Neumann was accurate in identifying weather simulation as a worthy challenge for the computer, and over the next thirty years, both the power of computers and the collection of weather data, especially via satellite, increased beyond anything he might have predicted. Supercomputers now work with thousands of variables in less time than Lorenz's machine took to handle only twelve, and the results are translated into three-dimensional graphic images in full color. Although the reliability of predictions diminishes rapidly for periods longer than a week, meteorologists can and do use their simulations to predict the weather reasonably accurately for several days in advance anywhere on the globe.

Meanwhile, Lorenz and others turned their attention to a careful exploration of the mathematics of sensitive dependence on initial conditions, which has led to the fields now known as *chaos theory* and *fractal geometry*. This exploration could not have been made without the computer. Some of the seminal work in this area had been begun early in this century, by Gaston Julia and Pierre Fatou, but they were hampered by the monumental amount of relatively simple computation needed to produce any intelligible results. The computer permitted a productive reconsideration of their work, and the expansion of that work to far greater detail and accuracy. As researchers used the computer screen to represent pictorially the results of millions of numerical calculations, they occasionally found themselves looking at oddly familiar shapes — the outline of a leaf, a meandering river, a jagged mountain landscape. They had stumbled upon a simulation of real-world phenomena, suggesting the possibility that mathematics might provide a theoretical explanation for these seemingly random shapes. Although Von Neumann's dream of precise long-term weather prediction was not fulfilled, the computer-simulated search has led to a whole new branch of mathematics and an unexpected ability to simulate and perhaps explain parts of nature previously thought to be too chaotic ever to be captured by a mathematical description.

Topics for Papers – Chapter 5

1. Research one of the following characters in the history of computer science and write a biographical account of the person you choose, concentrating on his/her contribution to computers:

    - Charles Babbage
    - Ada Lovelace
    - John von Neumann
    - Alan Turing
    - Norbert Wiener

    (Be sure to list the source(s) you used in writing this paper, and be careful that you do not simply copy or paraphrase the material from your sources.)

2. Write a paper of no more than two pages, using your current knowledge and opinions, on the question, "Can a machine think?" Then read Alan Turing's essay of that title. (It originally appeared in *Mind* in 1950; reprinted in *The World of Mathematics*, James R. Newman, ed. New York: Simon & Schuster, Inc., 1956, 1988.) Write a second paper, as a companion to and commentary on your own original paper, covering the following points:

- What portions of your original paper would you leave unchanged? What would you change?
- What issues, if any, did Turing raise that you did not deal with in your paper? Would you characterize these issues as an oversight on your part? Are they points of disagreement with Turing?

3. This is a combined exercise in programming and documentation. Write a BASIC program that performs some user-initiated procedure of your choice. (It can be a mathematical computation, a simple game, a visual display of data, or whatever else you think might interest your reader.) The program must be at least 25 lines long and must contain the following ingredients:

   (a) some kind of user interaction (e.g., data entry, keyboard "triggering," question response, etc.);
   (b) a loop;
   (c) a trap;
   (d) a graphic display of some sort.

   Supply full documentation for your program, written primarily for a potential user who has no knowledge of BASIC, as follows:

   - Write an opening paragraph explaining what the program is designed to accomplish.
   - Give step-by-step instructions that guide a first-time user through the process of executing the program and understanding its output.
   - As a technical appendix, describe (to a reader who understands BASIC) the purpose and behavior of the loop and the trap that you built into your program.

For Further Reading

1. Gersting, Judith L., and Michael C. Gemignani. *The Computer: History, Workings, Uses & Limitations.* New York: Ardsley House, Publishers, Inc., 1988.

2. Gleick, James. *Chaos, Making a New Science.* New York: Penguin Books, 1987.

3. Goldstine, Herman. *The Computer from Pascal to Von Neumann.* Princeton: Princeton University Press, 1972.

4. McCorduck, Pamela. *Machines Who Think.* San Francisco: W. H. Freeman and Company, 1979.

5. Peterson, Ivars. *The Mathematical Tourist.* San Francisco: W. H. Freeman and Company, 1988.

6. Poirot, James, and David Groves. *Computers and Mathematics.* Manchaca, TX: Sterling Swift Publishing Company, 1979.

7. Shelly, Gary B., and Thomas J. Cashman. *Introduction to BASIC Programming.* Brea, CA: Anaheim Publishing Company, 1982.

8. Silver, Gerald A., and Myrna L. Silver. *Simplified BASIC Programming for IBM PCs, PS/2s, Compatibles & Clones*, 2nd ed. New York: Ardsley House, Publishers, Inc., 1988.

CHAPTER

MATHEMATICS OF INFINITY: CANTOR'S THEORY OF SETS

6.1   What Is Set Theory?

Mathematics is many things to many people. To some it is just a routine tool, to others, a convenient language, to still others, a strict science. And to a surprisingly large number of people, mathematics is mainly an art, pursued for its own sake out of curiosity and with an appreciation of abstract beauty, much as when a chess grandmaster seeks an elegant checkmate of a respected opponent. This chapter treats some mathematics developed from that last point of view.

As soon as it was introduced by the German mathematician Georg Cantor in 1872, set theory caused intense controversy in mathematical, philosophical, and theological circles. That controversy contributed to Cantor's eventual mental breakdown, caused severe divisions of opinion among European theologians and mathematicians, and resulted in a continuing three-way philosophical split in regard to the foundations of mathematics. But Cantor's work also affected mathematics in a decidedly positive way. His basic set theory provided a simple *unifying* approach to many different areas of mathematics. For example, the study of probability and statistics [Chapter 4] now begins with set theory, as does modern geometry [Chapter 3] and abstract algebra [Chapter 7]. Moreover, the strange paradoxes encountered in some early extensions of his work encouraged mathematicians to put their logical house in order, so to speak. Their careful examination of the logical foundations of mathematics led to many new results in that area and paved the way for even more abstract unifying ideas.

The key to the universality of Cantor's work is the simplicity of its starting point. This fact is especially convenient for us because it allows us to get a good look at some of his most important results without requiring much preliminary material. In fact, we need assume only a few elementary ideas from your prior mathematical experience. Specifically, we assume you know (or are willing to believe) that two points determine exactly one line, and that each point of a line can be matched with exactly one real number, and vice versa. (A more detailed description of the real numbers appears in Section 6.4.) We also assume you have a little experience with various kinds of numbers — integers, fractions, decimals — and are somewhat familiar with the elementary language of sets, including *subset*, *proper subset*, *equal sets*, and *Cartesian product*. For convenience, we list here a brief summary of these essential concepts. (A more detailed description of the language of sets may be found in Section 4.2. If you have not seen this language before or if you feel unsure of it, we suggest you study that section before proceeding further.)

- Any collection of objects whatsoever is called a **set**; the objects themselves are called **elements** of that set. Sets are usually denoted by capital English letters, such as A or B or S; elements often are denoted by small (often italicized) letters, like a, b, c, etc. If a set is described by listing its elements, that list is enclosed in braces. Thus, the set of the first four letters of the alphabet is written {a, b, c, d}. The set of all letters of the alphabet may be written {a, b, c, ..., z}, where the ellipsis "..." is an abbreviated way of showing that all the letters between c and z are included. In general, the ellipsis is used to indicate that a pattern is continued. If it is followed by a specific number, as in

$$\{-1, -2, -3, \ldots, -25\}$$

the ellipsis indicates that the pattern ends with that final number. If it is not followed by a number, as in

$$\{-1, -2, -3, \ldots\}$$

it indicates that the pattern continues without end. Often letters or other symbols are used with ellipses to help describe the pattern, as in

$$\{-1, -2, -3, \ldots, -n, \ldots\}$$

where n is understood to be any natural number.

- Sometimes it is inconvenient or impossible to list all the elements in a set, or to establish enough of a pattern so that the ellipsis notation can be used without danger of confusion. The natural thing to do then is to specify some property or properties that describe all the elements in the set and nothing else. In that way it becomes possible to "build" the set from its description. In such cases we use a standard form of notation:

$$\{x \mid x \text{ has a certain property}\}$$

This is read "the set of all x such that x has a certain property." Such a set is **well-defined** if the statement "x has a certain property" always is unambiguously true or false, no matter what x is.

- If the defining condition for a set is *always* false regardless of what is substituted for the variable, then the set has no elements in it; but it is considered to be a set, just the same. The set containing no elements is called the **empty set** (or **null set**) and is denoted by $\emptyset$.

- A set A is a **subset** of a set B if every element of A is also an element of B. We write this as "$A \subseteq B$." Two sets A and B are **equal** if $A \subseteq B$ and $B \subseteq A$; in this case we write "$A = B$." A set A is a **proper subset** of a set B if A is a subset of B but A does not equal B; here we write "$A \subset B$."

- Sometimes it is necessary to specify the set of all elements that are considered appropriate for a particular discussion. This set is called the **universal set** for the discussion, and is usually denoted by $\mathcal{U}$. The set of all elements in the universal set that are not in a particular set A is called the **complement** of A, and is denoted by A'.

- The set of all elements in both a set A and a set B is called the **intersection** of A and B, and is symbolized by $A \cap B$. If A and B are sets such that $A \cap B = \emptyset$, we say that A and B are **disjoint**. The set of all elements that are either in a set A or in a set B, possibly in both, is called the **union** of A and B, and is symbolized by $A \cup B$.

- Another way of combining sets is based on the concept of an **ordered pair**, which is simply a pair of elements in a specified order. We write the ordered pair with first element a and second element b as (a, b). The **Cartesian product** of two sets A and B is the set of all ordered pairs whose first elements are from A and whose second elements are from B, and is denoted by $A \times B$.

Listed here for reference are some infinite sets of numbers that will be used frequently throughout the chapter. The first few are familiar to you, no doubt. A fuller description of the others, as well as a discussion of what is meant by saying they are "infinite," will appear shortly.

N $= \{1, 2, 3, 4, 5, \ldots\}$, the **natural numbers**

E $= \{2, 4, 6, 8, 10, \ldots\}$, the **even natural numbers**

D $= \{1, 3, 5, 7, 9, \ldots\}$, the **odd natural numbers**

$I = \{\ldots, -3, -2, -1, 0, 1, 2, 3, \ldots\}$, the **integers**

Q denotes the **rational numbers**, the set of all numbers that can be written as fractions with integer numerators and nonzero integer denominators.

R denotes the **real numbers**. It is the set of all finite or infinite decimals, and it contains **Q** as a subset. (An explanation of finite and infinite decimals appears in Section 6.4.) This set can be used to represent all the points on a single line.

$[0, 1]$ denotes the set of all real numbers between 0 and 1, inclusive. It is called the **unit interval**.

$[a, b]$ denotes the set of all real numbers between a and b, inclusive. It is called the **(closed) interval** a, b. (The term "closed" refers to the fact that the endpoints a and b are included in the set, which is also denoted by the use of the square brackets. If it is clear from context that the endpoints are included, the word "closed" is often omitted.)

R × **R** denotes the set of all ordered pairs of real numbers, which can be used to represent the set of all points of a plane.

EXERCISES 6.1

All of these exercises refer to the sets described at the end of this section. Answer *true* or *false*:

1. **E** is a subset of **N**.

2. **D** is a proper subset of **N**.

3. **E** is a subset of **D**.

4. **E** equals **D**.

5. **E** and **D** are disjoint.

6. **E** ∪ **D** = **N**

7. **E** ∪ **D** is a subset of **N**.

8. **E** ∪ **D** is a proper subset of **N**.

9. **E** and **N** are disjoint.

10. **E** ∪ **N** = **N**

11. **E** ∪ **N** is a subset of **N**.

12. **E** ∪ **N** is a proper subset of **N**.

13. **E** ∩ **D** = **N**

14. **E** ∩ **D** is a subset of **N**.

15. **E** ∩ **N** = **N**

16. **E** ∩ **D** is a proper subset of **N**.

17. **E** ∩ **N** is a subset of **N**.

18. **E** ∩ **N** is a proper subset of **N**.

19. **N** × **N** is a subset of **N**.

20. **N** × **N** and **N** are disjoint.

21. **I** is a subset of **N**.

22. **N** is a subset of **I**.

23. **Q** is a proper subset of **I**.

24. **I** is a proper subset of **Q**.

25. **Q** is a subset of **Q**.

26. $[0,1]$ is a subset of **R**.

27. $[0,1]$ is a subset of $[-1,5]$.

28. $[0,1]$ equals $[1,2]$.

29. $[0,1]$ is a subset of $[2,4]$.

30. $[0,1] \cup [1,2] = [0,2]$

31. $[0,1] \cap [1,2] = [0,2]$

32. .75 is an element of $[0,1]$.

33. 1.02 is an element of $[1,2]$.

34. $\frac{1}{3}$ is an element of $[0,1]$.

35. $\frac{3}{5}$ is an element of $[3,5]$.

36. The ordered pair $(3,5)$ is an element of **R** × **R**.

37. $[3,5]$ is a subset of **R** × **R**.

38. $[0,1] \times [0,1]$ is a subset of **R** × **R**.

WRITING EXERCISES

1. Compare the definitions of *union*, *intersection*, and *Cartesian product* as they are given in this section with the symbolic versions of these definitions as they appear in Section 4.2. Which form do you prefer, and why?

2. What do you think we should mean by the "size" of a set? Can you think of more than one way to define this concept? Use some specific sets, including some of the sets of numbers listed in this section, to illustrate your definition(s)?

3. How is the ellipsis ("...") used in ordinary English? How is that similar to its mathematical usage. How is it different?

6.2   Infinite Sets

In its most literal sense, infinite means "unbounded, endless, without limits of any kind"; but these definitions are not much more informative than the word itself. The confusion is compounded by the fact that nothing in the world around us seems to possess this property, with the possible exceptions of space and time, which appear to be limitless, but certainly are finite as far as the experience of any individual is concerned. Even a quantity as vast as the total number of electrons and protons in the universe can be calculated, at least approximately, and is thereby bounded by some large but finite number. (In 1938, A. S. Eddington proposed a number for this: 10^{79}.)

The first outright encounter with infinity usually comes from a consideration of numbers themselves. Very early in their mathematical experience, children discover that there is no largest number because 1 can be added to any number to get a larger one. The natural numbers are thus envisioned as an endless chain stretching out "to infinity," a place just beyond the farthest cloud and off limits to sane, stable people. This pathway to infinity has a definite starting point and is composed of discrete numerical steps, but it operates like a treadmill; you can walk as far as you like, but your goal will be as far away when you stop as it was when you began.

Contrasted with this discrete, step-by-step idea of infinity is the predominantly geometric idea of continuous infinity. Although a straight line can be shown to be infinite because one can "step off" intervals in either direction as far as one likes, there is no need to "step" at all. Because there are no gaps in the line, it is possible to proceed as far as is desired without singling out specific points along the way.

"But," you say, "this is really the same kind of process. One simply glides from one point to the next, thus taking very small but definite steps."

The Greeks considered a line (and time) to be composed of a succession of adjacent points (or moments) until Zeno showed how that assumption led to absurd results. (See Appendix Section B.3.) There is no "next" point to any given point and there are no gaps anywhere, so the set of points that make up a line appears to constitute a kind of infinity that is somehow different from that of the natural numbers. One might even speculate that these two infinite sets are of different "sizes," so to speak.

If, just for the sake of argument, we suppose that there are indeed at least two different sizes of infinity, are there more? What about the set I of integers or the set Q of rational numbers? I contains no smallest number, so the infinity appears two-sided, in some sense. Between any two rational numbers there are infinitely many others. Does that mean I and/or Q somehow represent larger sizes of infinity than N? If infinity is truly limitless, can there be "larger" or "smaller" infinite sets, or different sizes in any sense? To answer these questions (and others that probably are starting to occur to you) we need to define precisely what we mean by an "infinite set," and we must define "same size."

For instance, consider the sets

$$A = \{a, b, c, d\} \quad \text{and} \quad B = \left\{ \square, \triangle, \bigcirc, \rectangle \right\}$$

Are these two sets the same size? Before this question can be answered reasonably, we must specify what we mean by "same size." Clearly, if we are talking about the area of the page each occupies, the answer is *No*. However, the size question as it concerns us in this chapter is more accurately translated by asking if A and B have the same number of elements, and in this case the answer is *Yes*. But what does "same number" mean, especially in the case of infinite sets? Must we have particular numbers of elements before we can decide whether or not two sets are the same size in this sense? We certainly can use numbers to count the elements of A and B; but saying that A and B are the same size just means we ended up with the same number in each case. Now, counting is nothing more than matching things up with part of a known set (the natural numbers 1, 2, 3, ...). If we are going to match up sets, there is no need to use numbers at all; we can directly match the sets we want to compare. Thus, we can pair off the elements of A and B as in Figure 6.1 and conclude that A and B are the same size.

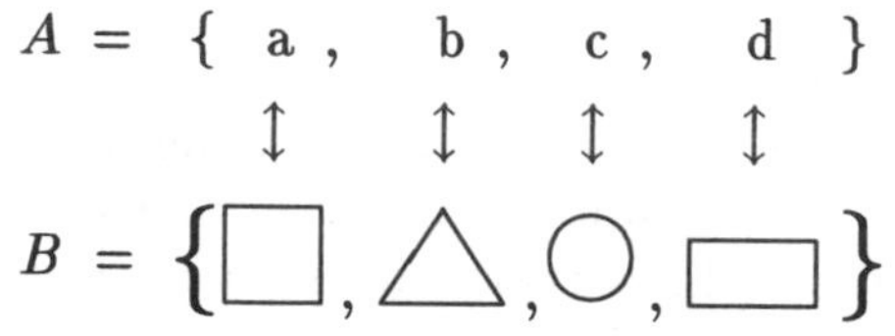

$$A = \{ \ a \ , \quad b \ , \quad c \ , \quad d \ \}$$

$$\updownarrow \quad\ \updownarrow \quad\ \updownarrow \quad\ \updownarrow$$

$$B = \{\square, \triangle, \bigcirc, \square\}$$

Figure 6.1   *Corresponding elements in sets A and B.*

As we prepare to investigate infinite sets, where counting all the elements is a hopeless task, it is useful to have a formal definition of this simpler comparison process.

DEFINITION   A **one-to-one** (or **1-1**) **correspondence** between two sets A and B is any rule or process by which each element of A is associated with exactly one element of B and each element of B is associated with exactly one element of A.

DEFINITION   Two sets A and B are **equivalent** if there exists a one-to-one correspondence between them. We shall write this as $A \leftrightarrow B$.

When we say that two sets *have the same size*, we mean that they are equivalent.

Example 6.1   There are many different 1-1 correspondences between the sets

$$A = \{a, b, c, d\} \quad \text{and} \quad B = \{w, x, y, z\}$$

Two of them are

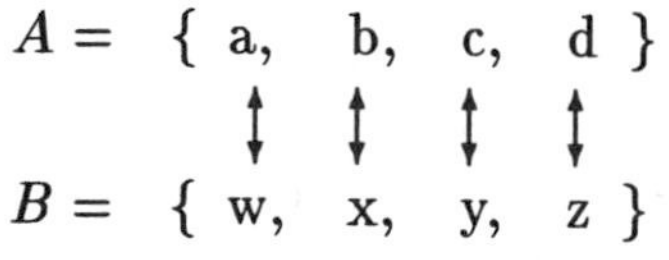

$$A = \ \{ \ a, \quad b, \quad c, \quad d \ \} \qquad\qquad A = \ \{ \ a, \quad b, \quad c, \quad d \ \}$$
$$\updownarrow \ \ \updownarrow \ \ \updownarrow \ \ \updownarrow \qquad \text{and}$$
$$B = \ \{ \ w, \quad x, \quad y, \quad z \ \} \qquad\qquad B = \ \{ \ w, \quad x, \quad y, \quad z \ \} \qquad \square$$

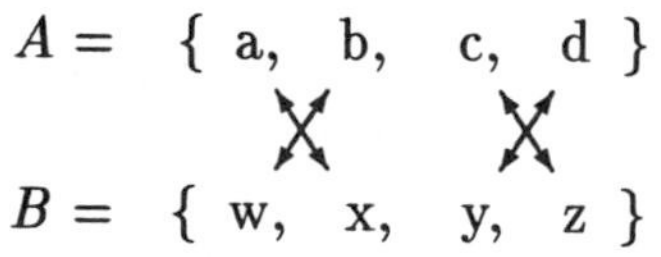

Example 6.2   The matching

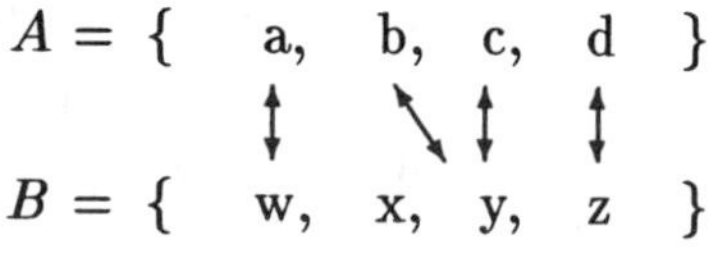

$$A = \{ \quad a, \quad b, \quad c, \quad d \quad \}$$
$$\updownarrow \ \ \ \diagdown\ \updownarrow \ \ \updownarrow$$
$$B = \{ \quad w, \quad x, \quad y, \quad z \quad \}$$

is not a 1-1 correspondence (even though the two sets are equivalent). Although each element of A corresponds to exactly one element of B, the element x in B is not matched with anything in A, and y is matched with two elements of A.   □

Example 6.3 There is no way to put the sets {a, b, c, d} and {x, y, z} in 1-1 correspondence. Therefore, these two sets are not equivalent.   □

Example 6.4 A 1-1 correspondence between the set of all natural numbers and the set of all negative integers is given by

$$\{ \quad 1, \quad 2, \quad 3, \quad 4, \quad \ldots, \quad n, \quad \ldots \quad \}$$
$$\updownarrow \quad \updownarrow \quad \updownarrow \quad \updownarrow \qquad \updownarrow$$
$$\{ \quad -1, \quad -2, \quad -3, \quad -4, \quad \ldots, \quad -n, \quad \ldots \quad \}$$

We cannot list every pairing in the correspondence, but the pattern given allows us to determine specifically how each element in either set is matched with an element of the other. For instance:

The natural number 37 is matched with the negative integer -37;
the negative integer -45 is matched with 45; etc.   □

DEFINITION A set A is **finite** if it is empty or if there is a natural number n such that A is equivalent to $\{1, 2, 3, \ldots, n\}$. A set is **infinite** if it is not finite.

This definition says that a set is infinite if its elements cannot be counted completely, no matter how fast we count or how much time we take. It fits well with the intuitive sense of infinity described at the beginning of this section and at the same time it makes good mathematical sense. So, let us move on to the first major question we face:

Are all infinite sets equivalent?

In other words, can two infinite sets *always* be put in 1-1 correspondence? If so, then all infinite sets are the same size (in this sense). If not, then there are actually different sizes of infinity! Intuitive reactions ("gut feelings") about that question vary widely. Since we human beings have never encountered actually infinite collections of things in our material experience, all of our attempts to deal with them must involve projecting our finite experience into an area in which its applicability is unknown. Therefore, we must rely on logical reasoning to guarantee the validity of any statements we make about infinity. In particular, we must apply the formal definitions carefully, and then be prepared to accept the consequences of our reasoning, regardless of whether or not they conform to our intuitive feelings.

If we try to apply the definition of *infinite set* to a specific set, a difficulty arises almost immediately. To verify that a set is infinite by that definition we must prove that no natural number will determine an equivalent set. The logical argument required for such a proof is often delicate and somewhat obscure. Fortunately, there is another property of infinite sets that serves to distinguish them from finite sets. Before giving its general description, let us consider an impossible, but informative, example from the folklore of mathematics.

An owner of a 100-room motel in a busy convention center is concerned because he often is completely booked up and has to turn away customers. In planning to enlarge his establishment, this innkeeper faces a dilemma — if he doesn't make it large enough, he will still have to turn away business, but if he makes it too big he will lose money on the vacant rooms. Based on past experience he decides that even on the slowest days he can fill half the rooms. After pondering the problem awhile, he decides that the perfect solution is to build a motel with infinitely many rooms, which he will number 1, 2, 3, ... He reasons as follows:

"On the nights when I have filled every room and an extra customer arrives, I can still accommodate him. I'll give him Room 1 and move the guest in Room 1 to Room 2, the guest in Room 2 to Room 3, the guest in Room 3 to Room 4, and so on. Because there is no last room, everyone will have a place to stay. On the slow nights, when only 'half' the rooms are filled, I'll just make sure I've assigned all the even-numbered rooms first. Then, before everyone settles in for the night, I'll move all the guests to the rooms whose numbers are half the numbers of the rooms they were originally assigned — 2 will move to 1, 4 will move to 2, 6 will move to 3, and so on. Then *all* the rooms will be filled for the night!"

We might suspect that this innkeeper is a better mathematician than a businessman. Although he might never finish building his infinite motel, his mathematical conclusions about it are correct, and they illustrate a characteristic property of infinite sets. Since all infinite sets have this property, and no finite sets do, we have a useful alternative definition:

DEFINITION   A set is **infinite** if it is equivalent to a proper subset of itself. A set is **finite** if it is not infinite.

This definition of infinite set and the one given earlier both yield the same results — a set that satisfies either definition also satisfies the other.

(The proof of this fact is somewhat difficult and is therefore omitted.) Hence, the two definitions may be used interchangeably, according to which is more convenient in a given situation.

Example 6.5   We can easily show that the set **N** of natural numbers is infinite by using the second definition of infinite set. Indeed, **E** is a proper subset of **N**, and we can establish the equivalence of the two sets by the 1-1 correspondence

$$\mathbf{N} = \{ \quad 1, \quad 2, \quad 3, \quad 4, \quad \ldots, \quad n, \quad \ldots \quad \}$$
$$\updownarrow \quad \updownarrow \quad \updownarrow \quad \updownarrow \qquad \updownarrow$$
$$\mathbf{E} = \{ \quad 2, \quad 4, \quad 6, \quad 8, \quad \ldots, \quad 2n, \quad \ldots \quad \} \qquad \square$$

Example 6.6   The easiest way to show that {a, b, c, d, e} is *finite* is to count its elements; that is, to apply the first definition of finite set:

$$\{ \quad a, \quad b, \quad c, \quad d, \quad e \quad \}$$
$$\updownarrow \quad \updownarrow \quad \updownarrow \quad \updownarrow \quad \updownarrow$$
$$\{ \quad 1, \quad 2, \quad 3, \quad 4, \quad 5 \quad \} \qquad \square$$

EXERCISES 6.2

1. Find three more 1-1 correspondences between {a, b, c, d} and {w, x, y, z}. How many such correspondences are there? Why?

2. Find three more 1-1 correspondences between the set of all natural numbers and the set of all negative integers.

3. Find three sets that can be put in 1-1 correspondence with {0, 1, 2, 3, 4}, and find two sets that cannot be.

4. Are the sets
   {1, 2, 3, ..., 50}   and   {1, 2, 3, 4, ...}
   equivalent? Are they equal? Why or why not?

5. Are the sets
   {1, 2, 3, ...}   and   {1, 2, 3, 4, ...}
   equivalent? Are they equal? Why or why not?

6. If two sets are equivalent, must they be equal? Why or why not?

7. If two sets are equal, must they be equivalent? Why or why not?

8. Show that **E** is equivalent to **D**.

9. Show that **E** is equivalent to
   {5, 10, 15, ..., 5n, ...}

10. If two sets are both equivalent to a third set, must they be equivalent to each other? Why or why not? Find an example to illustrate your answer.

In Exercises 11–14, find specific examples of finite sets of the indicated sizes. How would you show that each of these sets is finite?

11. 7 elements   12. 10 elements

13. more than 100 elements

14. more than 1000 elements

For Exercises 15–19, let $A = \{a, b, c, d, e\}$, and recall that **N** denotes the set of all natural numbers. Find an example of each of the following:

15. A set that is not equal to A, but that is equivalent to A.

16. A set that is not equal to **N**, but that is equivalent to **N**.

17. A proper subset of A that cannot be put in 1-1 correspondence with A.

18. A proper subset of **N** that cannot be put in 1-1 correspondence with **N**.

19. A proper subset of **N** that can be put in 1-1 correspondence with **N**.

In Exercises 20–23, find an example of each of the following:

20. An infinite set of positive numbers all less than 5.

21. An infinite set of numbers between 2 and 3.

22. An infinite set containing the alphabet as a proper subset.

23. A 1-1 correspondence between **E** and one of its proper subsets. (By doing this you are proving that **E** is an infinite set.)

For each of the sets given in Exercises 24–29, find a proper subset that is equivalent to it; then find one that is not.

24. **D** (the set of odd natural numbers)

25. $\{3, 6, 9, 12, \ldots, 3n, \ldots\}$

26. $\{1, \frac{1}{2}, \frac{1}{3}, \frac{1}{4}, \ldots, \frac{1}{n}, \ldots\}$

27. $\{5, 6, 7, 8, \ldots, n+4, \ldots\}$

28. $\{-2, -4, -6, -8, \ldots, -2n, \ldots\}$

29. $\{1, 4, 9, 16, \ldots, n^2, \ldots\}$

30. Use the second definition of infinite set to prove that the set **I** of integers is infinite.

WRITING EXERCISES

1. Do infinite sets actually exist in the physical world? In the real world? Are your answers to these two questions different? Why or why not?

2. Describe at least two other ways, besides 1-1 correspondence, of interpreting the statement that two things are "the same size." Can either of these other ways be applied generally to "things" that are sets? If so, how; if not, why not?

3. Using complete sentences, write a half-page summary that identifies the main theme of this section and outlines how that theme is carried out.

6.3   The Size of N

One way to rephrase the second definition of infinite set is to say that a set is infinite if we can throw away some of it and still be left with a set of the same size. Example 6.5 of Section 6.2 showed that **N** and **E** are the same size, even though **E** was obtained by throwing away "half" of **N** (that is, by throwing away all the odd natural numbers). It is also easy to see that **D**, the set of odd natural numbers, is equivalent to **N**. These two correspondences can be used to extend one step further our size comparison of the sets listed in Section 6.1.

Saying that some infinite sets can be put in 1-1 correspondence with the set of natural numbers is the same as saying that these sets can be ordered sequentially, starting with a first element, then a second element, then a third, and so on. Now, when we deal with sets of numbers — like **I** or **Q** or **R** — we tend to think of them as being arranged in their usual size order, as if they were on a number line. However, they are still the same sets even if we jumble the order or rearrange them to fit some other pattern. (If a box of computer registration cards for everyone in this class, arranged in alphabetical order, were spilled on the floor and then picked up in haste, the cards probably would not remain in the same order, but they would still be the same collection of cards. If they were rearranged according to the social security numbers on them, they would also still be the same set of cards.)

If we ignore the usual ordering of the integers, then, we can show that **I** is equivalent to **N** by choosing a first element of **I**, then a second element, then a third, and so on. The only tricky part is that we must be sure that every integer gets matched with some natural number. The matching between **N** and **E** in Example 6.5 provides a clue for an easy way to match **I** with **N**; it tells us we can match up all the positive integers with just the even natural numbers, as follows:

$$\mathbf{N} = \{ \quad 1, \quad 2, \quad 3, \quad 4, \quad 5, \quad 6, \quad 7, \quad 8, \quad 9, \quad \ldots \}$$
$$\mathbf{I}^+ = \{ \qquad 1, \qquad 2, \qquad 3, \qquad 4, \qquad \ldots \}$$

The odd natural numbers are then still available to be matched up with zero and the negative integers, as follows:

$$\mathbf{N} = \{ \quad 1, \quad 2, \quad 3, \quad 4, \quad 5, \quad 6, \quad 7, \quad 8, \quad 9, \quad \ldots \}$$
$$\mathbf{I}^- \cup \{0\} = \{ \quad 0 \qquad -1, \qquad -2, \qquad -3, \qquad -4, \ \ldots \}$$

Combining both match-ups, we obtain:

$$\mathbf{N} = \{ \quad 1, \quad 2, \quad 3, \quad 4, \quad 5, \quad 6, \quad 7, \quad 8, \quad 9, \quad \ldots \}$$
$$\mathbf{I} = \{ \quad 0 \quad 1 \quad -1, \quad 2 \quad -2, \quad 3 \quad -3, \quad 4 \quad -4, \ \ldots \}$$

It should be easy to see that the pattern established by the first few terms of this correspondence can be continued as far as we please. For instance, the positive integer 10 corresponds to the natural number 20, the natural number

51 corresponds to the negative integer −25, etc. A general description of this matching is:

Any even natural number n corresponds to the integer $\frac{n}{2}$;

any odd natural number m corresponds to the integer $\frac{1-m}{2}$.

Thus, we have shown that **N** and **I** are the same size.

At this point it is convenient to make a useful observation:

(6.1) | For any sets A, B, and C, if A is equivalent to B and B is equivalent to C, then A and C are also equivalent.

The required 1-1 correspondence between A and C may be obtained by "patching together" the A-to-B correspondence with the B-to-C correspondence. For example, if the sets

$$A = \{x, y, z\}, \ B = \{p, q, r\}, \text{ and } C = \{5, 6, 7\}$$

are matched as in Figure 6.2, then we automatically have a 1-1 correspondence between A and C that matches x with 6 (by way of q), y with 7 (by way of p), and z with 5 (by way of r), as shown in Figure 6.2.

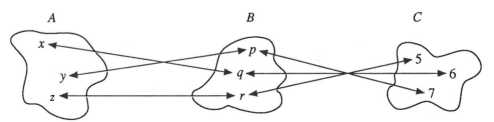

Figure 6.2 *A is in 1-1 correspondence with B and B is in 1-1 correspondence with C. Thus, A is in 1-1 correspondence with C.*

So far, our work has shown that the first four infinite sets in the list of Section 6.1 are all the same size. Now let us deal with **Q**, the set of all rational numbers. By again ignoring the usual ordering of the numbers, we can devise a method for putting the positive integers and the positive rationals in 1-1 correspondence. Arrange all the positive rationals in an infinite rectangular array, listing all the fractions with numerator 1 in the first row, all the fractions with numerator 2 in the second row, all the fractions with numerator 3 in the third row, and so on, as shown in Figure 6.3. Then, starting at the upper-left corner and omitting all fractions that are not in lowest terms, take the positive rationals in a zigzag diagonal order and match them with successive positive integers.

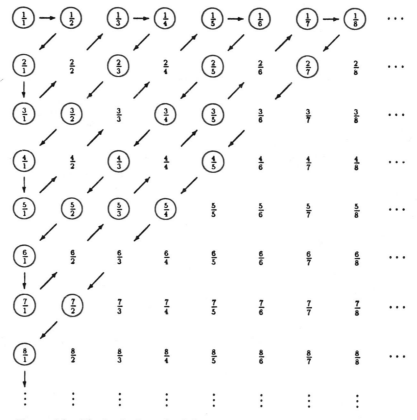

Figure 6.3   *The beginning of a 1-1 correspondence between the positive integers and the positive rationals.*

As the arrows in Figure 6.3 indicate, this 1-1 correspondence between the positive integers and the positive rationals begins

$$\mathbf{Q^+} = \{\ \ 1, \quad \tfrac{1}{2}, \quad 2, \quad 3, \quad \tfrac{1}{3}, \quad \tfrac{1}{4}, \quad \tfrac{2}{3}, \quad \tfrac{3}{2}, \quad 4, \quad 5, \quad \tfrac{1}{5}, \quad \dots\ \}$$

$$\mathbf{I^+} = \{\ \ 1, \quad 2, \quad 3, \quad 4, \quad 5, \quad 6, \quad 7, \quad 8, \quad 9, \quad 10, \quad 11, \quad \dots\ \}$$

The same process also gives us a 1-1 correspondence between all the negative integers and all the negative rationals, just by labeling all the numbers in both sets negative and using the same matching. The 1-1 correspondence between **I** and **Q** is completed by assigning the integer 0 to the rational number 0. Thus, the sets **I** and **Q** are the same size!

EXERCISES 6.3

In Exercises 1–5, reorder each set in three different ways.

1. $\{1, 2, 3, 4, 5\}$

2. $\{a, b, c, d, \ldots, y, z\}$

3. **N**, the set of natural numbers

4. **I**, the set of integers

5. **Q**, the set of rational numbers

6. Set up a 1-1 correspondence that proves that the set **D** of all odd natural numbers is equivalent to **N**. Try to write a general formula for that correspondence; then use it to find which natural numbers correspond to the odd numbers 99 and 243.

Use the 1-1 correspondence between **N** and **I** given in this section to answer Exercises 7–12.

7. Which integer is matched with the natural number 100?

8. Which natural number is matched with the integer −100?

9. Which integer is matched with the natural number 75?

10. Which natural number is matched with the integer −30?

11. If p denotes a positive integer, what is its corresponding natural number?

12. If $-p$ denotes a negative integer, what is its corresponding natural number?

13. Set up a 1-1 correspondence between **N** and the set
$$\{30, 60, 90, 120, \ldots\}$$

Try to write a general formula for this correspondence; then use it to find which numbers correspond to the natural numbers 17 and 50.

14. Use your answers to Exercises 6 and 13 to set up a 1-1 correspondence between **D** and $\{30, 60, 90, 120, 150, \ldots\}$. Try to write a general formula for this correspondence; then use it to find which numbers correspond to the odd numbers 17, 99, and 243.

Use the correspondence between **I** and **Q** described in this section to answer Exercises 15–23.

15. Which rational number corresponds to the integer 15?

16. Which rational number corresponds to the integer 30?

17. Which rational number corresponds to the integer −15?

18. Which rational number corresponds to the integer −30?

19. Which integer corresponds to the rational number $\frac{5}{2}$?

20. Which integer corresponds to the rational number $\frac{4}{7}$?

21. Which integer corresponds to the rational number $\frac{-5}{3}$?

22. Which integer corresponds to the rational number $\frac{-3}{8}$?

23. Why were the fractions that are not in lowest terms omitted in the correspondence process?

24. Prove that the set $\{\frac{2}{1}, \frac{2}{3}, \frac{2}{5}, \frac{2}{7}, \ldots\}$ is equivalent to the set **N** of natural numbers.

WRITING EXERCISES

1. Does every infinite set have a subset that can be put in 1-1 correspondence with the set **N** of natural numbers? Explain.

2. Describe how to set up a 1-1 correspondence between **N** and the set of all rational numbers whose denominators are powers of 10. If you find this question difficult, describe how you might use at least two of the dozen problem-solving tactics of Section 1.2 to help you answer it.

6.4 Rational and Irrational Numbers

Note

For those already familiar with the characterization of rational and irrational numbers by means of their decimal expansions, this section may be omitted without loss of continuity.

How large is the set **R** of all real numbers, both rational and irrational? Before we can answer this, we must learn a little more about rational and irrational numbers. To begin with, the set **R** can be described as the set of all decimals, provided that we allow decimals to have infinitely many digits. For brevity, we shall call decimals with infinitely many digits **infinite decimals** and decimals with only finitely many digits **finite decimals**. Now, every rational number can easily be converted to a decimal by dividing its numerator by its denominator. Some rational numbers, such as

$$\frac{1}{3} = .33333\ldots$$

require infinitely many digits in their decimal expansions. If we agree to attach an infinite "tail" of zeros to an ordinary decimal, then every rational number can be written as an infinite decimal. For example,

$$\frac{3}{4} = .75000\ldots$$
$$\frac{7}{3} = 2.33333\ldots$$
$$\frac{13}{8} = 1.62500\ldots$$
$$\frac{2}{11} = .18181\ldots$$

and so forth. In other words, **Q** is a subset of **R**.

However, not every infinite decimal represents a rational number. To distinguish those decimals that represent rationals from those that do not, consider first an example of the division process just mentioned.

Example 6.7   To convert $\frac{47}{22}$ to a decimal, we divide 47 by 22 as follows:

$$
\begin{array}{r}
2.136 \\
22\,\overline{)47.0000}\quad\ldots \\
\underline{44} \\
3\,0 \\
\underline{2\,2} \\
80\quad\longleftarrow \\
\underline{66} \\
140 \\
\underline{132} \\
8\quad\longleftarrow
\end{array}
$$

The remainder, 80, at the fourth step is the same as the one at the second step, so the third and fourth digits of the quotient will repeat again and again from there on, without interruption. Thus,

$$\frac{47}{22} = 2.1363636\ldots \qquad\qquad \square$$

Must such a repetition of remainders happen with any division example, or is this just a carefully chosen special case? The answer is *Yes* to both parts of that question. A repetition must always occur *eventually*, but the fact that it happens so conveniently soon in this example is the result of careful choice.

Any case in which the numerator is larger than the denominator can be simplified by first taking care of the integer part of the quotient. For instance, we could have begun Example 6.7 by writing

$$\tfrac{47}{22} = 2 + \tfrac{3}{22}$$

then converting $\frac{3}{22}$ to the decimal .1363636... afterwards. In this way we can safely confine our attention to fractions in which the numerator is smaller than the denominator.

In general, when one integer is divided by another (larger) one using long division, the remainder at any step must be less than the divisor. Thus, if we are converting the rational number $\frac{p}{q}$ to a decimal, the largest possible remainder is $q - 1$; that is,

(6.2)
> The number of different nonzero remainders in the division problem $p \div q$ can be no larger than $q - 1$.

This means that, once the nonzero digits of p have all been used, after at most q more steps in this division problem the remainder *must* repeat an earlier one. Therefore, the digits in the quotient that occur in the steps between the two equal remainders will repeat over and over from there on without interruption. Sometimes this repetition occurs early, as in Example 6.7. Sometimes all possible nonzero remainders are used; for instance,

$$\frac{4}{7} = .571428\,571428\,571428\ldots$$

Thus, we have established an important fact:

(6.3)
> Every rational number can be expressed as an infinite decimal in which, from some specific digit on, a finite sequence of digits repeats again and again in the same order, without interruption.

Such decimals are called **repeating decimals.** To denote the repeating sequence of digits without having to write it several times over, we just put a bar over it. For instance, the results of the preceding two examples are written

$$\frac{47}{22} = 2.1\overline{36} \quad \text{and} \quad \frac{4}{7} = .\overline{571428}$$

A finite decimal such as $\frac{2}{5} = .4000\ldots$ can be written $.4\overline{0}$, but usually the repeated zeros are just omitted; in this case, we would write $\frac{2}{5} = .4$.

Two natural questions arise at this point:

1. Does *every* repeating decimal represent a rational number?

2. Are there really any (or many) such things as infinite *non*repeating decimals?

It is surprisingly easy to show that the answer to the first question is *Yes.* To do this we use a simple method for converting any repeating decimal to fraction form. The next two examples illustrate how this method is used to convert repeating decimals to fractions.

Example 6.8 Suppose $d = .\overline{35}$. Since there are two digits in the repeating sequence, we multiply d by 100 and then subtract d from that product:

$$
\begin{aligned}
100d &= 35.353535\ldots \\
-\ d &= .353535\ldots \\
\hline
99d &= 35.000000\ldots \\
d &= \frac{35}{99}
\end{aligned}
$$

□

Example 6.9   Suppose $d = 2.8\overline{473}$. Since there are three digits in the repeating sequence, we multiply d by 1000 and then subtract d from that product:

$$
\begin{aligned}
1000d &= 2847.3473473473\ldots \\
- \quad d &= 2.8473473473\ldots \\
\hline
999d &= 2844.5000000000\ldots \\
d &= \frac{2844.5}{999}
\end{aligned}
$$

Strictly speaking, a rational number is the quotient of two *integers*; to put the answer in appropriate form, then, we should eliminate the decimal point from the numerator. This is easily done by multiplying both numerator and denominator by 10. This answer can be reduced to lowest terms, if desired, by eliminating all common factors of the numerator and denominator. Thus:

$$
d = \frac{28{,}445}{9990} = \frac{5689}{1998} \qquad \square
$$

Example 6.10   A finite decimal, such as .375, can be handled in the same way by treating it as $.375\overline{0}$, a repeating decimal with a single-digit sequence. (Multiply by 10, subtract, and divide.) It is *much* simpler, however, to observe that a finite decimal is just another notation for a fraction whose denominator is a power of 10 and to write the fraction form immediately:

$$
.375 = \frac{375}{1000} = \frac{3}{8} \qquad \square
$$

The general method for converting infinite decimals to fractions may be described like this:

> Suppose d is a repeating decimal with n digits in its repeating sequence. If we multiply d by 10^n and then subtract d from $10^n \cdot d$, we will get a *finite* decimal that equals $(10^n - 1) \cdot d$. Then d is the fraction obtained by dividing that finite decimal by $10^n - 1$. To obtain an integer numerator, multiply both numerator and denominator by the same power of ten, if necessary.

Perhaps this description reads more like a magic incantation than a logical procedure. It was stated in general terms to show that there is a procedure that works all the time. Look back at the preceding examples, then reread the general method; you should find it much clearer.

Since every rational number can be expressed as a repeating decimal, and vice versa, we have just characterized all those real numbers (in decimal form) that are rational — they are just the repeating decimals. The other real numbers are called **irrational numbers**; they are the infinite nonrepeating decimals. But are there any such numbers? Certainly — here is one:

$$.101001000100001\ldots$$

where the ellipsis ("...") implies that the continuing pattern requires one more 0 after each successive 1. Because there is no finite sequence of digits that repeats again and again in the same order without interruption, this is not a repeating decimal.

To construct other examples of an infinite nonrepeating decimals, we can choose the sequence of successive multiples of some natural number, such as

$$3, 6, 9, 12, 15, 18, 21, 24, \ldots$$

and form decimals using these digits in order:

$$.3691215182124\ldots, \quad 3.691215182124\ldots, \quad 36.91215182124\ldots$$

and so forth. (Can you convince yourself that the digit sequence in such cases is nonrepeating?) These examples should suggest how to construct many other irrational numbers.

Sometimes irrational numbers arise in other contexts, and when they do, it often is difficult to prove that they are irrational. Some familiar irrational numbers are $\sqrt{2}$ (the length of the diagonal of a unit square), π (the ratio of the circumference of a circle to its diameter), and e (the base of the natural logarithms). The proof that $\sqrt{2}$ cannot be expressed as a rational number was known to the early Greeks; a version of it appears in Appendix Section B.3. The proofs that π and e are irrational are much more difficult; they were only established within the last 200 years.

The decimal characterization of rational and irrational numbers is important and useful; we summarize it here for emphasis:

(6.4)

> The real numbers that are rational are the repeating decimals; the real numbers that are irrational are the (infinite) nonrepeating decimals.

Note

There is a small problem of ambiguity of representation here. Decimals ending in repeated 9s represent the same numbers as decimals ending in repeated 0s. For example, $.4999\ldots = .5000\ldots$. (To see that this is true, convert these two decimals to fractional form.) If we discard one or the other of these representations in each case, then there is a 1-1 correspondence between the set of all repeating decimals and the set of all rational numbers. For our purposes, it suffices to assume that this has been done in a way that allows us to use whichever form is more useful in each particular situation.

EXERCISES 6.4

Before doing the divisions in Exercises 1–9, indicate the maximum number of different nonzero remainders that can occur. Then do the computations to find out how many different remainders actually do occur. Is the number of different remainders the same as the number of digits in the repeating sequence?

1. $3 \div 5$
2. $2 \div 3$
3. $11 \div 37$
4. $7 \div 10$
5. $3 \div 7$
6. $14 \div 6$
7. $70 \div 14$
8. $451 \div 999$
9. $7832 \div 9990$

Convert each fraction in Exercises 10–21 to a repeating decimal. (Find the entire repeating sequence of digits.)

10. $\frac{1}{9}$
11. $\frac{2}{5}$
12. $\frac{13}{6}$
13. $\frac{15}{22}$
14. $\frac{5}{11}$
15. $\frac{2}{13}$
16. $\frac{21}{9}$
17. $\frac{9}{21}$
18. $\frac{10}{7}$
19. $\frac{22}{7}$
20. $\frac{112}{37}$
21. $\frac{5}{111}$

Convert each decimal in Exercises 22–33 to fractional form.

22. $1.\overline{4}$
23. $.0\overline{2}$
24. $.\overline{9}$
25. $.\overline{02}$
26. $.\overline{15}$
27. $.1\overline{5}$
28. $.0\overline{15}$
29. $62.1\overline{75}$
30. $62.\overline{175}$

31. $.\overline{40}$
32. $6.85\overline{14}$
33. $6.8\overline{514}$

34. Write five irrational numbers in infinite decimal form.

WRITING EXERCISES

1. Describe at least two ways in which the term *repeating decimal* can be misleading if not understood carefully.

2. The word *numerator* literally means "a person or thing that numbers or counts something." The word *denominator* literally means "namer"; it is closely related to *denomination* — the name of a specific kind of thing — as in a religious denomination or a denomination of currency. With this in mind, explain why it is appropriate to use the words *numerator* and *denominator* to refer to the top and bottom numbers of a fraction, respectively.

3. Discuss the relative merits of using the upper-bar notation versus the ellipsis ("...") notation to represent repeating decimals.

4. Do you *really* believe that $.\overline{9}$ is actually equal to 1? Do you really believe that $.\overline{3}$ is actually equal to $\frac{1}{3}$? Discuss.

6.5   A Different Size

The equivalence of such different sets as **I** (the integers) and **Q** (the rationals) suggests that any two infinite sets may be equivalent. Let us assume for the moment that this assumption is true and see where it leads us. In particular, let us suppose that **N**, the set of natural numbers, is equivalent to [0, 1], the set of all real numbers between 0 and 1, inclusive. Because the real numbers are all the infinite decimals, the real numbers between 0 and 1 can be described as all infinite decimals that have 0 to the left of the decimal point. (The number 1 is represented by the infinite decimal .999... in this case.) Then any 1-1 correspondence between **N** and [0, 1] will be a sequential listing of these infinite decimals, one for each natural number. For example, such a correspondence might begin as illustrated in Figure 6.4.

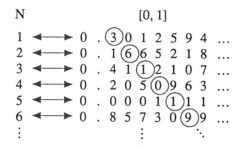

Figure 6.4   *The beginning of a proposed 1-1 correspondence
between* **N** *and [0,1].*

In any such 1-1 correspondence there is an infinite decimal corresponding to each natural number. Let us use this correspondence to define a new infinite decimal, according to the following rule:

> For each natural number n, look at the nth digit of its corre-
> sponding decimal. If that digit is 1, let 2 be the nth digit of the
> new decimal; otherwise, let this digit be 1.

Looking at the encircled digits in Figure 6.4, we see that the new decimal would begin .112121.... Notice that this new number differs from the first number in the [0, 1] listing in at least the first decimal place; it differs from the second number in at least the second place; it differs from the third number in at least the third place; and so on. Because this new number differs from each number of [0, 1] in the list, it does not correspond to any natural number; yet it is clearly a real number between 0 and 1. In other words, we have found a number in $[0, 1]$ that was left out of the alleged 1-1 correspondence between all of $\mathbf{N}$ and all of $[0, 1]$.

Therefore, the proposed correspondence does *not* verify the equivalence of $\mathbf{N}$ and $[0, 1]$.

But *any* 1-1 correspondence between $\mathbf{N}$ and $[0, 1]$ must be of the same form as the one we just saw, and, despite the specific choice of the first few decimals to illustrate the correspondence, the method given for constructing the "new" decimal number is general enough to apply *to any proposed listing*. Thus, the process just described — called **Cantor's diagonalization process** — shows that

> any proposed 1-1 correspondence between $\mathbf{N}$ and $[0, 1]$ must necessarily leave out at least one element of $[0, 1]$; therefore, $\mathbf{N}$ and $[0, 1]$ cannot be equivalent.

(In fact, any attempted 1-1 correspondence between $\mathbf{N}$ and $[0, 1]$ actually will leave out many, many of the numbers in $[0, 1]$; but to prove the correspondence invalid, all we need to show is that at least one number in $[0, 1]$ is skipped.) This means that the infinite sets $\mathbf{N}$ and $[0, 1]$ are *not* the same size. Therefore, *there are at least two different sizes of infinity*!

Having used the unit interval $[0, 1]$ to discover that there are different sizes of infinite sets, it seems natural to ask whether the length of an interval is related to its size as a set of points. For example, can we get an infinite set that is not equivalent to $[0, 1]$ by choosing a longer interval, such as $[0, 5]$ or $[-36, 247]$, or a shorter interval, such as $[\frac{1}{2}, \frac{3}{4}]$ or $[.00001, .00002]$? It would seem reasonable to assume that an interval of length 5 or 50 or 1000 should contain too many points to be equivalent to an interval of length 1. However, if $a < b$, a surprisingly simple geometric argument proves that:

(6.5) | Any interval $[a, b]$ of any length can be put in 1-1 correspondence with $[0, 1]$.

To show how this can be done, place the interval $[0, 1]$ perpendicular to $[a, b]$, with point 0 lying on point a. Draw the line determined by the two points 1 and b, and choose a point P somewhere on the extension of that line from b past the point 1, as shown in Figure 6.5. Now, if we choose any point x in $[0, 1]$, the line through P and x must intersect $[a, b]$ at some point y. Also, if we choose a point y' in $[a, b]$, the line through P and y' necessarily intersects $[0, 1]$ at some point x'. This process gives us a correspondence between the points of $[0, 1]$ and the points of $[a, b]$.

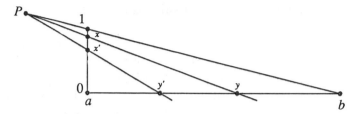

Figure 6.5 *A 1-1 correspondence be-*
tween [0,1] and [a,b].

Notice that different points of $[0,1]$ determine different lines through P, so these lines must intersect $[a,b]$ at different points, and vice versa (because two points determine exactly one line). Therefore, the correspondence is one-to-one, and consequently $[0,1]$ and $[a,b]$ are equivalent sets. In fact, the presence or absence of endpoints does not affect this result, as Exercise 34 shows. This means that *all* intervals, regardless of length, are equivalent to $[0,1]$ and hence to each other. Thus, *the length of an interval does not affect its size as a set of points!*

Similarly, it can be shown that the interval $[0,1]$ is equivalent to the entire number line $\mathbf{R}$. (We omit the details of this argument.) Coupling this information with the first result of this section, we arrive at a startling conclusion about the number line and its intervals, *considered as sets of points*:

(6.6) | $\mathbf{R}$ is the same size as $[0,1]$ or any other interval!

That is, the *length* of a line or line segment and its *size* as a set of points are totally different mathematical concepts.

What if we change dimension? Do we get a different-sized set? Conceivably, it seems plausible that the set of all points in a plane, for example, is far too large to be matched up with the set of points on a single line. Once again, however, we are in for a surprise; a 1-1 correspondence between these two sets is possible. We shall look at a special case — a correspondence between the unit interval $[0,1]$ and the set of all points in a 1-by-1 square.

First, recall from high-school algebra that a **coordinate system** can be placed on the plane by using two copies of the real line $\mathbf{R}$, one horizontal and one vertical, intersecting at their respective zero points. Each of these two lines is called a **coordinate axis**. Using this system, any point P on the plane can be labeled by an ordered pair of real numbers (x,y), where x

is the number on the horizontal axis directly above or below P and y is the number on the vertical axis directly to the left or right of P. The numbers x and y are called the **coordinates** of the point P. (If P is on an axis, one of its coordinates is 0.) Thus, the set of all points in the plane can be treated as the set $\mathbf{R} \times \mathbf{R}$ of all ordered pairs of real numbers. If we take all the points whose coordinates lie within the $[0, 1]$ interval on each axis, we have a square 1 unit on a side. Set up a 1-1 correspondence between this square and a single unit interval roughly as follows:[1] Each point in the square has coordinates of the form

$$(.x_1 x_2 x_3 \ldots, .y_1 y_2 y_3 \ldots)$$

where $.x_1 x_2 x_3 \ldots$ and $.y_1 y_2 y_3 \ldots$ are decimals between 0 and 1. Match each such point with the decimal formed by "interweaving" the x-digits and y-digits in an alternating pattern, starting with x_1:

$$.x_1 y_1 x_2 y_2 x_3 y_3 \ldots$$

This number is also a decimal between 0 and 1. For example, the point $(\frac{1}{3}, \frac{2}{3})$ in the square, which can be written

$$(.333\ldots, .666\ldots)$$

corresponds to the point $.363636\ldots$ in the unit interval. Conversely, the point $.\overline{724}$ ($= .724724\ldots$) in the unit interval corresponds to the point $(.\overline{742}, .\overline{274})$ in the square. (See Figure 6.6.)

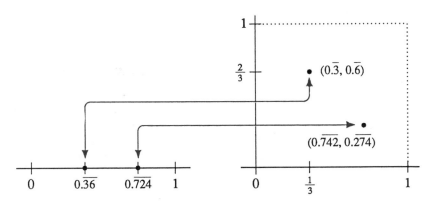

Figure 6.6 A 1-1 correspondence between the unit interval and the unit square.

---

[1]We say "roughly" because the repeated-0/repeated-9 ambiguity described in the *Note* at the end of Section 6.4 causes some trouble. Certain decimal forms have to be omitted initially, then "patched" back in by an argument similar to the one needed for Exercise 34. Nevertheless, the main argument works as outlined here.

The construction of a 1-1 correspondence between $\mathbf{R}$ and $\mathbf{R} \times \mathbf{R}$ is an extension of the technique shown here. We omit its details as well as the proof that this matching leads to a 1-1 correspondence. The fact that such a correspondence *can* be constructed is the main point here because it establishes the equivalence of $\mathbf{R}$ and $\mathbf{R} \times \mathbf{R}$. If we disregard the spatial properties of lines and planes and consider them just as sets of points, this equivalence means:

(6.7)   The set of all points in a plane (a two-dimensional geometric object) is the same size as the set of all points on a single line (a one-dimensional geometric object).

The trick of interweaving the digits of decimal representations of numbers can also be used for three-dimensional space. For instance, each point in the "unit cube" $[0,1] \times [0,1] \times [0,1]$ is determined by three coordinates of the form

$$(.x_1x_2x_3\ldots, \ .y_1y_2y_3\ldots, \ .z_1z_2z_3\ldots)$$

Each such point can be matched with the single infinite decimal

$$.x_1y_1z_1x_2y_2z_2x_3y_3z_3\ldots$$

in $[0,1]$. The extension of this technique to all triples of real numbers can be used to establish the equivalence of the sets $\mathbf{R}$ and $\mathbf{R} \times \mathbf{R} \times \mathbf{R}$. Thus:

(6.8)   The set of all points on a single line is the same size as the set of all points in three-dimensional space!

EXERCISES 6.5

1. Find five rational numbers between 0 and 1.

2. Find five irrational numbers between 0 and 1.

3. Find five rational numbers between .5 and .53.

4. Find five irrational numbers between .5 and .53.

5. Suppose that an alleged 1-1 correspondence

between $\mathbf{N}$ and $[0,1]$ begins as follows:

N		[0, 1]
1	$\leftrightarrow$	.040404...
2	$\leftrightarrow$	.111111...
3	$\leftrightarrow$	.257346...
4	$\leftrightarrow$	.101001...
5	$\leftrightarrow$	.763988...
6	$\leftrightarrow$	.121121...
$\vdots$		$\vdots$

(a) Construct the first six decimal places of a number in $[0, 1]$ that cannot be matched with any natural number in this correspondence.

(b) Construct three more such numbers, all different from each other and from the number you found in Part (a). (*Hint*: There is nothing special about the digits 1 and 2 in Cantor's diagonalization process.)

6. Answer Parts (a) and (b) of Exercise 5 for the following proposed correspondence:

N		$[0, 1]$
1	$\leftrightarrow$	.111111...
2	$\leftrightarrow$	.101010...
3	$\leftrightarrow$	.321321...
4	$\leftrightarrow$	.765432...
5	$\leftrightarrow$	.151617...
6	$\leftrightarrow$	.222222...
$\vdots$		$\vdots$

Using a ruler, draw a 2-inch line segment T perpendicular to a 5-inch line segment F, with an endpoint of one segment coinciding with an endpoint of the other. Then, mimicking the argument given in conjunction with Figure 6.5, answer Exercises 7–16 as precisely as you can by measuring.

7. Which point of F corresponds to the point at $1''$ on T?

8. Which point of F corresponds to the point at $\frac{1}{2}''$ on T?

9. Which point of F corresponds to the point at $\frac{3}{2}''$ on T?

10. Which point of T corresponds to the point at $1''$ on F?

11. Which point of T corresponds to the point at $2''$ on F?

12. Which point of T corresponds to the point at $3''$ on F?

13. Which point of T corresponds to the point at $4''$ on F?

14. Which point of T corresponds to the point at $5''$ on F?

15. Which point of T corresponds to the point at $0''$ on F?

16. Which point of F corresponds to the point at $2''$ on T?

In Exercises 17–24, decide whether or not the two sets given are equivalent. Give reasons to justify your answers.

17. **I** and the unit interval, $[0, 1]$

18. $[0, 1]$ and **Q**

19. $[0, 1]$ and the set of all points in a plane

20. $\mathbf{R} \times \mathbf{R}$ and $\mathbf{R} \times \mathbf{R} \times \mathbf{R}$

21. $[0, 1]$ and the set of all points in three-dimensional space

22. **Q** and the set of all points in a plane

23. **I** and the set of all points in three-dimensional space

24. The interval $[3, 8]$ and the set of all points in three-dimensional space

The 1-1 correspondence between $[0, 1] \times [0, 1]$ and $[0, 1]$ described in this section matches ordered pairs of real numbers with single real numbers by "interweaving" decimals. In Exercises 25–32, a point in one of these two sets is given; find the corresponding point in the other set.

25. $(0.555\ldots, 0.111\ldots)$

26. $(0.85, 0.222\ldots)$

27. $\left(\frac{3}{4}, \frac{1}{3}\right)$ 28. $(0.\overline{43}, 0.\overline{918})$

29. $0.777\ldots$ 30. $\frac{2}{3}$

31. $\frac{1}{2}$ 32. $.\overline{73125}$

33. A 1-1 correspondence between two intervals of different lengths can be defined without pictures by using a little elementary algebra. For instance, the interval $[0, 1]$ can be matched with $[0, 3]$ by "stretching" it according to the formula $y = 3x$. That is, each point x in $[0, 1]$ is matched with the point $3x$ in $[0, 3]$.

(a) By this correspondence, which points of $[0,3]$ are matched with the following points of $[0,1]$?

$$\frac{1}{4}, \ \frac{1}{2}, \ \frac{3}{4}, \ \frac{1}{3}, \ .4, \ 0, \ 1$$

(b) By this correspondence, which points of $[0,1]$ are matched with the following points of $[0,3]$?

$$0, \ 1, \ 2, \ 3, \ \frac{1}{2}, \ \frac{1}{3}, \ .243, \ \frac{5}{2}, \ 2.7$$

34. Show that the unit interval $[0,1]$ is equivalent to the "open" unit interval

$$(0,1) = \{x \mid x \in \mathbf{R} \text{ and } 0 < x < 1\}$$

(*Hint*: Is the set $\{0, 1, \frac{1}{2}, \frac{1}{3}, \frac{1}{4}, \ldots\}$ equivalent to the set $\{\frac{1}{2}, \frac{1}{3}, \frac{1}{4}, \ldots\}$?)

WRITING EXERCISES

1. Describe how at least three of the dozen problem-solving tactics of Section 1.2 can be used to help in the solution of Exercise 34. (Even if you can't complete Exercise 34, you should be able to do this writing exercise.)

2. Consider the subset H of $[0,1]$ consisting of all decimals that are one hundred digits long.

   (a) Is H a finite set or an infinite set? Justify your answer.

   (b) If you tried to apply Cantor's diagonalization process to H, what would go wrong? Be specific.

3. Consider the subset V of $[0,1]$ consisting of all the infinite decimals that have only even digits. Does Cantor's diagonalization process still work for this set? Explain.

6.6 Cardinal Numbers

We have now examined all the sets listed in Section 6.1 and have found that they can be separated into two distinct sizes — those that are equivalent to **N** and those that are equivalent to **R**. It is convenient to have names for these sizes, so we shall adopt for them the names Cantor used.

DEFINITION The collection of all sets that are equivalent to **N** is denoted by $\aleph_0$, and is read **aleph null**. ($\aleph$ is the first letter of the Hebrew alphabet; the significance of the subscript 0 will appear later.)

DEFINITION The collection of all sets that are equivalent to **R** is denoted by c, and is called the **cardinality of the continuum**. ("Continuum" is the name chosen by mathematicians to convey the idea of an unbroken line.)

The sizes $\aleph_0$ and c can be regarded as numbers, much like the usual counting numbers 1, 2, 3, 4, 5, ... The number 3, for example, can be considered as the common property possessed by all sets that can be put in 1-1 correspondence with $\{a, b, c\}$. But the idea of "common property" (or "three-ness," if you prefer) is not precise enough to work with. Instead, we consider 3 to be the collection of all sets that are equivalent to $\{a, b, c\}$. In general:

DEFINITION A **cardinal number** is the collection of all sets that are equivalent to a particular set. Any particular set used to describe a cardinal number is called a **reference set** for that number. The cardinal number to which a set belongs is called the **cardinality** of that set.

Thus, 3 is a cardinal number and $\{a, b, c\}$ is a reference set for it. Of course, we could use $\{\star, \Diamond, \heartsuit\}$ or any other three-element set as a reference set for 3 just as well. Similarly, the definitions of $\aleph_0$ and c indicate that they are also cardinal numbers. Those definitions also specify the most commonly used reference sets for these two cardinal numbers —

$$\mathbf{N} \text{ for } \aleph_0 \quad \text{and} \quad \mathbf{R} \text{ for } c$$

DEFINITION The **finite** cardinal numbers are those with finite reference sets; the **transfinite** cardinal numbers are those with infinite reference sets.

The whole numbers $0, 1, 2, 3, \ldots, n, \ldots$ are the finite cardinals; $\aleph_0$ and c are transfinite cardinals.

Some natural questions arise at this point. One that fairly cries out for an answer is: Are there any other transfinite cardinal numbers? Another is: Does it make sense to ask which of the numbers $\aleph_0$ and c is the bigger one? Let us take on the second question first, using our experience with finite numbers as a helpful (if perhaps unreliable) guide.

What does it mean to say that 5 is larger than 3? Because we are dealing with cardinal numbers, the behavior of their reference sets should somehow determine the behavior of the numbers. In this sense, we would say that 5 is larger than 3 because any attempted 1-1 correspondence between a reference set for 5 and a reference set for 3 results in having some elements of 5's reference set "left over." For example, the letters d and e are not used in the following correspondence:

$$\{ \quad a, \quad b, \quad c, \quad d, \quad e \quad \}$$
$$\updownarrow \quad \updownarrow \quad \updownarrow$$
$$\{ \quad x, \quad y, \quad z \quad \}$$

This observation can be generalized to provide a definition of *larger than* that applies to any cardinal numbers:

DEFINITION Let $\mathcal{A}$ and $\mathcal{B}$ be cardinal numbers, with reference sets A and B, respectively. We say the number $\mathcal{A}$ is **larger than** the number $\mathcal{B}$ (or $\mathcal{B}$ is **smaller than** $\mathcal{A}$) if

1. there exists a 1-1 correspondence between all of set B and a proper subset of set A, and
2. there does not exist a 1-1 correspondence between B and all of A.

We symbolize "A is larger than B" by writing $A > B$ (or $B < A$).

Condition 2 might seem redundant at first, and in fact for finite numbers it is, but it is necessary in order for the definition to make sense when applied to transfinite numbers. For example, consider two different reference sets for $\aleph_0$, namely **N** and **E**. There obviously is a 1-1 correspondence between **E** and a proper subset of **N** (**E** itself), so Condition 1 of the definition is satisfied. Thus, if we were to use Condition 1 alone as the definition of *larger than*, we would find ourselves in the ridiculous position of having to say that $\aleph_0$ is larger than itself! Condition 2 allows us to avoid this absurdity because there is also a 1-1 correspondence between **E** and all of **N**. Therefore, the given definition of *larger than* does not apply to $\aleph_0$ and itself.

Now that we have a workable definition of *larger than*, it is easy to prove that c is larger than $\aleph_0$ by using the results of our previous work. Choose **N** as the reference set for $\aleph_0$ and $[0, 1]$ as the reference set for c. We have already proved that there is no 1-1 correspondence between **N** and all of $[0, 1]$, so Condition 2 of the definition is satisfied. To verify Condition 1, observe that

$$\{1, \tfrac{1}{2}, \tfrac{1}{3}, \tfrac{1}{4}, \ldots, \tfrac{1}{n}, \ldots\}$$

is a proper subset of $[0, 1]$ and it can be put in 1-1 correspondence with **N** by matching each fraction of the form $\frac{1}{n}$ with the natural number n that is its denominator. Thus, we have shown that

(6.9)

$$\boxed{\aleph_0 < c}$$

EXERCISES 6.6

1. Find three different sets that can be used as reference sets for the cardinal number 5.

2. Find three different sets that can be used as reference sets for the cardinal number 8.

3. $\{a, b, c, \ldots, y, z\}$ is a reference set for which cardinal number?

4. Determine a reference set for the cardinal number 0.

5. Set up a 1-1 correspondence between
    $$\{\spadesuit, \heartsuit, \diamondsuit, \clubsuit\}$$
    and a proper subset of
    $$\{a, b, c, d, e, f, g\}$$
    Is there more than one way to do this?

6. Let A be a cardinal number with reference set $\{x, y, z\}$. Which specific number is A? Why?

7. Suppose that B is a cardinal number with reference set $\{p, q\}$ and that C is a cardinal number with reference set $\{t, y\}$. What can you say about B and C? Why?

8. Suppose that A is a cardinal number with reference set A and that B is a cardinal number with reference set B. If A can be put in 1-1 correspondence with B, how are the numbers A and B related? Why?

9. If two cardinal numbers are equal, must a randomly chosen reference set for one be equal to a randomly chosen reference set for the other? Must these reference sets be equivalent? Why?

10. Use the definition given in this section to prove that 7 is larger than 4.

11. Prove that $\aleph_0$ is larger than 3.

12. Prove that c is larger than 5.

13. Prove that $\aleph_0$ is larger than any whole number n.

14. Find an example to illustrate the following general fact:

    For any three cardinal numbers A, B, and C, if $A < B$ and $B < C$, then $A < C$.

15. Use the general fact given in Exercise 14 and the statement of Exercise 13 to show that c is larger than any whole number n.

WRITING EXERCISES

1. (a) What is the distinction between *cardinal numbers* and *ordinal numbers*? (You might find it helpful to consult a dictionary.)

    (b) Write a paragraph defending the use of the adjective "cardinal" in this numerical sense as compatible with its nonmathematical usage in English.

2. Suppose there were a cardinal number X strictly between $\aleph_0$ and c; that is,
    $$\aleph_0 < X < c$$
    What information would this give you about sets? Write a paragraph or two "thinking with your pen" about this question, just writing down what comes to mind without worrying about paragraph structure or organization.

6.7   Cantor's Theorem

Up to now, our investigation of particular infinite sets has led us to classify all the infinite sets we have seen into two sizes: The "discretely" infinite size $\aleph_0$ of the set of natural numbers, and the "continuously" infinite size c of the real number line. It is tempting, especially after finding that even a change of dimension does not produce a set of larger size than c, to guess that there are no other infinite sizes. But this is a deceptive temptation. In proving his most important general statement, the theorem that bears his name, Georg Cantor showed that no matter what size we consider, there is a way to find a set of a larger size. Thus, there are *infinitely many* different sizes of infinity!

Cantor's Theorem is based on a simple, almost trivial, observation about finite sets — a set has more subsets than it has elements. The surprise is that, unlike many other observations about finite sets, this one also holds for infinite sets. We look first at a finite example. The three-element set $\{1,2,3\}$ has eight subsets:

$$\emptyset, \{1\}, \{2\}, \{3\}, \{1,2\}, \{1,3\}, \{2,3\}, \{1,2,3\}$$

This collection of subsets cannot be put in 1-1 correspondence with the original set $\{1,2,3\}$, but there are many ways to put $\{1,2,3\}$ in 1-1 correspondence with some subcollection of its subsets. One obvious way is to match each number with the single-element subset containing it:

$$
\begin{array}{cccc}
\{ & 1, & 2, & 3 \quad \} \\
   & \updownarrow & \updownarrow & \updownarrow \\
\{ & \{1\}, & \{2\}, & \{3\} \quad \}
\end{array}
$$

Clearly, such a correspondence can be established between *any* set and all its single-element subsets. For instance, the infinite set $\mathbf{N}$ of all natural numbers can be matched with its one-element subsets in this way:

$$
\begin{array}{ccccccc}
\{ & 1, & 2, & 3, & 4, & \ldots, & n, & \ldots \quad \} \\
   & \updownarrow & \updownarrow & \updownarrow & \updownarrow & & \updownarrow & \\
\{ & \{1\}, & \{2\}, & \{3\}, & \{4\}, & \ldots, & \{n\}, & \ldots \quad \}
\end{array}
$$

Thus, $\mathbf{N}$ can be matched with a proper subset of the set of all its subsets.

It is not obvious, however, that an infinite set *cannot* be put in 1-1 correspondence with the set of all its subsets. That was the ingenious part of Cantor's proof. He used a method analogous to his diagonalization process to show that no such matching is ever possible. In particular, he showed that, given any alleged 1-1 correspondence between an infinite set and all of its subsets, one could construct a specific subset, step by step, so that it could not possibly be matched with any element of the original set. The details of his proof are given below. The steps of that proof are not difficult, but you may have to ponder them awhile to understand how they establish a contradiction, and why this contradiction is the key to the entire logical argument.

(6.10) | **Cantor's Theorem:** Let $\mathcal{A}$ be any cardinal number, with reference set A, and let S be the set of all subsets of A. If $\mathcal{S}$ is the cardinal number that represents the size of S, then $\mathcal{A}$ is smaller than $\mathcal{S}$.
|
| (Informally, this says that the size of any set is smaller than the size of the set of all its subsets.)

Proof: [*optional*] As stated in the theorem, A is a reference set for the cardinal number $\mathcal{A}$, and S (the set of all subsets of A) is a reference set for the cardinal number $\mathcal{S}$. By the definition of *larger than*, to prove that $\mathcal{S}$ is larger than $\mathcal{A}$ we must show that

1. A can be put in 1-1 correspondence with a proper subset of S, and
2. A cannot be put in 1-1 correspondence with all of S.

The first part is easy; we shall dispose of it quickly. The second part is a bit more complicated, but a little concentrated study should give you not only an understanding of the logical steps, but an appreciation of the elegance of Cantor's work, as well.

1. (For simplicity, we assume $A \neq \emptyset$; the special case where $A = \emptyset$ is handled by a logical "trick" which need not concern us here.) Among all the subsets of A are the single-element subsets; that is, for each element x in A there is the subset $\{x\}$ in S. The matching of each x with $\{x\}$ is obviously a 1-1 correspondence between all the elements of A and some, but not all, of the elements of S. (In particular, $\emptyset$ is an element of S not used in this correspondence.) Thus, Condition 1 is satisfied.

2. To prove that there cannot exist a 1-1 correspondence between A and all of S, we use an indirect argument. We assume that such a correspondence does exist and derive a contradiction from this assumption.[2]

    Thus, assume that each subset of A corresponds to exactly one element of A, and vice versa. Because each element of A is matched with a subset of A by this presumed correspondence, it makes sense to ask whether or not an element is actually contained in the set with which it is matched. We select each element of A that is *not* in the set with which it is matched, and

---

[2] This indirect method of proof is quite common in mathematics and in classical philosophy. It is called *proof by contradiction*, or *reductio ad absurdum*. See the end of Appendix A for a brief description of its logical structure.

denote the set of all such elements by W. Clearly, W is a subset of A, and hence it is an element of S.

Because A and S are in 1-1 correspondence, there must be some element z in A that is matched with the set W. Now, either z is contained in W or it is not; let us examine both of these alternatives:

- If z is contained in W, then z is contained in the set with which it is matched. But then, the definition of W implies that z *cannot* be contained in W, which is an outright contradiction.

- If z is not contained in W, then z is not contained in the set with which it is matched. Then the definition of W implies that z *must* be in W, and again we have a contradiction.

There are no other alternatives, so we are forced to conclude that our argument contains a flaw somewhere. But each step follows logically from the one before it, except for our initial supposition that a 1-1 correspondence between A and S exists. Hence, this supposition must be false, so Condition 2 of the definition of *larger than* is satisfied.

Thus, we have proved that S, the size of the set of all subsets of A, is larger than A, the size of A.

EXERCISES 6.7

1. Write out the set of all subsets of $\{a, b, c\}$.

2. Write out the set of all subsets of $\{2, 4, 6, 8\}$.

In Exercises 3–6, find a proper subset of the set **N** of natural numbers that has size:

3. 3   **4.** 10   **5.** 500   **6.** $\aleph_0$

In Exercises 7–10, find a proper subset of the interval $[0, 1]$ that has size:

7. 3   **8.** 100   **9.** $\aleph_0$ **10.** c

11. Let $A = \{a, b\}$. Write out the set S of all subsets of A; then write out the set T of all subsets of S. What size is S? What size is T?

12. Let $A = \{p, q, r, s, t\}$ and let S be the set of all subsets of A.

(a) Find a three-element subset of A.

(b) Find two elements of S.

(c) Find a three-element subset of S.

(d) What cardinal number represents the size of S? Why?

13. (a) Let $A = \{u, v, w, x, y, z\}$. Set up a 1-1 correspondence between A and a set consisting of six of its subsets. (You may choose any six subsets.)

(b) In the proof of Cantor's Theorem we constructed a set W of all elements of A that are not in their corresponding subsets. Using the set A and your correspondence from Part (a) of this exercise, list the elements of such a set W.

(c) Do Parts (a) and (b) again, choosing a different collection of six subsets.

14. An alleged 1-1 correspondence between **N** and its set of subsets begins:

$$1 \leftrightarrow \emptyset$$
$$2 \leftrightarrow \mathbf{D}$$
$$3 \leftrightarrow \{p \mid p \text{ is prime}\}$$
$$4 \leftrightarrow \mathbf{E}$$
$$5 \leftrightarrow \{1, 4, 7, 10, 13, 16, \ldots\}$$
$$6 \leftrightarrow \{2, 4\}$$
$$\vdots \qquad \vdots$$

(a) Describe briefly how the "contradiction" set W of Cantor's Theorem is constructed.

(b) Following the process you described in Part (a), which of the numbers 1, 2, 3, 4, 5, 6 are in W?

WRITING EXERCISES

1. Referring to the statement of Cantor's Theorem, discuss the notational distinctions between the script capital letters ($\mathcal{A}$ and $\mathcal{S}$) and the italic capital letters (A and S). Why is this distinction made? Is the distinction necessary? Do you think this notation is helpful or confusing in conveying its message? How would you improve upon it? (In what way is your approach better than the one in the text?)

2. Reread the proof of Cantor's Theorem. What is the *first* thing in the proof that you don't understand? Write out two questions about it (in complete sentences), as specifically as you can.

3. Explore the analogy between the proof of Cantor's Theorem and Cantor's diagonalization process. How are they similar? How are they different?

6.8   The Continuum Hypothesis

Successive applications of Cantor's Theorem, first to a set A, then to the set S of all subsets of A, then to the set of all subsets of S, and so on, provides a way of getting larger and larger sizes of infinite sets, thus guaranteeing an infinite string of different transfinite numbers. However, the theorem does *not* guarantee that the numbers obtained in this way are successive; that is, if a set A represents a particular cardinal number, Cantor's Theorem does not tell us whether the set of all subsets of A represents the *next* cardinal number, or even whether the idea of "next" makes sense in this context. We do not get successive numbers in the finite case (except for 0 and 1) — a 2-element set has 4 subsets, a 4-element set has 16 subsets, and so on. However, our previous experience with basing expectations for infinite sets on finite examples should warn us against assuming the situations are alike.

Basing his work on some commonly accepted axioms,[3] Cantor established several useful facts about transfinite numbers; we state them here without proof:

---

[3] One of Cantor's implicit assumptions was the Axiom of Choice, which was not recognized explicitly as an axiom until about 25 years later. See page 519 for a brief description of the Axiom of Choice.

- $\aleph_0$ is the smallest transfinite cardinal number.

- Each of the transfinite cardinal numbers has an immediate successor, a "next larger" cardinal number. Thus, we may begin to list the transfinite cardinals as

$$\aleph_0 < \aleph_1 < \aleph_2 < \ldots < \aleph_n < \ldots$$

  with no other cardinal numbers in between.

- c is the size of the set of all subsets of the natural numbers.

All three of these statements are interesting and important to mathematicians, but one of Cantor's most intriguing statements about transfinite numbers is an assertion he could *not* prove. Having observed that a set of the smallest infinite size, $\aleph_0$, has a set of all subsets whose size is c, Cantor conjectured that c is the next smallest infinite size; that is, he guessed that $c = \aleph_1$. However, he was unable to prove his conjecture, and by 1900 this question had become one of the most famous unsolved mathematical problems. Known as the "Continuum Hypothesis," it may be stated more formally as follows:

(6.11)

> **Continuum Hypothesis:** If a set has size $\aleph_0$, then the set of all its subsets has size $\aleph_1$. More generally, if a set has size $\aleph_n$ for some n, then the set of all its subsets has size $\aleph_{n+1}$.

(Properly speaking, the Continuum Hypothesis is just the first of these two sentences. The second sentence is known, naturally enough, as the *Generalized Continuum Hypothesis*.)

During the first half of this century there were many fruitless efforts to prove the Continuum Hypothesis. The elusiveness of this problem was made even more frustrating by the fact that set theory had rapidly become respectable in most areas of mathematics and by 1930 had been put on a solid axiomatic foundation. Many people felt that such a rigorously logical approach to set theory would quickly provide a solution for the Continuum Hypothesis. But the problem was more difficult than they anticipated, and the eventual solution came in a surprising form.

In 1940, Kurt Gödel, an Austrian logician, proved that the Continuum Hypothesis is consistent with the axioms for set theory. In other words, he proved that the *assumption* that the Continuum Hypothesis is *true* will not lead to any contradictions within set theory. Of course, this did not prove the Continuum Hypothesis itself; rather, it showed only that the Continuum Hypothesis *could not be proven false* by a logical argument based on any of the commonly accepted axiom systems for set theory.

The problem remained unresolved until 1963, when Paul Cohen, of Stanford University, proved that the *assumption* that the Continuum Hypothesis is *false* also will not lead to any contradictions within set theory! This means that the Continuum Hypothesis can neither be proved nor disproved within set theory. In other words, we may treat it as a separate axiom that provides information not found in the other axioms of set theory. If we assume it to be true, we get one kind of set theory; if we assume it to be false, we get another kind of set theory, different from the first, but equally valid. (Compare this with the discussion of Euclid's Parallel Postulate in Chapter 3, especially in Section 3.6.)

This discovery provides a striking illustration to support the modern view of mathematics as a study that is independent of the physical world. If there were a single "true" theory of sets waiting for discovery, we could not have conflicting, but equally consistent, theories of sets. It appears, then, that mathematics is invented by human beings and is then applied to the world around them. This imposes on the universe a convenient order useful for explaining observed phenomena but is not determined by those phenomena. In the words of Cantor himself,

> "mathematics is entirely free in its development and its concepts are restricted only by the necessity of being noncontradictory and coordinated to concepts previously introduced by precise definitions.... The essence of mathematics lies in its freedom."[4]

EXERCISES 6.8

For the purpose of these exercises, assume that the general form of the Continuum Hypothesis is true.

1. (a) Give an example of a set A of size $\aleph_1$.
   (b) List three specific elements of A.
   (c) Find a subset of A that has size 5.
   (d) Find a subset of A that has size $\aleph_0$.

2. (a) Give an example of a set B of size $\aleph_2$.
   (b) List three specific elements of B.
   (c) Find a subset of B that has size 5.
   (d) Find a subset of B that has size $\aleph_0$.
   (e) Find a subset of B that has size $\aleph_1$.

In Exercises 3–8, let K be a set of size $\aleph_5$ and let S be the set of all its subsets.

3. What is the size of S?

4. Is K a subset of S or is it an element of S?

5. Is $\{K\}$ a subset of S or is it an element of S? What is its size?

6. What is the size of the set of all subsets of S?

---

[4]Georg Cantor, 1883, as quoted on page 1031 of [7] in the list *For Further Reading* at the end of this chapter.

7. Find a subset of S that can be put in 1-1 correspondence with K.

8. Find a subset of S that is infinite but that cannot be put in 1-1 correspondence with K.

9. How many transfinite cardinal numbers are smaller than $\aleph_3$? Give an example of a set of each of those sizes.

WRITING EXERCISES

1. What idea in this section do you find most confusing or difficult to understand? If, after rereading the section, you still do not understand it, write three specific questions that you think might help you. If you do understand it, write a brief explanation that you think might help a classmate who is having trouble with the idea.

2. Assume that the (first part of the) Continuum Hypothesis is false. In this case, what can be said about sets of size $\aleph_1$? Write as many characteristics of these sets as you can find.

3. If, as this section suggests, mathematical truth is not determined by reality but, rather, simply depends on assumptions you choose to begin with, how is it that mathematics works so well in describing the physical world? (After all, mathematics is fundamental to the construction and operation of almost all the modern machinery we have — automobiles, televisions, airplanes, satellites, CD players, etc.) Think about this for a little while; then summarize your thoughts in one or two paragraphs.

6.9   The Foundations of Mathematics

Although Cantor's theory of sets was well received in many parts of the mathematical community, acceptance was by no means universal. Cantor's set-theoretic treatment of infinity generated heated opposition from some of his foremost contemporaries, notably Leopold Kronecker, a prominent mathematics professor at the University of Berlin. Kronecker based his approach to mathematics on the premise that a mathematical entity does not exist unless it is actually constructible in a finite number of steps. From this point of view, infinite sets do not exist because it is clearly impossible to construct infinitely many elements in a finite number of steps. The natural numbers are "infinite" only in the sense that the finite collection of natural numbers constructed to date may be extended as far as we please, but "the set of all natural numbers" is not a legitimate mathematical concept. To Kronecker and those who shared his views, Cantor's work was a dangerous mixture of heresy and alchemy that introduced potentially lethal dosages of fantasy into the bloodstream of mathematics.

Kronecker's fears for the safety of mathematical consistency were at least partially justified by the appearance of several paradoxes in set theory. Among the most renowned of these is the self-contradictory notion of the "set of all sets." Cantor's concept of set was extremely general:

> By a set we are to understand any collection into a whole of definite and separate objects of our intuition or our thought.

Now, a set is itself a "definite and separate object of our thought," so it would seem sensible to consider the set S of all sets. By its very nature, S would have to contain at least as many elements as any other set; that is, the cardinal number of S would have to be greater than or equal to the cardinal number of any other set. But Cantor's Theorem states that there is no greatest cardinal number, so there must be a set whose cardinal number is greater than that of S (namely, the set of all subsets of S).

Perhaps the most famous set-theoretic paradox of all was formulated by Bertrand Russell in 1902. It depends solely on the notion of set, thus striking at the very heart of set theory. The paradox begins by observing that all sets may be classified according to whether or not they are elements of themselves. For example, the set of abstract ideas is an abstract idea and hence is an element of itself, but the set of all elephants is hardly an elephant. Let us call a set that is not an element of itself *normal*, and consider the set N of all normal sets. In symbols,

$$N = \{S \mid S \notin S\}$$

Question: Is N normal?

If we answer *Yes*, then N is in the set of all normal sets; that is, N is an element of itself, implying that it is not normal. If we answer *No*, then N is an element of itself, and hence must be normal because it is thereby contained in the set of all normal sets. Thus, either choice leads to a contradiction.[5]

Dilemmas such as this, resulting from an unrestricted use of the seemingly harmless concept of set, forced mathematicians of the late 19th and early 20th centuries to undertake a thorough reappraisal of the foundations of mathematics in an attempt to free it from the dangers of self-contradiction. This, in turn, led to the formulation of several different philosophies of mathematics.

A philosophy of mathematics may be described as a viewpoint from which the various bits and pieces of mathematics can be organized and unified by some basic principles. There have been many philosophies of mathematics throughout history. As the body of mathematical knowledge grew and was changed by the results of new investigations, its philosophies underwent similar mutations. The advent of set theory in all its unifying simplicity and then the discovery of serious flaws in its fundamental struc-

---

[5]There are many popularized versions of this paradox. Russell himself gave one in 1919: A (beardless) barber in a certain village claims that he shaves all those villagers who do not shave themselves, and that he shaves no one else. If his claim is true, who shaves the barber?

The Beardless Barber. He shaves all those who do not shave themselves. Who shaves the barber?

Picture Collection, The Branch Libraries, The New York Public Library

ture, all within less than half a century, brought about a violent upheaval in mathematical philosophy. Its development may be separated into three branches, each attempting in its own way to safeguard mathematics from internal contradictions. In the brief space remaining we summarize a few of the basic tenets that characterize each of these schools of thought.

Logicism regards mathematics as a branch of logic, claiming that all mathematical principles are completely reducible to logical principles. Several attempts have been made to reduce all of mathematics to a symbolic logical system, culminating in Bertrand Russell and Alfred North Whitehead's *Principia Mathematica* (1910). The *Principia* bases all of mathematics on a logical system derivable from five primitive logical statements whose truth is founded on basic intuition. Refinements of Russell and Whitehead's work, made by a number of people during the first half of this century, have succeeded in ironing out many of the minor difficulties in the *Principia*.

Nevertheless, there are some fundamental objections to the logistic viewpoint as a whole. It is claimed that some primitive mathematical ideas must be used to develop the system in an orderly fashion, and thus the system is not completely self-contained. Some have also objected that logicism implies that all mathematical ideas are contained in the five initial statements and that the rest of mathematics is merely an exercise in redundancy, a formalized restatement of these five principles, involving no additional information at all.

Intuitionism proceeds from the premise that mathematics must be based solely on the intuitively given notion of a succession of things, exemplified by the sequence of natural numbers. The intuitionists claim that mathematics is dependent neither on language nor on classical logic. The symbols in mathematics are used for communication only and they are incidental, because mathematics is essentially an individual matter and need not be communicated in order to exist. Moreover, the rules of mathematical reasoning are arrived at intuitively from a logic system that differs from classical logic and is applicable only to mathematics. Like Kronecker, the intuitionists reject the idea that infinitely many elements can be treated as a single thing (a set). They define a set as a *law* that generates a succession of elements. Thus, the "law"

$$\frac{1}{n} \text{ for any } n \in \mathbf{N}$$

is a set whose elements are 1, $\frac{1}{2}$, $\frac{1}{3}$, ..., but the collection of its elements cannot be completed by proceeding in this way.

Intuitionist logic does not accept the Law of the Excluded Middle (the principle that a statement that is not false must be true). Thus, for them, proof by contradiction is an invalid procedure. The existence of a mathematical object can be proven only by establishing a method for constructing it in a finite number of steps. These restrictions successfully eliminate the contradictions stemming from set theory, but they also eliminate sizable portions of generally accepted classical mathematics. It remains to be seen whether the intuitionists can refine their approach sufficiently to obtain all of classical mathematics while retaining their advantage of working in a contradiction-free system.

Formalism claims that mathematics is concerned solely with the development of systems of symbols. A formal mathematical system is a collection of abstract statements expressing relationships among undefined terms that are subject to a variety of interpretations. Because these systems have no necessary relation to reality, the formalists must guard against potential contradictions by proving that their systems are internally consistent.

Attempts were made to find one provably consistent axiom system for all of mathematics, but in 1931 Kurt Gödel proved this goal to be unattainable. Formalism is thus restricted to a piecemeal verification of the consistency of separate parts of mathematics,[6] and is forced to be content with the tentative assurance that any one mathematical system is as likely to be consistent as any other.

---

[6]In many instances, a mathematical system is proved to be consistent relative to the *assumed* consistency of some other, more familiar system, such as the arithmetic of the integers.

It would be inaccurate and misleading to suggest that all or even most mathematicians adhere rigidly to one of these three mathematical philosophies. The fact is that most mathematicians work in their respective fields "doing mathematics" and concern themselves very little with questions of philosophy. They have formulated opinions about what constitutes mathematics that are sufficient to guide them in their research, and these opinions are often mixtures of the viewpoints expressed here. Even those who do concern themselves directly with mathematical philosophy seldom agree in every detail. The classifications that we have described represent general trends of thought regarding the foundations of mathematics, especially the mathematical treatment of infinity. However, they are at best incomplete views of a subject far more complex than any of these three philosophies seems willing to admit. A truly adequate philosophy of mathematics, if such is possible, invites the efforts of a future generation.

EXERCISES 6.9

WRITING EXERCISES

1. Summarize Russell's Paradox in your own words, without using *any* symbols. Try to capture the main idea as briefly as you can.

2. (a) Rewrite this section's description of Logicism in a different style; that is, rewrite it in such a way that someone reading the book's description and yours side by side would see all the same main ideas, but would not think that the two descriptions had been written by the same person.

   (b) Do the same as in Part (a) for the description of Intuitionism.

   (c) Do the same as in Part (a) for the description of Formalism.

3. Reflect on what you know about mathematics so far; then decide which of the three philosophies of mathematics described in this section best fits your sense of what mathematics is. State and explain your choice. What do you find wrong or uncomfortable about the other two? How (if at all) does the one you chose *fail* to match what you think?

4. What is mathematics?

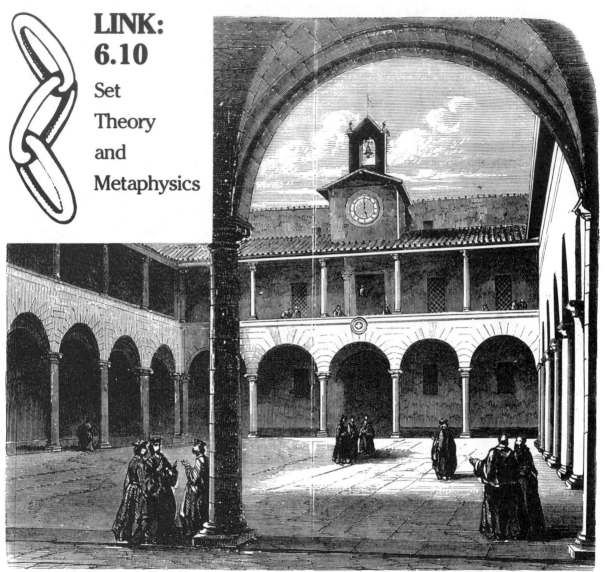

LINK: 6.10

Set Theory and Metaphysics

Picture Collection, The Branch Libraries, The New York Public Library

Neo-Thomism at Catholic Universities. Some Neo-Thomists adopted Cantor's theory to defend the existence of infinities. Pictured is the University of Pisa in the latter part of the 19th century.

In very general terms, terms that might well make a professional philosopher bubble over with disclaimers and qualifying phrases, philosophy is the search for a true understanding of reality by means of logical reasoning.

Traditionally, it has been broken down into a number of major subfields, including

- logic — the study of reasoning processes
- aesthetics — the study of beauty
- ethics — the study of the principles of human conduct
- politics — the study of the principles of social organization
- epistemology — the study of the limits and validity of human knowledge
- metaphysics — the study of being and reality.

Philosophy as a study distinguishable from religious activity began in Greece in the sixth century B.C., along with deductive mathematics. In fact, sometimes it was impossible to distinguish mathematical questions from philosophical ones, especially when it came to ideas such as motion, time and infinity. (See, for example, the material on the Pythagoreans and on Zeno's Paradoxes in Appendix Section B.3.)

Bertrand Russell has described philosophy as lying somewhere between the dogmatic pronouncements of religion and the experimental verifications of science. This distinction, while largely accurate today, has not always been valid. In the past, philosophy has often been regarded as encompassing both religion and science, borrowing from or lending to those fields the authority to defend the correctness of some particular theory or thesis. One example of this attempted merger of philosophy, theology, and science occurred in the late nineteenth century, just as Cantor's set theory was emerging into the intellectual daylight.

In 1879, Pope Leo XIII issued the encyclical *Aeterni Patris*, in which he instructed the Catholic Church to revitalize its study of Scholastic philosophy. This type of philosophy, also called Thomism, is based on the *Summa Theologica* of Saint Thomas Aquinas (c. 1225–1274), a truly monumental work that attempted to integrate traditional Roman Catholic doctrine with Aristotle's metaphysics. *Aeterni Patris*, along with an earlier encyclical, gave rise to neo-Thomism, a school of philosophical thought that proceeded in part from the view that atheism and materialism result from an incorrect philosophical interpretation of modern scientific achievements and that Scholastic philosophy is the key to a "proper" (i.e., orthodox Catholic) understanding of science. With the urging and guidance of the Vatican and the educational influence of various teaching orders of priests (notably the Jesuits), neo-Thomism soon became a major form of philosophy taught and studied in Catholic colleges and universities throughout the world.

When Cantor's work on the mathematics of infinity became known in the 1880s, it generated considerable interest among neo-Thomist philosophers.

There was some initial concern that the assertion of the actual existence of infinite sets would lead logically to the heresy of pantheism. Cantor, a devoutly religious man, nevertheless maintained the view that his mathematics of infinite sets indeed dealt with actual reality. For instance, he stated in an 1896 letter:

> The establishing of the principles of mathematics and the natural sciences is the responsibility of metaphysics.... The general theory of sets...belongs entirely to metaphysics.[7]

Thus, to Cantor, actually infinite collections of numbers had a real (though not necessarily material) existence. In patient, persistent correspondence with some of the leading Catholic theologians, he successfully distinguished his position from heretical views and gained semiofficial acceptance. Some neo-Thomistic philosophers in Germany even adopted his theories to defend the existence of actual infinities. For example, they argued that, because the Mind of God is all-knowing, It must know *all* numbers; hence, not only do all the natural numbers actually exist in the Mind of God, but so do all rationals, all infinite decimals, and so forth.

The most important effect of set theory in philosophy goes far beyond the arguments of the neo-Thomists, however. The attempts of Cantor and his successors to rid set theory of contradictions and thereby make it metaphysically sound led to profound investigations into the foundations of mathematics. Those introspective investigations early in this century that led to the three major schools of mathematical thought described in Section 6.9 have also led to clarifications of logical forms, methods of proof and errors of syntax, which have, in turn, been used to refine the arguments of philosophy. Modern mathematics has provided philosophy with some explicit, formal guidelines for admissible kinds of reasoning and possible logical constructions. The boundaries between religion, philosophy, and science have been brought into sharper focus as a result.

In particular, a major effect of these efforts has been the *removal* of mathematics from the realm of metaphysics! Thus, mathematics is now seen as a subject that is independent of reality. In the words of David Hilbert:

> It is not truth, but only 'certainty,' which is at stake.[8]

Thus, the saga of set theory has a final ironic twist: Cantor, working

---

[7]Quoted in Meschkowski, Herbert. *Ways of Thought of the Great Mathematicians* (trans. by John Dyer-Bennet). San Francisco: Holden-Day, Inc., 1964.

[8]ibid., p. 129.

to establish the metaphysical reality of set theory, began the process that has freed all of mathematics from metaphysics. Perhaps Cantor would have seen this as a tragedy; most modern mathematicians and philosophers see it as a giant forward stride in the progress of human thought.

Topics for Papers – Chapter 6

General Instructions: In doing these papers, you may use any outside sources you like. However, please remember that you must properly credit ideas and statements drawn from those other sources. Be careful that you do not simply copy or paraphrase the material from your sources. Any material not explicitly quoted should have been mentally digested by you, so that your own words describe what really are your own ideas when you write them down.

1. This paper assignment asks you to compare and contrast the mathematical idea of infinity with the idea of infinity as it occurs in some other area or as it is seen from some other point of view. Your title should be something like

    "Infinity in Mathematics and Infinity in _____,"

    or

    "Mathematical Infinity vs. _____ Infinity."

    Consider the reader of your paper to be an anonymous student in your class — someone who knows the material of Chapter 6 fairly well, but who knows little or nothing about the other area you choose to discuss.

    (a) Choose some area of thought or activity, other than mathematics, in which some idea of infinity arises. Define the term *infinite* as it is used in that area. (Consider: Is the term formally defined in that area? If not, then how do we arrive at an understanding of it? From where does it derive its meaning?)
    Note: Before going further, it might be wise to check with your instructor to be sure that the area and idea you have chosen are suitable for this project.

    (b) Give an example to illustrate the definition you stated in Part (a). (Consider: What is the connection between this example and your abstract definition or description?)

(c) Compare and contrast the idea of infinity you have chosen with the mathematical idea of infinity described in this chapter. How are they similar? How are they different? Apply these questions specifically to the definitions and examples in both areas; then try to formulate some general assessment of the similarities and differences.

(d) How do the similarities and differences you described in Part (c) reflect more general differences between mathematics and the other field you have chosen to discuss?

2. The 20th-century distinction in mathematics between truth and consistency, noted at the end of this chapter's LINK section, raises many paradoxical questions. Consider these statements:

- As we saw in Section 6.10, Georg Cantor said: "The establishing of the principles of mathematics and the natural sciences is the responsibility of metaphysics...." However, he also said: "The essence of mathematics lies in its freedom."

- Albert Einstein said: As far as the laws of mathematics refer to reality, they are not certain; and as far as they are certain, they do not refer to reality." Nevertheless, scholars in many areas of thought — from physics and chemistry to psychology, sociology, medicine, and even literary criticism — work hard to make their areas more mathematical in order that their conclusions will be accepted as reliably true.

- David Hilbert said about mathematics: "It is not truth, but only 'certainty,' which is at stake." But many people indicate their conviction that something is *true* by saying that it is "as sure as $1 + 1 = 2$," or words to that effect.

Pretend that you are writing an article about truth and consistency for the *New York Times* Magazine Section. (That is, think of your reader as a fairly intelligent person who is widely read, but who has no special background in mathematics.) Write a carefully organized paper in which you —

(a) describe clearly the distinction between truth and consistency;

(b) analyze each of the three given statements with respect to the truth/consistency distinction;

(c) use your analysis as the basis for a general discussion of the relationship between mathematics and reality;

(d) apply your general discussion to the concept of mathematical infinity as it has been described in this chapter. (If you covered

Chapter 3, you might also find it worthwhile to make connections with some of the ideas presented there.)

You may include your own opinions, but you must give arguments to justify any position you take.

For Further Reading

1. Bell, E. T. *Men of Mathematics*. New York: Simon and Schuster, Inc., 1937, Chapter 29.

2. Dauben, Joseph Warren. *Georg Cantor*. Cambridge, MA: Harvard University Press, 1979.

3. Davis, Philip J., and Reuben Hersh. *The Mathematical Experience*. Boston: Birkhäuser Boston, Inc., 1981, Part 7.

4. Devlin, Keith. *Mathematics: The New Golden Age*. London: Penguin Books, 1988, Chapter 2.

5. Dunham, William. *Journey Through Genius: The Great Theorems of Mathematics*. New York: John Wiley and Sons, Inc., 1990, Chapters 11 and 12.

6. Eves, Howard, and Carroll V. Newsom. *An Introduction to the Foundations and Fundamental Concepts of Mathematics*. New York: Holt, Rinehart and Winston, 1965, Chapters 8 and 9.

7. Kline, Morris. *Mathematical Thought from Ancient to Modern Times*. New York: Oxford University Press, 1972, Chapters 41, 43, and 51.

8. Newman, James R., ed. *The World of Mathematics*, Vol. 3. New York: Simon and Schuster, Inc., 1956, Part X.

9. Reid, Constance. *From Zero to Infinity*, 3rd Ed. New York: Thomas Y. Crowell Company, 1964, Chapter ... .

CHAPTER

MATHEMATICS OF SYMMETRY: FINITE GROUPS

7.1   What Is Group Theory?

The historical roots of group theory lie in early 19th-century Europe. At that time mathematicians in several different countries made largely independent efforts to free algebra from its dependence on numbers. As they worked to strip their algebraic tools down to the essential underlying ideas, they were led to remarkably similar patterns and concepts. This process of stripping down an idea to its essential features is called "abstraction"; it will be discussed more a little later in this section. For now, let us just say that the abstraction process is reminiscent of the Cheshire Cat of Wonderland, who faded slowly away before Alice's very eyes, until only the grin remained.

The grin that remained in algebra after the numbers disappeared is a mathematical theory of symmetry known as *group theory*. Symmetry is one of the most basic concepts of human thought. We observe balance and proportion in nature and we try to capture it in art; we observe its structure in science and repeat its harmony in music. Group theory's importance to us today stems from the fact that it is the mathematical tool used to describe symmetry in all these areas. Groups occur in the geometry of artistic perspective, in the design of wallpaper patterns, in the theory of modern computer codes, and in the patterns of traditional bell ringers. Chemists use group theory to classify and analyze molecules somewhat as they use the periodic table to relate the chemical properties of the elements. Physicists use group theory as a basis for theories of nuclear structure. Mathematical group theory is even beginning to turn up in sociology and psychology. And, despite all these applications to the "real world," some mathematicians study group theory for its own sake, to extend their understanding and appreciation of its internal symmetries and harmonies.

The theory of groups is part of a mathematical field called *abstract algebra*. Each of these two words has been known to cause discomfort; together they can be positively terrifying. Let us put these fears to rest by examining exactly what this phrase means. Many people consider the word *abstract* to be a synonym for "vague" or "hard to understand"; but, in fact, its meaning is quite different. Its literal meaning is derived from Latin and is better reflected in our use of the verb *to abstract*, which means "to pull out of" or "to separate from." (It has the same root as "tractor.") Thus, the adjective *abstract* is used to indicate some property of a thing that is considered apart from that thing's other characteristics. The abstraction process is a way of simplifying a situation by focusing directly on a specific aspect of it, separating its essential features from other facts that might confuse the issue.

Let us turn to arithmetic for an example. Among the many common arithmetic facts at our disposal, we know that

$$6 \cdot 4 = 24, \ 7 \cdot 3 = 21, \ 8 \cdot 2 = 16, \ 9 \cdot 1 = 9, \ 10 \cdot 0 = 0$$

This list of data has many features, some interesting from one point of view, some from another. For instance, we might observe that every digit except 5 was used, or that the products are alternately even and odd, or that all the numerals are printed with the same-size type, or that the two factors add up to 10 in each case. All these observations are true, but they may distract us from some more important ones. In this example, let us rearrange the information given, by starting with the observation that in each case 5 is "in the middle of" the two numbers being multiplied:

$$
\begin{aligned}
(5+1)(5-1) &= 6 \cdot 4 = 24 = 25 - 1 \\
(5+2)(5-2) &= 7 \cdot 3 = 21 = 25 - 4 \\
(5+3)(5-3) &= 8 \cdot 2 = 16 = 25 - 9 \\
(5+4)(5-4) &= 9 \cdot 1 = 9 = 25 - 16 \\
(5+5)(5-5) &= 10 \cdot 0 = 0 = 25 - 25
\end{aligned}
$$

A close look at these facts in this form should suggest a general observation about the behavior of numbers, which can be written in the following *abstract* form:

> If two numbers are equidistant from 5 on opposite sides, then their product equals 25 minus the square of their distance from 5; that is,

$$(5 + d) \cdot (5 - d) = 5^2 - d^2$$

The use of the letter *d* here in place of the various specific distances is an essential abstraction step; it unifies the five specific cases of the one pattern and brings the pattern itself into focus. Moreover, this last equation suggests that the number 5 might not be crucial to the behavior of such products. This leads us to an even more general form, which you may remember from high school:

$$\text{For any numbers } x \text{ and } y, \ (x + y)(x - y) = x^2 - y^2$$

This example typifies much of what is called *algebra* in high school. High-school algebra, and indeed, the only algebra studied and practiced by mathematicians until fairly modern times, is little more than symbolized arithmetic. Various methods for manipulating numbers and the basic arithmetic operations $+$, $-$, $\times$, and $\div$ are abstracted (pulled out) by using letters to stand for numbers. This makes it easier for us to focus on general rules that describe the behavior of our number system. That area of mathematics is sometimes called "classical algebra"; it might appropriately be called "abstract arithmetic."

Abstract algebra takes a further step in generality (and thus also in simplicity). Rather than confining itself to the numbers and operations of our usual number system, this form of algebra looks at various collections (*sets*) of objects and discusses operations on them. Because the objects considered need not be numbers (for instance, they could be letters, or motions, or just about anything), the usual operations of arithmetic may not apply to them. Hence, the fundamental properties of the "operation" concept must be abstracted from their numerical settings and put in a general form to permit application to other collections of things. This shift of attention from the concept of number to the concept of operation is what distinguishes modern algebra from classical algebra.

The conceptual unity that results from the abstraction process is one of the most significant features of modern mathematics as a whole. Let us look at a simple example of this unifying effect. Consider the following three situations:

1. A 3-way light has four switch positions: *low, medium, high,* and *off*. Each time the light switch is turned, the light goes from one of these settings to the next, in order. Each "click" of the switch signals a movement from one level of brightness to the next; with the fourth click the light is *off*, with the fifth it is at *low* again, and so on.

2. The duty periods for crew members aboard ship are called *watches*. By tradition, dating back to the early days of sailing ships, each watch is four hours long and the passing of every hour is marked by pairs of bells — 2 bells for the first hour, 4 for the second, 6 for the third, and 8 bells to signal the end of a watch (and the beginning of another).

3. A square tile works loose from its position in a bathroom floor. Its shape allows it to be put back in any of four ways: exactly as it

Since the days of sailing ships, a watch has been four hours long.

was originally, or rotated (clockwise) 90°, or 180°, or 270°. (A 360° rotation puts it back to its original position.) Even if it has been kicked around for a while, whatever way it is put back *must* correspond to one of these four "quarter turns" relative to its original (but perhaps forgotten) position.

Let us put the abstraction process to work. Some common features of these three very different situations are obvious: Each involves four "things" — brightness levels, watch hours, tile positions — and we can go from each to the next in succession until we return to a "starting point" — *off*, beginning of watch, original position. Now, if we focus not on the things themselves but on the way of getting from one to another, a common structure emerges. In each case, we go from one "thing" to another by steps in a cyclic pattern; moreover, if we take more than 3 steps we will have "gone around" the cycle completely and will have started over. Thus, we might as well ignore all "strings" of four successive steps. For example, 5 switch clicks from *off* gets us to *low* again, so it's the same as 1 click; 6 hours from the beginning of a watch is 2 hours into a watch (the next one, of course); 7 quarter turns of the tile puts it in the same position as 3 quarter turns, and so on.

*The square shape of a loosened tile allows
it to be put back in any of four ways.*

The number of steps that takes us to the different positions in each
situation, then, is just the set {0, 1, 2, 3}, and if we follow any of these
numbers of steps by any other, we get the same result as some one of these
four numbers (2 steps following 3 steps is equivalent to 1 step, etc.). If we
denote "following" by f, we can tabulate all possible combinations of these
four step numbers, as in Table 7.1.

$$0\ f\ 0 = 0 \quad 0\ f\ 1 = 1 \quad 0\ f\ 2 = 2 \quad 0\ f\ 3 = 3$$
$$1\ f\ 0 = 1 \quad 1\ f\ 1 = 2 \quad 1\ f\ 2 = 3 \quad 1\ f\ 3 = 0$$
$$2\ f\ 0 = 2 \quad 2\ f\ 1 = 3 \quad 2\ f\ 2 = 0 \quad 2\ f\ 3 = 1$$
$$3\ f\ 0 = 3 \quad 3\ f\ 1 = 0 \quad 3\ f\ 2 = 1 \quad 3\ f\ 3 = 2$$

Table 7.1   Combinations of the four steps.

If we further simplify this tabulation by listing just the four columns of answers, we obtain a table whose patterns and symmetry almost beg for further exploration! (See Table 7.2.)

$$
\begin{array}{cccc}
0 & 1 & 2 & 3 \\
1 & 2 & 3 & 0 \\
2 & 3 & 0 & 1 \\
3 & 0 & 1 & 2
\end{array}
$$

Table 7.2   The common pattern of Situations 1, 2, and 3.

Further examination of this pattern is best left until we get some basic definitions and structural ideas under our belts. For now, it suffices to say that this pattern is an example of something called a *group,* and this particular group describes the operational behavior of the three different situations that we have described. From this viewpoint, then, the three situations have been unified into one, and *the study of the single table representing this one operational pattern tells us about all three of the situations it came from.*

This example illustrates in a simple way the value of abstraction. The abstraction process unifies seemingly different situations by focusing on (abstracting) some particular properties those situations have in common, so that a single study of the abstract properties can give us information about all the "different" situations at the same time. Thus, abstraction is a powerful tool in applied science; it reduces the amount of effort required to analyze things with similar characteristics and sometimes suggests unsuspected connections among seemingly different phenomena.

EXERCISES 7.1

For Exercises 1–4, write three specific instances of the given abstract numerical formula. (Assume x, y, and z represent numbers of any kind.) Work out the arithmetic to verify that your examples are true statements.

1. $3x + 3y = 3(x + y)$

2. $(x + y) + (x - y) = 2x$

3. $(x + y) + z = x + (y + z)$

4. $(-x)(-y)(-z) = (-x)yz$

For Exercises 5–8, write a general expression using letters such as x, y, and z to abstract a general form from the given set of numerical data.

5. $2 \cdot 3 = 3 + 3$
   $2 \cdot 5 = 5 + 5$
   $2 \cdot 14 = 14 + 14$
   $2 \cdot 27 = 27 + 27$

6. $2 + 3 = 3 + 2$
   $5 + 8 = 8 + 5$
   $7 + 19 = 19 + 7$
   $43 + 88 = 88 + 43$

7. $2 \cdot 3 + 2 \cdot 5 = 2 \cdot 8$
   $7 \cdot 12 + 7 \cdot 8 = 7 \cdot 20$
   $5 \cdot 40 + 5 \cdot 2 = 5 \cdot 42$
   $3 \cdot 1 + 3 \cdot 1 = 3 \cdot 2$

8. $25 - 9 = 8 \cdot 2$
   $36 - 25 = 11 \cdot 1$
   $100 - 4 = 12 \cdot 8$
   $81 - 49 = 16 \cdot 2$

WRITING EXERCISES

1. Examine Table 7.2 and describe patterns (of any sort) that you see in it. Explain how each pattern you find might be translated to say something about each of the three different situations the table describes.

2. Find and describe another real-world situation (besides the three given in the text) that fits the pattern summarized by Table 7.2.

3. Table 7.2 expresses an underlying structure that is common to three real-world situations, each of which was decribed by a short paragraph. In like manner, find and describe three real-world situations with a common underlying structure that is described by Table 7.3.

0	1
1	0

Table 7.3

7.2   Operations

The first step in our investigation of groups will be to isolate the "operation" idea in some general form. If we look at the familiar arithmetic operations of addition, subtraction, multiplication, and division, we see that they share a common characteristic: They all are ways of assigning an "answer" number to two numbers. For instance, addition assigns 8 to 3 and 5 (because $3 + 5 = 8$), whereas multiplication assigns 15 to 3 and 5 (because $3 \cdot 5 = 15$). To abstract this idea from its numerical setting, we begin with the idea of an **ordered pair** — a pair of elements (of any sort) in which the first element is distinguishable from the second. We write (x, y) to denote the ordered pair in which x is the first element and y is the second. If x and y are different, the ordered pairs (x, y) and (y, x) are different. (By way of contrast, the set $\{x, y\}$ is considered to be the same as the set $\{y, x\}$ because both sets contain exactly the same elements.) Now it becomes easy to state a useful general definition of operation.

DEFINITION   An **operation** on a set S is any rule or method that assigns to each ordered pair of elements of S exactly one element of that set S.[1]

---

[1]Strictly speaking, such operations are usually called *binary operations* because they combine two elements at a time. This is the only kind of operation we shall consider, so the word *binary* has been omitted for simplicity.

NOTATION   We shall use ∗ as the general symbol for an operation on a set. The element of the set that ∗ assigns to the ordered pair (x, y) will be written $x ∗ y$.

Example 7.1   As noted above, the four arithmetic operations are examples of operations. The first three are defined on the set of all real (or complex) numbers; division is an operation on the set of all nonzero real (or complex) numbers.

- Addition assigns to each ordered pair (x, y) the sum $x + y$;
- subtraction assigns to (x, y) the difference $x - y$;
- multiplication assigns to (x, y) the product $x \cdot y$;
- division assigns to (x, y) the quotient $x \div y$.   □

Example 7.2   The tabulation at the end of Section 7.1 defines an operation "f" on the set $\{0, 1, 2, 3\}$ because it assigns to each of the 16 ordered pairs of those four numbers some number from that set — $(2, 1)$ is assigned 3, $(1, 3)$ is assigned 0, $(0, 1)$ is assigned 1, and so forth.   □

Example 7.3   As we saw in Section 7.1, a particular case of the previous example is the set of the four rotations of a square — rotations of 0°, 90°, 180°, and 270°. Because any of these rotations followed by any other gives the same result as if one of the original four rotations had occurred, the *follows* process is an operation on this set of rotations.

A similar operation can be defined for rotations of any regular polygon, such as an equilateral triangle or a regular pentagon or hexagon. To find the numbers of degrees for the appropriate rotations, just divide 360 by the number of sides of the figure; then take successive multiples of that number until you get back to 360. For instance, the appropriate rotations for a triangle are

0°, 120°, and 240°

For a hexagon they are

0°, 60°, 120°, 180°, 240°, and 300°

In each case the result of following one of these rotations by another gives the same result as one of the original rotations in the set. We will have more to say about movements of geometric figures in a later section.   □

Example 7.4   Let S be the set of all students in a class. If we assign to each ordered pair of students the taller one, we have an operation on S. If we pick out a particular student and assign that student to each ordered pair, we have another operation on S (even though in this case all the "answers" are the same). However, if we assign to each ordered pair of students in S the average of their grades on the last exam, we do *not* have an operation on S, because the "answers" are not in the set S of students.   □

Example 7.5   The process that assigns to each pair (x, y) of numbers the mean $(x + y) \div 2$ is an operation on the set of all rational numbers, but not on the set of whole numbers. (Why not?)   □

Example 7.6   Let S be the set of letters {n, d, q}, and define an operation ∗ on S by listing all possible ordered pairs and assigning elements to them at random. An example of this is shown in Table 7.4. Because every ordered pair is assigned exactly one element of S, this listing specifies an operation on S (even though there is no apparent pattern to the assignments).   □

$$n * n = q \qquad n * d = d \qquad n * q = n$$
$$d * n = d \qquad d * d = n \qquad d * q = n$$
$$q * n = n \qquad q * d = n \qquad q * q = d$$

Table 7.4

Listing the assignments for each ordered pair is sometimes the only way a particular operation can be written down, because it is not always possible to give a succinct general rule for it. We can make the job shorter and the result clearer, however, by organizing the information in an **operation table**. An operation table works like the grid used for reading a road map and is perhaps most easily explained by example. To put the operation of Example 7.6 into tabular form, we list the elements of S in a row and in a column as shown in Table 7.5. This arrangement creates a square array of nine boxes to be filled, as indicated by the dotted lines. If we "name" each box by the letter indicating its row and then the letter for its column, we have a box for each ordered pair of elements of S, and we can fill each one with the element assigned to that pair by the operation. The box filled in Table 7.5 shows the entry that corresponds to d ∗ q. Table 7.6 shows the complete table for this operation. The information supplied is the same as that given in Table 7.4, but it is displayed much more compactly. The symbol for the operation is noted in the upper-left corner for convenience.

	n	d	q
n			
d			n
q			

Table 7.5   The form of an operation table, showing the entry d ∗ q = n.

∗	n	d	q
n	q	d	n
d	d	n	n
q	n	n	d

Table 7.6   The information of Table 7.4 in more compact form.

Example 7.7 Let $S = \{a, b, c\}$, and define the operation $*$ on S by Table 7.7, which says
$a * a = b$, $a * b = a$, $a * c = c$, etc.   □

$*$	a	b	c
a	b	a	c
b	c	b	a
c	a	c	b

Table 7.7

Example 7.8 Tables 7.8 (a) and (b) define two operations, $*$ and $\circ$, on the set $\{0, 1\}$.
(How many other operations can you define on this set?)   □

$*$	0	1
0	0	1
1	1	0

(a)

$\circ$	0	1
0	0	0
1	0	1

(b)

Table 7.8

Let us denote the set $\{\ldots, -3, -2, -1, 0, 1, 2, 3, \ldots\}$ of all integers by **I**
and the set $\{1, 2, 3, \ldots\}$ of all positive integers by **I**$^+$. Ordinary addition
and multiplication are operations on both **I** and **I**$^+$. However, subtraction
is an operation on **I** but not on **I**$^+$ because, for example, subtraction assigns
to the ordered pair of positive integers $(3, 4)$ the number -1, which is in **I**
but not in **I**$^+$. (Recall that the definition of operation requires the assigned
elements to come from the original set.)

In general, an operation defined on a set S may not be an operation on
every subset[2] of S, because the operation on S may assign to an ordered
pair of subset elements some element of S that is not in the subset. If
the operation does assign to each pair of subset elements an element from
that subset, we say that the subset is **closed** under the operation (or the
operation is closed on the subset). This property is called **closure**.

Example 7.9 The even integers are closed under addition; that is, the sum of two even
integers is always even. The odd integers are not closed under addition; for
example, $3 + 5 = 8$, so the sum of two odd integers is not always (in fact,

---

[2]A subset of a set S is any set whose elements all are also elements of S.

is never) odd. Both the even integers and the odd integers are closed under multiplication. (Can you justify these statements?)   □

Example 7.10   Table 7.9 defines an operation ∗ on the set {a, b, c, d}. The subset {a, b} is closed under ∗, but the subsets {c, d} and {a, b, c} are not. (Can you justify these statements?)   □

∗	a	b	c	d
a	a	b	c	d
b	b	a	d	c
c	c	d	a	b
d	d	c	b	a

Table 7.9

EXERCISES 7.2

1. Referring to the operation ∗ defined by Table 7.9, find $a ∗ b$, $b ∗ c$, $d ∗ d$, and $c ∗ a$.

Table 7.10 defines an operation ◇ on the set

$$\{2, 3, 4, 5, 6\}$$

Use this table for Exercises 2–11.

◇	2	3	4	5	6
2	4	2	6	3	5
3	5	5	2	5	2
4	3	6	2	2	3
5	4	4	4	4	4
6	2	6	3	4	5

Table 7.10

2. $2 ◇ 4 =$

3. $3 ◇ 5 =$

4. $4 ◇ 2 =$

5. $6 ◇ 6 =$

6. $5 ◇ 5 =$

7. $6 ◇ 3 =$

8. $(2 ◇ 3) ◇ 4 =$

9. $2 ◇ (3 ◇ 4) =$

10. $(5 ◇ 5) ◇ 5 =$

11. $5 ◇ (5 ◇ 5) =$

12. Let $S = \{q, r, s, t\}$ and define an operation ∗ on S as follows:

   - If q is in a pair, that pair is assigned q;
   - If r is in a pair, that pair is assigned the other element in the pair, and $r ∗ r = r$;
   - $s ∗ t = t ∗ s = t$;
   - $s ∗ s = t ∗ t = s$.

   Describe ∗ by an operation table.

13. Let $S = \{1, 2\}$. Construct tables describing *all* possible operations on S. [This exercise is used in Exercise 2 of Section 7.3.]

For Exercises 14 and 15, let $S = \{1, 2, 3, 4, 5\}$ and let x and y represent arbitrary elements of S.

14. Define an operation ⋆ on S by $x ⋆ y = y$ for all elements x and y in S. Construct an operation table for ⋆ on S.

15. Define an operation ○ on S by:
   − If $x ≠ y$, then $x ○ y$ is the larger number;
   − if $x = y$, then $x ○ y = 1$.
   Construct an operation table for ○ on S.

16. Let $T = \{x, y, z\}$.

    (a) Construct five different operation tables for T.

    (b) How many operations are possible on T? Why?

    (c) Which of the operations that you defined in Part (a) are closed on the subset $\{y, z\}$ of T?

17. Find two examples to show that the odd integers are not closed under addition.

18. Find three examples to illustrate that the even integers are closed under addition.

19. Abstract from the examples given in Exercise 18 to make a general argument showing that the sum of two even integers must be even. (*Hint*: Any even integer is of the form $2n$, where n is an integer.)

20. Keeping in mind the discussion in Example 7.3, list the appropriate rotations for a regular pentagon, and show that any one of them followed by any other must give the same result as one of the original rotations. (*Hint*: There are five of them.)

WRITING EXERCISES

1. We have seen that an operation on a finite set can be described either by stating some sort of succinct procedural rule or by filling in a table; many operations can be described both ways. Discuss the relative advantages and disadvantages of each approach.

2. The footnote to the definition of *operation* observes that this definition actually only refers to *binary* operations because it is stated in terms of pairs.

    (a) By analogy, define "ternary operation."

    (b) Construct and describe at least one interesting example of such an operation. (An interesting example should not be simply the repeated application of a single binary operation.)

    (c) What would an operation table for a ternary operation look like? If possible, construct a table for a ternary operation on the set 0, 1 and describe how it works.

7.3   Some Properties of Operations

The abstract definition of an operation is so general that the list of all possible operations even on a comparatively small set would be far too long to permit a meaningful examination of each one. For instance, because any operation on a three-element set can be written as a 3-by-3 table and because each of the nine entries in that table can be any of the three elements in the set, there is a total of 3^9 such operation tables. Now, $3^9 = 19,683$, so on any three-element set we can define 19,683 different operations! To organize this bewildering variety of things in some manageable way, we look at several properties that make some operations "nicer" than others. The properties discussed here should already be familiar to you; they are just generalizations of arithmetic concepts.

The most obvious "nice" properties are suggested by a close look at the definition of operation itself. The definition says that the operation must deal with ordered pairs, implying that the element assigned to a pair in one order might differ from the one assigned to that pair taken in the reverse order. Indeed, this is what happens with the usual arithmetic operations of subtraction and division:

$$5 - 3 = 2 \qquad \text{but} \qquad 3 - 5 = -2$$
$$10 \div 2 = 5 \qquad \text{but} \qquad 2 \div 10 = \tfrac{1}{5}$$

However, addition and multiplication assign the same number to a pair of numbers in both orders:

$$5 + 3 = 8 \qquad \text{and} \qquad 3 + 5 = 8$$
$$10 \cdot 2 = 20 \qquad \text{and} \qquad 2 \cdot 10 = 20$$

Thus, addition and multiplication are easier to work with. This suggests the following general property:

DEFINITION   An operation $*$ on a set S is **commutative** if
$$a * b = b * a$$
for all elements a and b in S.

Example 7.11   The operation on the set of letters {a, b, c} defined in Table 7.7 of Section 7.2 is not commutative because, for example, $a * b = a$, but $b * a = c$. However, the operations given in Tables 7.6, 7.8, and 7.9 are all commutative.   □

An easy way to see whether or not a table defines a commutative operation is to imagine folding it along the diagonal from the upper-left corner of the table to its lower-right corner, thereby matching up each $a * b$ with $b * a$. If the table entries that match by this folding are the same, then the operation is commutative. Table 7.11 illustrates this.

Table 7.11   Commutativity in an operation table.

If an operation is not expressed by a table, then a proof that it is commutative might be difficult because a general argument is needed to show that *every* pair is assigned the same element in both orders. On the other hand, since the definition of commutativity requires the reversibility to hold in every case, to prove that an operation is *not* commutative, it suffices to find *one* pair that is assigned different elements in its two orders.

Returning to the definition of an operation, let us now focus on the requirement that operations assign elements to *pairs* of elements. It is natural to ask, then, how three or more elements can be combined. Any attempt to apply a (binary) operation to three elements must proceed two at a time, and (because not all operations are commutative) the order in which the elements are considered is important. However, even if the order of the elements is fixed, there are still two ways to apply the operation. If a, b, and c are three elements (taken in that order), we could

combine a and b, and then combine $a * b$ with c

or

combine b and c, and then combine a with $b * c$

In symbols, we could consider either $(a * b) * c$ or $a * (b * c)$. For instance, if asked to subtract the numbers 9, 5, and 2, we might compute

$$(9 - 5) - 2 = 4 - 2 = 2$$

or

$$9 - (5 - 2) = 9 - 3 = 6$$

On the other hand, adding 9, 5, and 2 is independent of this grouping process:

$$(9 + 5) + 2 = 14 + 2 = 16$$

and

$$9 + (5 + 2) = 9 + 7 = 16$$

This suggests another property that some operations possess.

DEFINITION   An operation $*$ on a set S is **associative** if

$$(a * b) * c = a * (b * c)$$

for all elements a, b, and c in S.

Example 7.12   The operation defined by Table 7.7 (in Section 7.2) is not associative because, for example,

$$(b * b) * c = b * c = a \quad \text{but} \quad b * (b * c) = b * a = c \qquad \square$$

Example 7.13   Both operations in Tables 7.8 (of Section 7.2) are associative, as is the operation defined by Table 7.9. Unfortunately, there is no easy way to observe this from the tables (as there is for commutativity); a proof of these facts would require checking all possible cases. We give here just one example for each operation from Tables 7.8; you may verify the other cases at your leisure.

$$(1 * 1) * 0 = 0 * 0 = 0 \quad \text{and} \quad 1 * (1 * 0) = 1 * 1 = 0$$
$$(1 \circ 1) \circ 0 = 1 \circ 0 = 0 \quad \text{and} \quad 1 \circ (1 \circ 0) = 1 \circ 0 = 0 \qquad \square$$

EXERCISES 7.3

1. Give an example that proves that ordinary division is not associative.

2. (This exercise refers back to Exercise 13 of Section 7.2.) In all the examples we have examined so far, operations either have been both associative and commutative or have had neither of these properties. It is natural to ask whether these properties always exist together. Show that such is not the case by finding among the 16 operations on the set $\{1, 2\}$ an example of

   (a) an operation that is commutative but not associative;

   (b) an operation that is associative but not commutative.

For Exercises 3–8: Is the operation defined by the given table commutative? Is it associative? Give reasons for your answers.

3.

	1	2	3
1	3	3	3
2	3	3	3
3	3	3	3

4.

	1	2	3
1	1	3	2
2	3	3	1
3	2	1	2

5.

	5	9
5	5	9
9	5	9

6.

	5	9
5	9	9
9	5	9

7.

	2	4	6
2	2	6	4
4	4	2	6
6	6	4	2

8.

	2	4	6
2	6	4	4
4	4	6	4
6	4	4	6

9. Construct an operation on the set $\{p, q, r\}$ that is both commutative and associative.

10. Give a general argument to prove that Table 7.12 defines an associative operation on the set $\{1, 2, 3, 4, 5\}$. (A case-by-case verification is a permissible but unattractive strategy; there are 125 cases to be checked.)

$*$	1	2	3	4	5
1	1	1	1	1	1
2	2	2	2	2	2
3	3	3	3	3	3
4	4	4	4	4	4
5	5	5	5	5	5

Table 7.12

WRITING EXERCISES

1. Do one of the following two parts:

(a) Explain why the words "commutative" and "associative" are appropriate choices to represent the properties described by their definitions. (Look at the nontechnical meanings of the roots of these two words.)

(b) If you do not think that "commutative" and "associative" are appropriate choices to represent the properties described by their definitions, suggest two words that you think would be better, and explain why.

2. Subtraction and division fail to be commutative in much the same way. That is, for nonzero rational numbers a and b,

$$a - b = -(b - a) \quad \text{and} \quad a \div b = (b \div a)^{-1}$$

(a) Write a definition for the general property suggested by these two examples; call it "anticommutativity."

(b) Find or construct another operation that is anticommutative. (You might try making up an operation table on some set.)

(c) Explain how you might tell from looking at an operation table whether or not the operation it describes is anticommutative.

(d) If an operation is both commutative and anticommutative, what can be said about inverses in that system? Explain.

7.4   The Definition of a Group

Of the four elementary arithmetic operations, the two "better-behaved" ones seem to be addition and multiplication because both of them are commutative and associative. Addition and multiplication are also well-behaved in other ways. For instance, if a and b are integers, then any equation of the form $a + x = b$ has an integer solution. For example,

$$\text{if } 7 + x = 2, \text{ then } x = -5$$

Similarly, if c and d are positive rational numbers, then any equation of the form $c \cdot x = d$ has a positive rational solution. For example,

$$\text{if } \tfrac{2}{3} \cdot x = \tfrac{1}{2}, \text{ then } x = \tfrac{3}{2} \cdot \tfrac{1}{2} = \tfrac{3}{4}$$

Now, if we examine the ways in which equations like these are solved, we find a common pattern that can be abstracted and made into a definition with surprisingly far-reaching consequences. Let us look at two examples, solving each equation in painfully inefficient, but enlightening, detail.

To solve $3 + x = 5$, we add -3 to both sides of the equation, regroup by associativity, and let -3 "cancel" the 3 on the left, leaving us with the desired solution:

$$\begin{aligned} -3 + (3 + x) &= -3 + 5 \\ (-3 + 3) + x &= 2 \\ 0 + x &= 2 \\ x &= 2 \end{aligned}$$

Similarly, to solve $3 \cdot x = 5$, we multiply both sides by $\frac{1}{3}$, regroup by associativity, and let $\frac{1}{3}$ "cancel" the 3 on the left, leaving us with the desired solution:

$$\begin{aligned} \tfrac{1}{3} \cdot (3 \cdot x) &= \tfrac{1}{3} \cdot 5 \\ (\tfrac{1}{3} \cdot 3) \cdot x &= \tfrac{5}{3} \\ 1 \cdot x &= \tfrac{5}{3} \\ x &= \tfrac{5}{3} \end{aligned}$$

In both cases we needed a number to "cancel" the number paired with x. Here, "cancel" means that, after regrouping by associativity, the resulting number, when combined with x, left x unchanged. (Notice that commutativity was *not* needed.) These observations can be translated into specific properties of an operation on a set that guarantee the solvability of simple equations like the ones shown here.

DEFINITION   Let G be a nonempty set and let $*$ be an operation on G with the following properties:

1. $*$ is associative on G;
2. there is an element z in G such that
$$z * a = a \qquad \text{and} \qquad a * z = a$$
   for any element a in G;
3. for each element a in G, there is an element a' in G such that
$$a * a' = z \qquad \text{and} \qquad a' * a = z$$

Then G together with the operation $*$, which we shall sometimes abbreviate as $(G, *)$, is called a **group**. The element z is called an **identity** element, and a' is called an **inverse** of a.

Example 7.14   As we have already implied, the system $(\mathbf{I}, +)$ of integers under addition is a group. Its identity element is 0, and the inverse of any integer a is $-a$. Similarly, the set of positive rational numbers under multiplication is a group. Its identity element is 1 and the inverse of any number a is its reciprocal, $\frac{1}{a}$.   □

Note
Tables 7.7, 7.8, and 7.9, cited in the following examples, are in Section 7.2.

Example 7.15   Table 7.8(a) represents a group. 0 is the identity, $0' = 0$, and $1' = 1$. Table 7.8(b) does not represent a group; there is an identity, 1, but there is no inverse for 0.   □

Example 7.16   Table 7.7 does *not* represent a group table because there is no identity element. (Can you see an easy way to recognize when there is an identity element in a table?)   □

Example 7.17   Table 7.9 represents a group. The identity element is a, and each element is its own inverse.   □

Note
As you can see from Example 7.17, an identity element is easy to recognize in a table because its row and column must be identical with the row at the top and the column at the side of the table.

Many other examples of groups will turn up in the course of our work, but for now, the foregoing examples should be enough to illustrate the definition. Notice that a group operation does not have to be commutative. A group with a commutative operation will be specifically called a **commutative group**.[3] So far, all the examples we have seen are commutative groups, but noncommutative ones will begin to appear shortly.

Groups are of fundamental importance in many areas of modern mathematics. Although the defining properties of a group are simple, mathematicians have unearthed a wealth of information in their exploration of group theory since its beginnings in the 19th century. This chapter is too brief to do more than scratch the surface of such a vast field of knowledge, but even here we shall see some of the power of the group axioms as we investigate some specific problems. The first of these problems can be stated immediately; it is the focus of our attention in the next section:

> We saw at the beginning of Section 7.3 that there are 19,683 possible operations on a three-element set. Which of these define groups?

---

[3]These groups are also called *abelian groups*. They are named after Neils Henrik Abel, a Norwegian mathematician of the early 19th century.

EXERCISES 7.4

Exercises 1–5 refer to Table 7.13.

*	2	4	6	8
2	4	8	2	6
4	8	6	4	2
6	2	4	6	8
8	6	2	8	4

Table 7.13

1. Determine $4 * 8$.

2. Is the set $\{2, 4, 6, 8\}$ closed under $*$? Why or why not?

3. Which element, if any, is the identity?

4. Check one case of the associative law for $*$.

5. Which element, if any, is the inverse of 8?

Exercises 6–18 are based on Tables 7.14 and 7.15.

o	p	q	r	s	t
p	s	r	t	p	q
q	t	s	p	q	r
r	q	t	s	r	p
s	p	q	r	s	t
t	r	p	q	t	s

Table 7.14

◇	0	1	2	3	4
0	0	1	2	3	4
1	1	2	3	4	0
2	2	3	4	0	1
3	3	4	0	1	2
4	4	0	1	2	3

Table 7.15

6. $t \circ r = $ ___

7. $3 \diamond 4 = $ ___

8. $(q \circ r) \circ p = $ ___

9. $1 \diamond (3 \diamond 2) = $ ___

10. Is there an identity element in Table 7.14? If so, what is it?

11. Is there an identity element in Table 7.15? If so, what is it?

12. Does q have an inverse in Table 7.14? If so, what is it?

13. Does 1 have an inverse in Table 7.15? If so, what is it?

14. Solve for x: $r \circ x = t$.

15. Solve for x: $2 \diamond x = 3$.

16. Find an example to show that one of these operations is not commutative.

17. Find an example to show that one of these operations is not associative.

18. One of these tables does not describe a group. Which is it, and which group properties fail?

For Exercises 19–25, let $S = \{a, b, c\}$. Construct an operation table for S that satisfies the given condition or explain why this is not possible.

19. S is commutative, but has no identity element.

20. S has an identity element, but not every element has an inverse.

21. S has an identity element, but no element has an inverse.

22. S is a commutative group.

23. S is a group that is not commutative.

24. Fill in Table 7.16 so that you have an operation $\diamond$ on the set $\{a, b, c, d\}$ with the following three properties:

   ▷ c is the identity element,
   ▷ b is the inverse of d, and
   ▷ $\diamond$ is commutative.

25. Fill in Table 7.17 so that you have an operation $*$ on the set $\{1, 2, 3, 4, 5\}$ with the following five properties:

   ▷ $4 * 3 = 1$,
   ▷ 2 is the identity element,
   ▷ 1 is the inverse of 5,
   ▷ 3 is its own inverse, and
   ▷ $*$ is not commutative.

26. Find an example of a commutative group

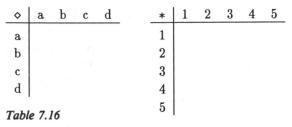

◇	a	b	c	d
a				
b				
c				
d				

Table 7.16

*	1	2	3	4	5
1					
2					
3					
4					
5					

Table 7.17

containing six elements. (Call this group *A* for future reference.)

WRITING EXERCISES

1. Évariste Galois, generally acknowledged to be the founder of group theory, was one of the more colorful, tragic figures of mathematical history. Read about him in at least two different mathematical history books; then write a short human-interest piece about some aspect of his life. (Be sure to list the sources you use, and be careful that you do not simply copy or paraphrase the material from your sources.)

2. Commutative groups are commonly called "Abelian" groups, in recognition of a mathematician whose last name was Abel. Write a paragraph that states who Abel was (what his full name was, when and where he lived, etc.) and explains why it is appropriate to name this kind of group after him. (List the sources you use, and be careful that you do not simply copy or paraphrase material.)

3. This section begins by observing that the defining properties of a group can be seen as an abstraction of the pattern for solving some simple equations.

   (a) Discuss the extent to which you think the ideas of this chapter *depend on* the standard algebraic notation for solving equations.

   (b) Appendix Section B.5 observes that the use of even such simple algebraic symbols as =, +, and − did not occur until the 16th century. With this in mind, expand on your answer to Part (a) to comment on whether or not group theory could have developed sooner than it did in our culture. (When and where did it start?)

7.5   Some Basic Properties of Groups

The problem we confront in this section is:

> Which of the 19,683 possible operations on a three-element set define groups?

It can be approached in many ways. The most obvious one is also the most tedious and the least enlightening: We could check each of the 19,683 different three-element tables to see if they had the properties required by the definition of group.

"There must be an easier way!" you say, and you are right. Instead of taking the direct, dull approach, let us first try to establish some easily recognizable properties that any group table must have, thereby eliminating many ineligible candidates at once.

The most easily recognizable group property in a table is the existence of an identity element. The row and column of an identity element must be exactly the same as the reference row and column given at the top and side of the table, respectively. For instance, an operation table on the set $\{1, 2, 3, 4\}$ with 3 as its identity element must fit the pattern of Table 7.18, regardless of how the operation behaves elsewhere on the set.

	1	2	3	4
1			1	
2			2	
3	1	2	3	4
4			4	

Table 7.18 *Recognizing an identity element.*

Moreover,

(7.1) | Any group has only one identity element. |

We prove this fact by showing that if y and z both represent identity elements in a group $(G, *)$, then they cannot be different. Consider the element $y * z$. If y is an identity, then

$$y * z = z$$

On the other hand, if z is an identity, then

$$y * z = y$$

Hence, we have

$$y = y * z = z$$

so y must equal z, and thus Property (7.1) has been proved.

Our experience with addition and multiplication of numbers suggests another useful general property. We know, for example, that for numbers x and y,

$$3 + x = 3 + y \quad \text{implies} \quad x = y$$

and also

$$3x = 3y \quad \text{implies} \quad x = y$$

In both cases the 3 can be "canceled." We abstract from such numerical examples a general property shared by all groups.

(7.2)

> **The Cancellation Laws:** Let a, b, and c be any elements of a group $(G, *)$.
>
> 1. If $a * b = a * c$, then $b = c$
> 2. If $b * a = c * a$, then $b = c$

We prove the first of these laws here; the proof of the second is almost exactly the same and is left as an exercise.

Because $(G, *)$ is a group, the element a has an inverse a' in G. Combining a' with both sides of the given equation, we obtain

$$a' * (a * b) \;=\; a' * (a * c)$$

and thus

$$(a' * a) * b \;=\; (a' * a) * c$$

by associativity. Denoting the identity of $(G, *)$ by z, we then have

$$z * b \;=\; z * c$$

by the definition of inverse, implying

$$b \;=\; c$$

by the definition of identity.

Cancellation is an important property of groups which, in turn, allows us to prove other basic properties. For example:

(7.3)

> Any element of a group has only one inverse.

As in the proof of Property (7.1), the approach here is to show that any two inverses of the same element must be equal. Let $(G, *)$ be a group with identity element z, and let g be an arbitrary element of G. Suppose that x and y both represent inverses of g. Then, by the definition of an inverse,

$$g * x = z \qquad \text{and} \qquad g * y = z$$

implying

$$g * x \;=\; g * y$$

Applying the first cancellation law, we get $x = y$, so Property (7.3) is proved.

Besides providing a powerful tool for proving other general facts about groups, the Cancellation Laws are easily recognizable properties of any operation expressed in table form. Suppose, for example, that Table 7.19 represents an operation * on the set $\{1, 2, 3, 4, 5\}$, and that two entries in the third row are the same, as shown. In this case we have

$$3 * 2 = 4 \quad \text{and} \quad 3 * 5 = 4$$

implying

$$3 * 2 = 3 * 5$$

But $2 \neq 5$, so Cancellation Law 1 is violated. Thus, Cancellation Law 1 implies that there can be no repetitions in any row.

*	1	2	3	4	5
1					
2					
3		4			4
4					
5					

Table 7.19 A violation of left cancellation.

Similarly, Cancellation Law 2 prohibits repetitions in any column. In other words:

(7.4)   In each row or column of any table that represents a group, every element of the set must appear exactly once.

Now we are ready to solve the problem of deciding which of the 19,683 three-element tables are groups. We shall use the general tools just developed to construct group tables for the set $\{1, 2, 3\}$. Notice first that every group must have an identity element, and that the choice of an element to play this role immediately determines five of the nine entries of a potential group table. For instance, the choice of 1 as identity requires that the row and column for 1 be identical with the reference row and column, as shown in Table 7.20.

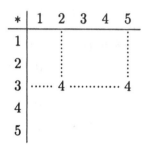

	1	2	3
1	1	2	3
2	2		
3	3		

Table 7.20 The identity is 1.

There are only four remaining entries, each of which can be chosen in three ways, so there are only 81 ($= 3^4$) ways to fill in the rest of the table. Now, there are only three possible identity elements and 81 ways to complete the table for each choice of identity, implying that there are only $3 \cdot 81$ ($= 243$) ways to construct a three-element table containing an identity element. Hence, the identity requirement alone reduces the number of possible three-element group tables from 19,683 to a mere 243.

The Cancellation Laws allow us to go even further. Referring to Table 7.20, it is not hard to see that the nonrepetition restriction on the rows and columns *requires* that the middle entry of the table be 3, since a 2 there would give us an obvious repetition in the second row and column, and a 1 there would lead to repeated 3s in the third row and column. This middle entry of 3, in turn, requires 1s at the end of the second row and column, and finally a 2 in the lower-right corner. That is, the *only way* to complete Table 7.20 without row or column repetitions is as shown in Table 7.21. Hence, there is *at most one* three-element group that has 1 as its identity element.

	1	2	3
1	1	2	3
2	2	3	1
3	3	1	2

Table 7.21 *The only possible group for this set with identity 1.*

Reasoning in precisely the same way, we can verify that there is at most one three-element group table with identity 2 and at most one with identity 3. We have now reduced the 243 possibilities of three-element group tables to only three.

Notice that we have not yet proved that these tables actually represent groups. There are operations that satisfy the Cancellation Laws but are not associative (see Exercise 18), so we still must check these tables for associativity. We do know, however, that no other three-element tables need to be checked because none of the others have identities and satisfy cancellation.

It is not hard to check each of the three tables directly for the group properties. For instance, referring to Table 7.21, it is easy to see that

- the operation is closed,
- 1 is the identity, and
- $1' = 1$, $2' = 3$, and $3' = 2$

Checking associativity is somewhat tedious (27 cases), but it can be done; in fact, associativity holds in every case. Hence, Table 7.21 is actually a group table. The group properties for the other two tables can be verified in exactly the same way. So the problem has been solved:

Of the 19,683 different three-element tables, exactly three yield groups!

EXERCISES 7.5

For Exercises 1–4, construct an operation table for the set $\{1, 2, 3, 4\}$ that satisfies the given condition.

1. both Cancellation Laws

2. Cancellation Law 1 but not 2

3. neither Cancellation Law

4. Cancellation Law 2 but not 1

5. Check that the operation shown in Table 7.21 is associative. (Work out all cases.)

6. Table 7.22 starts to describe an operation on the set $\{1, 2, 3, 4\}$ with identity 1. There are at least two different ways to complete the table so that the Cancellation Laws hold; complete it *in two ways*. Do both define group operations? Justify your answer.

	1	2	3	4
1	1	2	3	4
2	2			
3	3			
4	4			

Table 7.22

For Exercises 7–14, fill in each given table to define a group, if possible. If this is not possible, explain why not.

7.

	r	s
r	s	
s		

8.

	1	2	3
1	1	2	3
2	2	4	
3	3		

9.

	1	2	3
1	2		
2		2	
3			

10.

	2	5	8
2			5
5		5	
8	5		

11.

	2	4	6	8
2	4	2		
4	2	4	6	8
6		6		
8		8		6

12.

	a	b	c	d
a	b			
b		c		
c			d	
d				a

13.

	w	x	y	z
w	w			
x		x		
y			y	
z				z

14.

	0	1	2	3
0	0	1	2	3
1			0	
2				
3	0			

15. Supply a reason for each step in the proof of the statement:

If a and b are elements of a commutative group $(G, *)$ with identity z, then

$$\big((a * b) * a'\big)' = b'$$

Proof:
a' and b' exist in G.

――――――――――

Hence, $((a * b) * a') * b'$ is an element of G.

――――――――――

$$((a * b) * a') * b' = ((b * a) * a') * b'$$

$$= (b * (a * a')) * b'$$

$$= (b * z) * b'$$

$$= b * b'$$

$$= z$$

Therefore, $((a * b) * a')' = b'$.

	a	b	c	d	e
a	c	d	a	e	b
b	e	c	b	a	d
c	a	b	c	d	e
d	b	e	d	c	a
e	d	a	e	b	c

Table 7.23

16. Prove Cancellation Law 2.

17. Let G be a finite group. Prove that, if the number of elements contained in G is even, then at least one element besides the identity is its own inverse. (*This result is used later.*)

18. Table 7.23 satisfies the Cancellation Laws. (Why?) Moreover, there is an identity element (what is it?) and each element has an inverse (what are they?). Prove that this table does *not* define a group.

WRITING EXERCISES

1. Using complete sentences, write a half-page summary that identifies the main theme of this section and outlines how that theme is carried out.

2. Describe how at least three of the dozen problem-solving techniques of Section 1.2 can be used to help in the solution of Exercise 17. (Even if you can't complete the proof for Exercise 17, you should be able to do this writing exercise.)

3. Write two questions that extend the ideas of this section but are analogous in some way to questions answered in the section. Explain the analogy in each case.

7.6   Subgroups

One way to proceed with our investigation of finite groups would be to continue the approach of Section 7.5, trying to examine all four-element operation tables, then all five-element tables, and so on. But this route is tiresome and inefficient; the rapid growth of the number of possibilities supplies too many cases for even the powerful Cancellation Laws to dispose of. Checking associativity in a five-element table can be tedious, and by the time tables with ten elements come into view, we would be approaching the limits of human patience. (There are 1000 associativity cases to be checked for a ten-element table — not much for a computer, but a truly annoying chore if done by hand! Tables of several thousand elements involve billions of cases, very time-consuming even for computer-assisted checking.) We choose, instead, to seek another general insight into group theory which, like cancellation, will allow us to dispose of many particular cases all at once. This path will lead to a major result in group theory, Lagrange's Theorem. We begin by defining an important basic idea.

DEFINITION   Let G be a group with operation $*$. If S is a subset of G that is itself a group with respect to the same operation $*$ (restricted to the elements of S), then we call $(S, *)$ a **subgroup** of $(G, *)$.

Example 7.18   The integers form a group under addition. You know from elementary arithmetic that the sum of two integers is an integer and that addition is associative; the identity element is 0; the inverse of each element is its negative.

The set of even integers (under addition) forms a subgroup of of this group. The sum of two even integers is even; addition is (still) associative; 0 is an even number; the negative of any even number is even.   □

It is important to note that the definition of a subgroup requires that *the same operation* be used for the subset as was used for the original group. Otherwise, *any* subset could be made into a group under a totally unrelated operation; this would render the subgroup concept worthless. This necessary restriction simplifies the language and notation a little: Because the operation in question is necessarily the same, we can simply say that a *subset S* of $(G, *)$ is a subgroup, instead of having to specify that $(S, *)$ is a subgroup of $(G, *)$.

Moreover, because a group operation $*$ is necessarily associative and because the definition of associativity requires that

$$(a * b) * c \;=\; a * (b * c)$$

for *any* three elements of the group, this equality surely holds whenever a, b, and c all happen to be in a particular subset. Thus, a group operation is automatically associative on every subset. Hence, in checking to see if a subset is a subgroup, we may ignore the troublesome associative property. In other words:

(7.5)

> To verify that a subset S of a group $(G, *)$ is a subgroup, we need only check three properties:
> 1. S is closed under $*$.
> 2. S contains the identity element of $(G, *)$.
> 3. Each element of S has its inverse contained in S.

Example 7.19 Table 7.24 is a group with identity element 1. Let us find all its subgroups. Since 1 must be in every subgroup, we need to check only the subsets containing 1 to see if they are closed and if every element has an inverse. The subsets are:

	1	2	3	4
1	1	2	3	4
2	2	3	4	1
3	3	4	1	2
4	4	1	2	3

Table 7.24

$$\{1\}, \{1, 2\}, \{1, 3\}, \{1, 4\},$$
$$\{1, 2, 3\}, \{1, 2, 4\}, \{1, 3, 4\}$$

and, of course, the entire set $\{1, 2, 3, 4\}$, which is automatically a subgroup of itself.

- The set $\{1\}$ forms a subgroup because $1 * 1 = 1$; that is, $\{1\}$ is closed under $*$, and 1 is its own inverse.
- $\{1, 2\}$ is not closed because $2 * 2 = 3$, which is outside the subset.
- $\{1, 3\}$ is a subgroup; it is closed under $*$, $1' = 1$, and $3' = 3$.
- $\{1, 4\}$ is not closed because $4 * 4 = 3$, which is outside the subset.

(See Table 7.25.)

$*$	1
1	1

$*$	1	2
1	1	2
2	2	3

$*$	1	3
1	1	3
3	3	1

$*$	1	4
1	1	4
4	4	3

Table 7.25   Some subset tables.

We can dispose of the other cases similarly, or we can use the following slightly more efficient argument: By closure, any subgroup containing 2 or 4 must contain 3 $(= 2 * 2 = 4 * 4)$. But $2 * 3 = 4$ and $4 * 3 = 2$, so any group containing either 2 or 4 must contain the other three elements as well. Thus, the only subgroups of this group are

$$\{1\}, \; \{1, 3\}, \; \{1, 2, 3, 4\} \qquad \square$$

EXERCISES 7.6

1. Table 7.21 of Section 7.5 represents a three-element group with identity 1. Give a convincing argument to verify that this group has no two-element subgroups.

2. Table 7.26 describes a five-element group with identity 1. Find all of its subgroups.

	1	2	3	4	5
1	1	2	3	4	5
2	2	3	4	5	1
3	3	4	5	1	2
4	4	5	1	2	3
5	5	1	2	3	4

Table 7.26

3. Find all subgroups of the six-element group A that you constructed for Exercise 26 of Section 7.4.

4. Table 7.27 defines a group; call it $(B, \circ)$.

   (a) Find all of the subgroups of B.

   (b) Count the number of elements in each of the subgroups. How do these numbers relate to the total number of elements in B?

   (c) How does this group differ from your group A in Exercise 3?

$\circ$	j	f	g	h	i	k
j	j	f	g	h	i	k
f	f	j	i	k	g	h
g	g	h	j	f	k	i
h	h	g	k	i	j	f
i	i	k	f	j	h	g
k	k	i	h	g	f	j

Table 7.27

5. The solutions to the equation $x^2 + 1 = 0$ are the (complex) numbers whose squares equal -1. These numbers are usually denoted by i and $-i$; that is,

$$i = \sqrt{-1} \quad \text{and} \quad -i = -\sqrt{-1}$$

   (a) Make an operation table for the set
   $$S = \{1, i, -1, -i\}$$
   using ordinary multiplication and the fact that $i \cdot i = -1$.

   (b) Verify that $(S, \cdot)$ is a group by checking all the defining properties for a group. (You may remember or assume that this multiplication is associative.)

   (c) Find all the subgroups of $(S, \cdot)$.

   (d) What are the similarities between $(S, \cdot)$ and the group in Table 7.9 of Section 7.2? What are the differences?

   (e) What are the similarities between $(S, \cdot)$ and the group in Example 7.19? What are the differences?

For Exercises 6–11, let $\oplus$ be the operation defined on the set $C = \{0, 1, 2, 3, 4, 5, 6, 7, 8, 9, 10, 11\}$ by

$$x \oplus y = \text{the remainder when } x + y \text{ is divided by 12.}$$

6. $9 \oplus 7 = $ __

7. $(6 \oplus 8) \oplus 11 = $ __

8. $6 \oplus (8 \oplus 11) = $ __

9. It can be shown that $\oplus$ is an associative operation. Assuming this is true, verify the rest of the properties needed to make $(C, \oplus)$ a group.

10. Find all of the subgroups of C.

11. Count the number of elements in each of the subgroups. How do these numbers relate to the total number of elements in C?

12. Prove that the identity element of any subgroup must be the same as the identity element of the group containing it. (*This result is used later.*)

13. (a) Let G be a commutative group with the operation $*$. Prove that for any elements a and b in G,

   $$(a * b)' = a' * b'$$

   (b) In the noncommutative group $(B, \circ)$ defined by Table 7.27, find an instance of two elements for which the property stated in Part (a) does *not* hold.

(c) If $(G, *)$ is a group but is not commutative, how can you write $(a * b)'$ in terms of a' and b'? Justify your answer.

WRITING EXERCISES

1. In mathematical terminology, the prefix "sub-" is used to signify that a set or structure of some sort is included within another one of the same kind (as in *subset* and *subgroup*). How well does this conform to the use of "sub-" outside of mathematics? Explain, using appropriate examples from ordinary English.

2. Write three simpler questions that you think might provide helpful steppingstones for solving Exercise 12. Be as specific as possible.

3. Write three simpler questions that you think might provide helpful steppingstones for solving Exercise 13. Be as specific as possible.

4. Describe as specifically as you can an approach to one part of Exercise 13 that uses one (or more) of the problem-solving techniques listed in Section 1.2.

7.7   Lagrange's Theorem

To illustrate the variety of subgroups within a group, we consider the set of digits

$$D = \{0, 1, 2, 3, 4, 5, 6, 7, 8, 9\}$$

and define an operation $\oplus$ on D by

$$x \oplus y = \text{the remainder when } x + y \text{ is divided by 10}$$

Now, any number less than 10 when divided by 10 has quotient 0 and itself as remainder; so for pairs of numbers whose sum is less than 10 this operation is the same as addition. For pairs of numbers whose sum is greater than or equal to 10, however, the remainder feature guarantees that the results of the $\oplus$ process are always back in the set D. Thus, for example,

$$3 \oplus 5 = 8, \qquad 6 \oplus 8 = 4, \qquad 9 \oplus 2 = 1, \qquad 7 \oplus 3 = 0$$

It is not hard to verify that $(D, \oplus)$ is a group with identity element 0 and inverses as follows:

$$0' = 0 \quad 1' = 9 \quad 2' = 8 \quad 3' = 7 \quad 4' = 6$$
$$5' = 5 \quad 6' = 4 \quad 7' = 3 \quad 8' = 2 \quad 9' = 1$$

(Notice that x' is the inverse of x if $x + x' = 10$.)

Let us try to find all the subgroups of $(D, \oplus)$. We know from Subgroup Property 2 of (7.5) and Exercise 12, both in Section 7.6, that any subgroup of D must contain 0, and, in fact, $\{0\}$ itself is a subgroup. We can determine the other subgroups by trial and error, but a little ingenuity can save a lot

Joseph-Louis Lagrange. His work on the theory of equations led to basic theorems in group theory.

of effort. For example, any subgroup containing 1 must also contain $1 \oplus 1$, which is 2. But then it must also contain $2 \oplus 1 = 3$, and hence $3 \oplus 1 = 4$, $4 \oplus 1 = 5$, and so on. Thus, any subgroup of D containing 1 also contains every other element of D and hence must be D itself.

So far we have located a one-element subgroup and a ten-element subgroup (D itself). Clearly, $\{0\}$ is the only possible one-element subgroup. (Why?) Are there any two-element subgroups? If so, each must contain 0, which is its own inverse, so the other element must also be its own inverse. The list of inverses shows that 5 is the only element besides 0 that is its own inverse; thus, the only two-element subgroup of D is $\{0, 5\}$.

How about three-element subgroups? In this case, Subgroup Property 3 of (7.5) requires that the two nonzero elements be inverses of each other. Let us try the various combinations:

- $\{0, 1, 9\}$ is not a subgroup because it does not contain $1 \oplus 1 = 2$;
- $\{0, 2, 8\}$ does not contain $2 \oplus 2 = 4$;
- $\{0, 3, 7\}$ does not contain $3 \oplus 3 = 6$;
- $\{0, 4, 6\}$ does not contain $4 \oplus 4 = 8$.

Hence, there are no three-element subgroups of D.

These computations also ensure that there are no four-element subgroups because any such subgroup must contain 0 and 5 (by Exercise 17 of Section 7.5) and some other element with its inverse. Putting 5 into each of the three-element sets does not make them closed under $\oplus$.

A five-element subgroup must contain 0 and two pairs of elements that are inverses of each other (why?), and it cannot contain 1, as we have seen. Moreover, it cannot contain the pair consisting of 3 and 7 because $7 \oplus 7 \oplus 7$ equals 1. The only possibility, then, is

$$\{0, 2, 4, 6, 8\}$$

The sum of even numbers is even, and division by 10 preserves evenness in these cases; so this is indeed a subgroup.

Similar arguments may be used to rule out the possibility of six-, seven-, eight-, and nine-element subgroups; therefore, the only subgroups of D are:

$$\{0\}, \ \{0, 5\}, \ \{0, 2, 4, 6, 8\}, \ \text{and } D \text{ itself}$$

Notice that *the number of elements in each subgroup of this ten-element group is a divisor of the number 10.* Looking back at the results of Exercises 1 through 11 of Section 7.6, we can see the same phenomenon:

Exercise(s)	Group size	Subgroup sizes
1	3 elements	1, 3
2	5 elements	1, 5
3, 4	6 elements	1, 2, 3, 6
5	4 elements	1, 2, 4
6–11	12 elements	1, 2, 3, 4, 6, 12

This pattern is not just a curious coincidence; in fact, the assertion that subgroups are always related to their "parent" group in this way is the first major theorem of group theory:

(7.6) | **Lagrange's Theorem:**[4] The number of elements in any subgroup of a finite group must be a divisor of the total number of elements in the group.

(Recall that a whole number a is a **divisor** of a whole number b if there is a whole number c such that $a \cdot c = b$.)

---

[4] Named after Joseph-Louis Lagrange (1736–1813), an outstanding French mathematician and a friend of Napoleon.

The proof of Lagrange's Theorem is not trivial, but it is within the scope of the mathematical tools now at our disposal. A detailed explanation of that proof appears in the next (optional) section. The proof, however, is not necessary for an understanding of the role of Lagrange's Theorem in group theory. Hence, in the rest of this section we look at a few consequences of the theorem itself to illustrate its power both in handling specific groups and in proving other general results.

For a first instance of Lagrange's Theorem at work, we turn to an example just like the one that began this section. Consider the set

$$E = \{0, 1, 2, 3, 4, 5, 6, 7\}$$

and define an operation $\oplus$ on E by

$x \oplus y = $ the remainder when $x + y$ is divided by 8

$(E, \oplus)$ is a group with identity element 0 and inverses as follows:

$$0' = 0 \quad 1' = 7 \quad 2' = 6 \quad 3' = 5$$
$$4' = 4 \quad 5' = 3 \quad 6' = 2 \quad 7' = 1$$

Here, x' is the inverse of x if $x + x' = 8$. (Can you justify these statements about the identity and the inverses?)

Problem
Find all possible subgroups of $(E, \oplus)$.

Solution
Since E contains eight elements, Lagrange's Theorem tells us that it can only have subgroups containing one, two, four, or eight elements. The only one-element subgroup is $\{0\}$ because every subgroup must contain 0. Clearly, the only eight-element subgroup is E itself. Now, any two-element subgroup must contain a nonzero element that is its own inverse (by Exercise 17 of Section 7.5); we can see from the list of inverses that the only two-element group is $\{0, 4\}$. Finally, any four-element subgroup must contain 0, 4, and two elements that are inverses of each other. (Why?) Checking the available inverse pairs, we must discard 1 and 7 because $1 \oplus 1 = 2$, and closure would require a fifth element in the subgroup. Similarly, the pair consisting of 3 and 5 will not work because $3 \oplus 3 = 6$. However, 2 and 6 fit with 0 and 4 to form the only possible four-element subgroup. Thus, we have found *all* the subgroups of $(E, \oplus)$.

Compare this solution with the long arguments used in finding the subgroups of $(D, \oplus)$ earlier in this section. Lagrange's Theorem allowed us to avoid a lot of tedious work!

Here is an even more striking example of the same sort. Define the operation $\oplus$ on the set $F = \{0, 1, 2, \ldots, 96\}$ by

$$x \oplus y = \text{the remainder when } x + y \text{ is divided by } 97$$

Again, $(F, \oplus)$ is a group.

Problem
Find all subgroups of this 97-element group $(F, \oplus)$.

Solution
The only divisors of 97 are 1 and 97. Thus, Lagrange's Theorem tells us that the only possible subgroups are of these two sizes. But $\{0\}$ is the only possible one-element subgroup and F itself is the only possible 97-element subgroup. Therefore, we have found *all* subgroups of the group $(F, \oplus)$.

This last problem is a particular case of an obvious, but important, general consequence of Lagrange's Theorem.

(7.7)

> If the number of elements in a group is prime,[5] then there are only two subgroups — the one-element subgroup consisting of the identity alone, and the entire group itself.

We close this section with one more interesting corollary of Lagrange's Theorem. This result is a twin to the often used Exercise 17 of Section 7.5.

(7.8)

> Let G be a finite group. If the number of elements contained in G is odd, then no element besides the identity can be its own inverse.

To prove this, let z denote the identity element of G, and suppose there is some other element g in G that is its own inverse. Then $\{z, g\}$ is a two-element subgroup of G (why?); so, by Lagrange's Theorem, 2 must divide the number of elements in G. But this is impossible because the number of elements in G is odd. Therefore, there cannot be an element besides z that is its own inverse.

---

[5]A prime number is a whole number greater than 1 whose only divisors are 1 and the number itself. Thus, 97 is a prime number.

EXERCISES 7.7

Table 7.28 represents a group that we shall call G. Use this table to answer Exercises 1–10.

*	a	b	c	d	e	f
a	a	b	c	d	e	f
b	b	c	a	e	f	d
c	c	a	b	f	d	e
d	d	f	e	a	c	b
e	e	d	f	b	a	c
f	f	e	d	c	b	a

Table 7.28

1. What is the identity element of G?

2. List the inverse of each element of G.

3. Is {a, b} a subgroup of G? Why or why not?

4. Find a two-element subgroup of G.

5. How many two-element subgroups does G have? How do you know?

6. Is {a, b, c, d} a subgroup of G? Why or why not?

7. Is {d, e, f} a subgroup of G? Why or why not?

8. Is {a, b, c} a subgroup of G? Why or why not?

9. Is {a, c, e} a subgroup of G? Why or why not?

10. Is {a, d, f} a subgroup of G? Why or why not?

11. Let $S = \{0, 1, 2, 3, 4, 5, 6, 7, 8\}$ and define $\oplus$ on S by

   $x \oplus y =$ the remainder when $x + y$ is divided by 9

   This operation makes $(S, \oplus)$ a group. Find all its subgroups.

12. Tables 7.29 and 7.30 represent groups. Find all their subgroups.

	p	q	r	s	t	u	v
p	p	q	r	s	t	u	v
q	q	r	s	t	u	v	p
r	r	s	t	u	v	p	q
s	s	t	u	v	p	q	r
t	t	u	v	p	q	r	s
u	u	v	p	q	r	s	t
v	v	p	q	r	s	t	u

Table 7.29

	1	2	3	4	5	6	7	8
1	1	2	3	4	5	6	7	8
2	2	3	4	1	8	7	5	6
3	3	4	1	2	6	5	8	7
4	4	1	2	3	7	8	6	5
5	5	7	6	8	1	3	2	4
6	6	8	5	7	3	1	4	2
7	7	6	8	5	4	2	1	3
8	8	5	7	6	2	4	3	1

Table 7.30

13. Consider the following information:

   Group W contains 22 elements.
   Group X contains 23 elements.
   Group Y contains 24 elements.
   Group Z contains 25 elements.

   (a) Which group has the most different sizes of possible subgroups, and which has the fewest?

   (b) What sizes of subgroups are possible in group W? In group Z?

   (c) Which of the four groups must contain an element besides the identity that is its own inverse? Which cannot contain such an element?

14. Let G be a finite group. Prove that the only subgroup of G that contains more than half of the total number of elements in G is G itself.

15. Let $T = \{a,\ b,\ c,\ \ldots,\ t\}$, the set of the first 20 letters of the alphabet, and let $U = \{b,\ d,\ g,\ j\}$.

(a) Split T into a collection of subsets such that

  i. Every element of T is in exactly one subset.

  ii. U is one of the subsets in the collection.

  iii. All subsets in the collection contain the same number of elements.

(b) Is there more than one way to choose a collection of subsets as described in Part (a)? Why or why not?

(c) Find an example of a subset V of T such that the splitting-up process described in Part (a) *cannot* be carried out if V is substituted for U.

(d) Is it possible to find a set W containing a different number of elements than U such that the same kind of splitting-up process as described in Part (a) can be carried out if W is substituted for U? If such a set can be found, what are the possible numbers of elements it could contain? If such a set W cannot be found, why not?

WRITING EXERCISES

1. Explain how three of the dozen problem-solving techniques of Section 1.2 were used in this section to lead up to the statement of Lagrange's Theorem. (You may be able to see more than three techniques used here; just explain three of them.)

2. (A research problem in mathematical history.) Group theory is generally regarded as having its origins in a long letter from Évariste Galois to a friend, written in 1832 on the night before Galois was killed. However, Lagrange's Theorem, a fundamental result of group theory, is named after Joseph Louis Lagrange, who died in 1813 (when Galois was about two years old). Explain this historical paradox. Be sure to cite the sources you use.

7.8   Lagrange's Theorem Proved [optional][6]

The proof of Lagrange's Theorem gives us a chance to see how all the basic machinery developed so far interacts to produce a major mathematical result. It is hardly surprising that the argument is neither short nor trivial; rather, it is remarkable that such an elegant and far-reaching result can be proved at all from the few facts we have established! For an idea of how the proof will proceed, we begin by examining the general version of Exercise 15 of Section 7.7.

Suppose that a set G contains n elements and a subset H of G contains r elements. Suppose, further, that G can be split up into a collection of subsets such that:

---

[6]This section may be omitted without disturbing the continuity of the chapter.

(7.9)   1. Every element of G is in exactly one subset.

2. H is one of the subsets in the collection.

3. All subsets in the collection contain the same number of elements.

In such a case, conditions 2 and 3 imply that each subset contains r elements. Then, because each of the n elements of G is in exactly one subset, the total number of elements in G can be found by counting the number of subsets in the collection and multiplying that number by r. That is, r must be a divisor of n.

To prove Lagrange's Theorem, then, we need only show that, for any group G and any sub*group* H of G, there is a way of splitting G up into a collection of subsets satisfying conditions 1, 2, and 3 listed in (7.9). To do this, we need one new concept.

DEFINITION   Let H be a subgroup of a group G with operation $*$. If g is some fixed element of G, then the set of all elements $g * h$ for each element h of H is called a **coset** of H in G and is denoted by $g * H$.

(Notice that the fixed element g is always on the left when it is combined with the elements of H. It is important to keep this in mind when forming cosets because $*$ may not be a commutative operation. Such cosets are usually called *left cosets*, but since they are the only kind we shall use, we omit the word "left.")

Example 7.20
(a)

(This example is continued throughout the section to illustrate the proof.)

As we have seen before, one of the subgroups of the ten-element group $(D, \oplus)$ defined in Section 7.7 is $\{0, 2, 4, 6, 8\}$; call this subgroup S. Then the coset of S determined by 3 is

$$
\begin{aligned}
3 \oplus S &= \{3 \oplus 0, \, 3 \oplus 2, \, 3 \oplus 4, \, 3 \oplus 6, \, 3 \oplus 8\} \\
&= \{3, 5, 7, 9, 1\}
\end{aligned}
$$

Similarly, the coset of S determined by 4 is

$$
\begin{aligned}
4 \oplus S &= \{4 \oplus 0, \, 4 \oplus 2, \, 4 \oplus 4, \, 4 \oplus 6, \, 4 \oplus 8\} \\
&= \{4, 6, 8, 0, 2\}
\end{aligned}
$$

A complete list of cosets of S in D is as follows:

$$
\begin{array}{ll}
0 \oplus S = \{0, 2, 4, 6, 8\} & 1 \oplus S = \{1, 3, 5, 7, 9\} \\
2 \oplus S = \{2, 4, 6, 8, 0\} & 3 \oplus S = \{3, 5, 7, 9, 1\}
\end{array}
$$

$$4 \oplus S = \{4, 6, 8, 0, 2\} \qquad 5 \oplus S = \{5, 7, 9, 1, 3\}$$
$$6 \oplus S = \{6, 8, 0, 2, 4\} \qquad 7 \oplus S = \{7, 9, 1, 3, 5\}$$
$$8 \oplus S = \{8, 0, 2, 4, 6\} \qquad 9 \oplus S = \{9, 1, 3, 5, 7\}$$

Notice that all the sets in the left column are the same, as are all the sets in the right column. (Remember that two sets are equal if they contain the same elements, regardless of order.) Thus, there are only two different cosets of S in D.   □

Example 7.21 One of the subgroups of the group $(B, \circ)$ defined by Table 7.27 in the Exercises for Section 7.6 is $\{j, f\}$; call this subgroup J. Then the coset of J in B determined by the element h is

$$h \circ J = \{h \circ j, \ h \circ f\} = \{h, \ g\}$$

A complete list of cosets of J in B is:

$$j \circ J = \{j, f\} \qquad g \circ J = \{g, h\} \qquad i \circ J = \{i, k\}$$
$$f \circ J = \{f, j\} \qquad h \circ J = \{h, g\} \qquad k \circ J = \{k, i\}$$

Notice that there are only three different cosets of J in B.   □

A closer look at these two examples can provide more insight into the arguments at the beginning of this section. In Example 7.21, notice that the three different cosets $\{j, f\}$, $\{g, h\}$, and $\{i, k\}$ form a set consisting of subsets of B that satisfy Conditions 1, 2, and 3 of (7.9) with respect to the subgroup J. Notice further that the number, 2, of elements of the subgroup J multiplied by the number, 3, of different cosets gives us the total number of elements of the original group, 6.

The same observations can be made about Example 7.20. The two different cosets $\{0, 2, 4, 6, 8\}$ and $\{1, 3, 5, 7, 9\}$ form a set of subsets of D satisfying (7.9) with respect to the subgroup S, and the number, 5, of elements of S multiplied by the number, 2, of cosets is the total number, 10, of elements of D. It seems, then, that a reasonable way to proceed with proving Lagrange's Theorem is to show that these three conditions hold for the set of different cosets of *any* subgroup in *any* finite group. The rest of the section is devoted to a formal argument that does exactly this.

Proof of Lagrange's Theorem:

Let $(G, *)$ be a finite group containing n elements and let H be a subgroup of G containing r elements. We want to show that the set of all the different cosets of H in G satisfies conditions 1, 2, and 3 of (7.9).

1. The identity element, z, must be in the subgroup H, so each element g of G is in at least one coset, namely, the one it determines. (That is, g is in $g * H$ because z is in H and $g * z = g$.) To show that no element of G can be in two different cosets, we prove that any two cosets containing a common element are necessarily the same, as follows:

Suppose two cosets $x * H$ and $y * H$ both contain some particular element g. Then g in $x * H$ implies $g = x * h_1$ for some h_1 in H, and g in $y * H$ implies $g = y * h_2$ for some h_2 in H. Thus,

(a) $$x * h_1 = y * h_2$$

(because they both equal g). Since H is a group, the inverse of h_1 is in H; so we may combine h'_1 with both sides of Equation (a) to get

(b) $$(x * h_1) * h'_1 = (y * h_2) * h'_1$$

Now, by associativity, this equation becomes

(c) $$x * (h_1 * h'_1) = y * (h_2 * h'_1)$$

and because $h_1 * h'_1$ is the identity element (why?), we get

(d) $$x = y * (h_2 * h'_1)$$

We use Equation (d) to show that *every* element of $x * H$ must also be in $y * H$, as follows. Suppose $x * h$ is an arbitrary element of $x * H$. By Equation (d), we have

(e) $$x * h = (y * (h_2 * h'_1)) * h$$

and, by associativity, this can be rewritten as

(f) $$x * h = y * ((h_2 * h'_1) * h)$$

But h_2, h'_1, and h are all in H, and because H is a subgroup, it must be closed under the operation $*$. Hence, $(h_2 * h'_1) * h$ is an element of H. Thus, Equation (f) tells us that the element $x * h$ can be written as

$$y * (\text{some element of } H)$$

implying that $x * h$ is in $y * H$. Because this is true for *any* $x * h$ in the coset $x * H$, it follows that $x * H$ is a subset of $y * H$.

By an argument like the one we just did, except for the use of h'_2 instead of h'_1 in passing from Equation (a) to an equation like (b), we can show that $y = x * (h_1 * h'_2)$ and then that $y * H$ is a subset of $x * H$. Because each of the cosets $x * H$ and $y * H$ is a subset of the other, they must be equal (by the definition of equal sets), so Condition 1 has been verified. That is to say, every element of G is in *exactly one* of the different cosets of H in G. (Whew!)

Example 7.20 *(b)*

As an illustration of the main argument in Part 1 of the proof, consider the cosets $3 \oplus S$ and $7 \oplus S$, both of which contain 5 because $5 = 3 \oplus 2$ and $5 = 7 \oplus 8$. Thus,

$$3 \oplus 2 = 7 \oplus 8$$

The inverse of 2, which is 8, is also in S, so

$$
\begin{aligned}
(3 \oplus 2) \oplus 8 &= (7 \oplus 8) \oplus 8 \\
3 \oplus (2 \oplus 8) &= 7 \oplus (8 \oplus 8) \\
3 &= 7 \oplus 6 \qquad \text{(Why?)}
\end{aligned}
$$

Now, for any element in $3 \oplus S$, such as 1, for instance, we have

$$1 = 3 \oplus 8 = (7 \oplus 6) \oplus 8 = 7 \oplus (6 \oplus 8) = 7 \oplus 4$$

which is also an element of $7 \oplus S$.   □

Proof of Lagrange's Theorem, continued:

2. This part is easy. If we denote the identity element of G by z, then the coset $z * H$ is H itself (because $z * h = h$ for each element h in H). Therefore, H is one of the cosets, satisfying Condition 2.

Example 7.20 (c)

Since 0 is the identity of D,

$$
\begin{aligned}
0 \oplus S &= \{0 \oplus 0, 0 \oplus 2, 0 \oplus 4, 0 \oplus 6, 0 \oplus 8\} \\
&= \{0, 2, 4, 6, 8\} \\
&= S
\end{aligned}
$$

□

Proof of Lagrange's Theorem, continued:

3. To prove that all cosets have the same number of elements, we show that any coset of H in G contains the same number of elements as H itself. Consider any coset $g * H$. If we write out the elements of H and of $g * H$, there appears to be a natural matching between the two sets that shows they are the same size:

$$
\begin{array}{cccccc}
H = \{ & h_1, & h_2, & h_3, & \ldots, & h_r \quad \} \\
 & \updownarrow & \updownarrow & \updownarrow & & \updownarrow \\
g * H = \{ & g * h_1, & g * h_2, & g * h_3, & \ldots, & g * h_r \quad \}
\end{array}
$$

One thing must be checked, however, before this argument is complete: We must be sure that different h's are paired with *different* coset elements. That is, whenever we have two different elements of H, say

$h_i \neq h_j$, then we must be sure that

$$g * h_i \neq g * h_j$$

But this is easy to verify: If $g * h_i = g * h_j$, then the cancellation law would imply $h_i = h_j$, contrary to our hypothesis. Hence, each coset contains r elements, satisfying Condition 3.

Example 7.20 (d)

Here is an example of the 1-1 correspondence between the subgroup S and one of its cosets in D.

$$S = \{ \quad 0, \quad 2, \quad 4, \quad 6, \quad 8 \quad \}$$
$$\updownarrow \quad \updownarrow \quad \updownarrow \quad \updownarrow \quad \updownarrow$$
$$3 \oplus S = \{ \ 3 \oplus 0, \ 3 \oplus 2, \ 3 \oplus 4, \ 3 \oplus 6, \ 3 \oplus 8 \ \}$$

This correspondence is a matching between distinct elements of the subgroup S and the coset $3 \oplus S$ because $3 \oplus x$ can equal $3 \oplus y$ if and only if $x = y$ (by cancellation).   □

Proof of Lagrange's Theorem, continued:

At this point the hard work has been done. All that remains is to put the pieces together, as suggested earlier: There are only a finite number of different cosets; say there are k of them. Because each coset contains r elements and no two cosets have any elements in common, the total number of elements in the cosets (that is, in their union[7]) is $r \cdot k$. But each element of G must be in *some* coset, so $r \cdot k = n$. Therefore, r is a divisor of n, and Lagrange's Theorem is proved!

EXERCISES 7.8

Table 7.31 represents a six-element group with identity element 1. $S = \{1, 3, 5\}$ and $T = \{1, 4\}$ are subgroups. (Can you verify this?) Use this table to find the cosets for Exercises 1–6.

*	1	2	3	4	5	6
1	1	2	3	4	5	6
2	2	3	4	5	6	1
3	3	4	5	6	1	2
4	4	5	6	1	2	3
5	5	6	1	2	3	4
6	6	1	2	3	4	5

Table 7.31

1. $2 * S =$ **2.** $2 * T =$ **3.** $4 * S =$

4. $4 * T =$ **5.** $5 * S =$ **6.** $5 * T =$

Recall the 12-element group C defined for Exercises 6–11 of Section 7.6. Two of its subgroups

---

[7]The **union** of a collection of sets is the set of all elements that are in at least one of the sets.

are $S = \{0, 3, 6, 9\}$ and $T = \{0, 4, 8\}$. For Exercises 7–14, list the elements of the given coset in C.

7. $5 \oplus S$ **8.** $5 \oplus T$ **9.** $2 \oplus S$

10. $2 \oplus T$ **11.** $1 \oplus T$ **12.** $9 \oplus S$

13. $3 \oplus T$ **14.** $4 \oplus S$

15. Write out all the cosets of the subgroup $\{a, c\}$ in the group defined by Table 7.9.

16. Let K denote the system described by Table 7.32, and let $L = \{p, q, r\}$.

   (a) For each element x in K, write out the elements of the set

$$x \circ L = \{x \circ p, x \circ q, x \circ r\}$$

   (b) How many different sets did you get in Part (a)? Do all these sets contain the same number of elements?

   (c) Is $(K, \circ)$ a group? Justify your answer in terms of your answer to Part (b).

$\circ$	p	q	r	s	t	u
p	p	q	r	s	t	u
q	q	p	s	s	u	t
r	r	s	p	s	t	q
s	s	s	s	p	q	u
t	t	u	t	q	p	r
u	u	t	q	u	r	p

Table 7.32

17. Let $(G, *)$ be defined by Table 7.33.

   (a) Write a table describing this same operation $*$ on the subset $H = \{1, 4\}$.

   (b) Is $(H, *)$ a group? Why?

   (c) Write out the sets $1 * H$, $2 * H$, $3 * H$, $4 * H$, $5 * H$, and $6 * H$.

   (d) What information do your results supply about whether or not $(G, *)$ is a group? Why?

$*$	1	2	3	4	5	6
1	1	2	3	4	5	6
2	2	5	4	3	6	1
3	3	4	5	6	1	2
4	4	3	6	1	2	5
5	5	6	1	2	3	4
6	6	1	2	5	4	3

Table 7.33

WRITING EXERCISES

1. Does the mathematical notation used to present the proof of Lagrange's Theorem in this section help or hinder your understanding? Would it be easier or more difficult to present this proof without so much (or any) notation? Discuss.

2. In Section 7.7 we checked quite a few groups for which Lagrange's Theorem works; why is a proof of this theorem still necessary? What is the purpose of proof in mathematics?

3. Near the end of the proof of Part **1**, the text says, "By an argument like the one we just did, except for the use of h_2' instead of h_1' in passing from Equation (a) to an equation like (b), we can show that $y = x * (h_1 * h_2')$ and then that $y * H$ is a subset of $x * H$." Write out this argument in detail.

7.9   Groups of Symmetries

This section provides a quick look at how groups can be used to describe the symmetries of some simple planar shapes. Many current applications of groups are merely more complicated versions of these elementary ideas.

Consider a straight split log lying flat side down in sand on a beach. If the log is picked up, there are exactly two ways to replace it so that it coincides with its imprint in the sand (flat side down again, of course) – it can either be returned to its original position, or be reversed and then put down. The rigidity of the log ensures that we can describe each of these movements by the initial and final positions of its ends. Thus, if the ends are numbered 1 and 2, the log movements may be symbolized by

$$1 \longrightarrow 1 \qquad 1 \diagdown \nearrow 1$$
$$2 \longrightarrow 2 \qquad 2 \diagup \searrow 2$$

If we call these two movements of the log s (for *same*) and r (for *reverse*) and look at what happens when we follow one of these movements by another, we can see that

s followed by r is the same as r

r followed by s is the same as r

r followed by r is the same as s

s followed by s is the same as s

This information is summarized in Table 7.34, where o denotes "followed by." Notice that this is a group table. The log movements s and r are examples of **rigid motions** — geometric figure movements that do not allow any distortion of the figure. With this understanding, we could easily replace our log model by a strictly geometric one using a line segment.

o	s	r
s	s	r
r	r	s

Table 7.34   The rigid motions of a line segment.

A more interesting group can be obtained by considering the rigid motions of an equilateral triangle that returns to its original "imprint" in the plane. Such motions must take each vertex to some original vertex position (not necessarily its own), and, because the triangle cannot be distorted, the positions of the three vertices determine the position of the entire triangle. There are exactly six of these rigid motions for an equilateral triangle. (See Exercise 1.) Three are the rotations of 120°, 240°, and 360° (or 0°); the other three are reflections with respect to the three *medians*. A **median** of a triangle is a line through a vertex of the triangle and the midpoint of the opposite side; a **reflection** of the triangle with respect to a median can be pictured as holding the median fixed and flipping the triangle over, as shown in Figure 7.1.

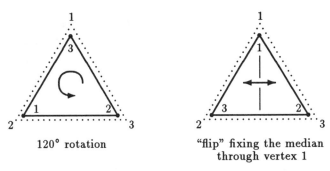

120° rotation   "flip" fixing the median
                         through vertex 1

Figure 7.1   *In each diagram, the posi-*
tion is indicated outside the
triangle, and the vertex in
that position is noted inside
the triangle.

If the vertices are numbered 1, 2, and 3 (counterclockwise), then any rigid motion can be represented by a chart indicating *how each vertex is matched with the original vertex positions.* For example, if the triangle is rotated 120° counterclockwise, vertex 1 goes to the place originally occupied by vertex 2, vertex 2 goes to the vertex 3's place, and vertex 3 goes to vertex 1's place. The six rigid motions of the triangle are as shown in Figure 7.2.

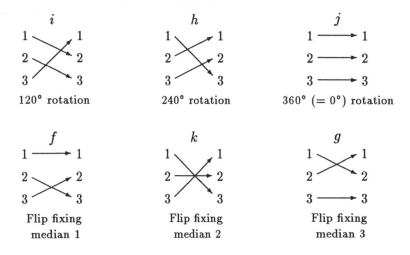

Figure 7.2   *The six rigid motions of an*
equilateral triangle.

Now let us consider what happens when we follow one of these rigid motions by another. For example, look at the 120° rotation followed by the flip fixing the original median 3. The vertices are moved around as follows:

- Vertex 1 is moved to position 2 by the rotation, and the flip fixing the original median 3 takes that position 2 to position 1; so vertex 1 ends up back where it started, in position 1.

- The rotation takes vertex 2 to position 3, and the flip leaves position 3 alone; so vertex 2 ends up in position 3.

- Vertex 3 is rotated to position 1, and then this vertex in position 1 is flipped over to position 2; so, vertex 3 ends up in position 2.

Looking at the result of all this, we see that the final position of the triangle is the same as if it had only been moved by the flip fixing median 1.

The process of following one motion by another is actually much easier than the foregoing description suggests, provided we adopt some convenient notation. Let us name the six rigid motions by the letters shown in Figure 7.2 and denote the *followed-by* process by "o." Then, using the diagrams of Figure 7.2, it is easy to observe the result of this process just by following arrows, as shown in Figure 7.3. Following the arrows from the left column to the right one, we see that $1 \longrightarrow 1, 2 \longrightarrow 3$, and $3 \longrightarrow 2$; that is, $i \circ g = f$.

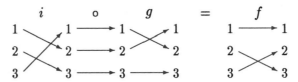

Figure 7.3 *i followed by g equals f.*

Because all possible rigid motions of the equilateral triangle are listed in Figure 7.2, any one of them followed by another must yield the same result as one of the original six motions; that is,

the set {i, h, j, f, k, g} of rigid motions is closed under the operation o.

In fact, if we compute all possible combinations of rigid motions (using the method described) and arrange the results in an operation table, we actually get the group shown in Table 7.27 of Section 7.6. Thus:

(7.10) | The set of all rigid motions of an equilateral triangle forms a six-element noncommutative group.

Similar groups may be formed by considering the rigid motions of various geometric figures. For instance, there are eight rigid motions of a square — four rotations (90°, 180°, 270°, 360°) and four reflections (holding fixed each of the four dotted lines in Figure 7.4.) The resulting eight-element group is called the **octic group**. This group and the rigid-motion groups for several other planar figures are explored in the exercises.

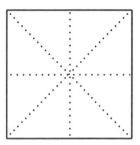

Figure 7.4   The dotted lines are the axes for the four reflections of a square.

EXERCISES 7.9

1. Cut an equilateral triangle out of cardboard and mark the vertices 1, 2, 3 on both sides. (Make sure each vertex has the same number on both sides of the cardboard.) Trace the outline of your triangle on a piece of paper and number the vertices (outside the traced figure). Then, by moving your triangle around, verify that there are only six different ways to replace the triangle in its tracing. These correspond to the six rigid motions listed in Figure 7.2.

2. (a) Compute the following combinations of rigid motions of an equilateral triangle according to the process illustrated by Figure 7.3:

   $$f \circ k, \; g \circ i, \; i \circ h, \; j \circ f, \; g \circ g, \; h \circ h$$

   (b) Compute the six combinations listed in Part (a) by actually moving around the triangle you constructed in Exercise 1.

3. Construct an operation table for the octic group (the group of rigid motions of a square), and verify that all the defining properties of a group are satisfied. (You may assume that associativity holds.) Is it commutative?

4. (a) Find all the rigid motions of a rectangle

that is not a square (there are 4), and make an operation table for them.

(b) Verify that this table satisfies all the defining properties of a group. Is it commutative?

(c) Relabel these four rigid motions using the letters a, b, c, and d so that your operation table matches Table 7.9.

(d) Find a subgroup of the octic group (Exercise 3) that matches this rectangle group.

(e) Repeat Parts (a)–(d) for a rhombus that is not a square.

5. (a) Find all the rigid motions of an isosceles right triangle and construct an operation table for them.

(b) Verify that this table describes a group. Is it commutative?

(c) Relabel these rigid motions so that your operation table matches one of the tables in Section 7.2.

(d) Can you find a subgroup of the equilateral-triangle group that matches this isosceles-right-triangle group? Why or why not?

6. (a) Find all the rigid motions of a regular pentagon and construct an operation table for them.

   (b) Assuming associativity, check the other properties needed to verify that this table describes a group. Is it commutative?

   (c) Find all the subgroups of this pentagon group.

   (d) Is this group like any group we have seen earlier in the chapter? If so, which one, and how are they similar?

WRITING EXERCISES

1. Look up the word "symmetry" in a dictionary and discuss how the definition(s) given there relate to the use of "symmetry" in this section.

2. Explain how the ideas of this section are related to the ideas presented in Section 7.1.

3. Does *every* polygon have a group of symmetries? Explain.

LINK: 7.10

Groups in Music and in Chemistry

Patterns in Music.

We end this chapter with two elementary examples of groups as they appear in other subject areas. The first one is in music, and it is really just an observation of a particular pattern, rather than a serious application of the theory. Nevertheless, formalizing observed harmonic and melodic patterns both in the theory of harmony and in the analysis of classical compositions has led to the inclusion of deliberate mathematical patterns in the works of some 20th-century composers, such as Schönberg and Webern. Thus, this sort of link between mathematics and music is not as far-fetched as it may initially appear.

Our musical example uses the twelve-tone chromatic scale. Imagine yourself sitting at a piano. If you play middle C and then continue up the keyboard playing every key, black or white, until you reach high C, you will have played the notes of the chromatic scale in the order

C	C#	D	D#	E	F	F#	G	G#	A	A#	B	(C)
0	1	2	3	4	5	6	7	8	9	10	11	(12)

If you were to continue playing notes beyond high C, you would just repeat the same scale notes an octave higher. Consider two notes to be the same if they have the same name, regardless of the octaves they are in. Thus, the high C in the scale shown here is in parentheses to indicate that it is to be considered as a copy of the first C in the scale. Now we define an operation $\oplus$ on the chromatic scale by using the numbers listed below the notes. To combine two notes, add their corresponding numbers, subtracting 12 if the sum is larger than 11; the result is the number of the "answer" note. For example,

$$G \oplus D\# = A\# \text{ because } 7 + 3 = 10$$

and

$$F \oplus A = D \text{ because } 5 + 9 = 14 \text{ and } 14 - 12 = 2$$

C is the identity note for this group, because 0 is the identity for addition. Each note has an inverse — the inverse of C# is B, the inverse of D is A#, and so forth. In other words, the notes of the twelve-tone scale form a group. (This group is a disguised version of the group $(C, \oplus)$ defined for Exercises 6–11 of Section 7.6. You might find it interesting to compare the two.)

This formal way of looking at the scale allows us to identify mathematically some items of musical interest. For instance, two subgroups of this 12-element group are musically significant. The (only) four-element subgroup is {C, D#, F#, A} (Why?), and these are precisely the notes of the "diminished seventh" chord in any of the keys represented by those four notes. The other two diminished-seventh chords,

$$\{C\#, E, G, A\#\} \text{ and } \{D, F, G\#, B\}$$

can be obtained by "adding" to each note of the original one any single note in the new chord. For instance,

$$\{D, F, G\#, B\} = \{D \oplus C, \ D \oplus D\#, \ D \oplus F\#, \ D \oplus A\}$$

(If you have read Section 7.8, you should recognize this as a *coset*. Using this terminology, we are saying that the three distinct cosets of the subgroup $\{C, D\#, F\#, A\}$ are the (only) three diminished-seventh chords.)

The (only) 3-element subgroup is $\{C, E, G\#\}$, the notes of the C-augmented chord (and also the E-augmented chord and the G#-augmented chord). Any of the other augmented chords —

$$\{C\#, F, A\}, \ \{D, F\#, A\#\}, \ \text{and} \ \{D\#, G, B\}$$

— may be obtained by "adding" any single note of the desired chord to the subgroup chord. Thus,

$$\{C\#, F, A\} = \{C\# \oplus C, \ C\# \oplus E, \ C\# \oplus G\#\}$$

(In the terminology of Section 7.9, the augmented chords are the distinct cosets of the 3-element subgroup chord.) We might observe in passing that the diminished-seventh and augmented chords are the only two common chord types that remain the same regardless of which of their notes is considered to be the root.

If you are interested in exploring group patterns in music, an excellent source of further information is Chapter 23 of [1] in the list *For Further Reading* at the end of this chapter. You will find in there a more detailed discussion of scales and harmony, and also an analysis of mathematical patterns in rounds, fugues, canons, and other musical forms.

Our second example of groups in other disciplines comes from the study of molecular structure, an important area of investigation in chemistry and other physical sciences. The shape of a molecule influences the way it behaves with respect to electromagnetic radiation. Because of this, a molecule's symmetric structure is closely related to its spectral properties. The symmetric structure of a molecule can be described as a group, known as the *point group* of the molecule. This name, which is also used in the study of crystalline symmetric structure, comes from the fact that any symmetry operation (rotation or reflection) that takes a molecule back into itself cannot move the point at its center of mass.

Let us look at a specific example of a point group. The methyl-chloride molecule can be diagrammed as in Figure 7.5, with three hydrogen atoms forming the equilateral base of a triangular pyramid that is topped by a carbon atom linked to a chlorine atom.

The requirement that any symmetric operation on this molecule must leave its center of mass fixed implies in this case that both the carbon and chlorine atoms must stay where they are. This means that all the symmetries of this molecule are determined by the rigid motions of the equilateral base triangle $H_1 H_2 H_3$. This group is easy to describe. In fact, we have already studied it in Section 7.9! It is the six-element group of rotations and reflections displayed in Figure 7.2.

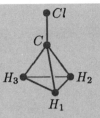

Figure 7.5   A methyl chloride molecule.

If you are interested in finding out more about molecular symmetry, Chapter 2 of [7] in the list *For Further Reading* is an excellent source. It provides a wealth of detail about the use of groups and other abstract mathematical systems in the study of molecular structure.

Topics for Papers – Chapter 7

General Instructions: Consider the reader of your paper to be an anonymous student in your class — someone who knows the material of Chapter 7 fairly well, but does not know anything about your particular topic. Make sure the paper introduces your topic clearly and takes the reader step by step through an orderly development of your ideas to your desired conclusion.

1. Refer to the twelve problem-solving techniques described in Section 1.2. Write a paper that treats Chapter 7 as an extended example of how these problem-solving techniques are used. In particular, find and describe one or two instances of each of these techniques as they occur in Chapter 7, and comment on how useful each instance seems to be. Conclude with some general opinions — and supporting arguments — about how effective and/or how distracting it is to focus on the problem-solving techniques when doing mathematics.

   Suggestion: The most annoying part of writing this paper may well be deciding how to organize it. Take a little extra time to think it through, particularly from the viewpoint of your reader. You might find it helpful to write a brief, specific outline first.

2. Think of an operation-table pattern that you might like to investigate. Here are some suggestions to prod your imagination, but feel free to make up something different, if you wish.

   • The same element appears everywhere in the upper-left/lower-right diagonal.

- The same element appears everywhere in the upper-right/lower-left diagonal.
- Two rows (and/or columns) are identical.
- Each row (or column) that appears also appears in reverse order.
- A particular pair of elements always appears next to each other.
- Only certain rearrangements of the order of the elements are allowable (as in the clock-arithmetic "cycles").

Choose a specific pattern — one of those listed here or one of your own — and get your instructor's approval to use it as a topic for your paper. The type of structure (set and operation) described by your pattern will be yours to name and investigate, using as many of the problem-solving techniques from Chapter 1 as you find helpful. Pay particular attention to the concepts described in this chapter — associativity, commutativity, cancellation, identity element, inverses, substructures — but don't feel limited to them. Your paper is to be a description of this investigation, reporting what specific questions you asked yourself, describing in detail the particular problem-solving tactics you pursued, showing where they led, etc. Begin with a clear definition of your kind of structure, along with two or three examples of tables that fit the definition, and end it with some questions that seem to be promising avenues for further investigation.

3. This is an open-ended exploration assignment. Since it is difficult to observe associativity from an operation table, the construction of a group table is often done as follows:

- Fill in the identity row and column.
- Choose an inverse for each element.
- Fill in the rest of the table so that there are no repetitions in any row or column. (That is, left and right cancellation hold.)
- Hope that the resulting table is associative; check (somehow).

For 2-, 3-, and 4-element sets, the first three of these steps actually yield groups; associativity follows automatically. However, for sets of 5 or more elements, an operation table constructed in this way may not be associative. What can be said about such tables? How are they like groups? How are they different? Investigate these questions and write a paper about your investigation. Begin with the following definition:

Definition: A **loop** is a set with an operation on it such that (i) there is an identity element, (ii) every element has an inverse, and (iii) left and right cancellation hold.[8]

Investigate loops in a way that parallels the development of groups in this chapter. Wherever possible, use the problem-solving techniques from Chapter 1, and describe how you are using them. Here are some particular questions you might consider:

(a) Can you construct some loops that are not groups?

(b) Is every loop commutative?

(c) Can you construct a loop of every size such that each element is its own inverse?

(d) How would you define "subloops"? What can be said about them?

(e) Does Lagrange's Theorem hold for loops?

Don't be limited by these questions; rather, use them to help you think of new ideas. If your explorations lead you in an interesting direction, follow it, even if you do not cover all the questions listed here.

Your paper is to be a description of *how you went about investigating loops.* Report what specific questions you asked, describe in detail the particular problem-solving techniques you pursued, show where they led, etc. (In this type of investigation, even a blind alley is something to report, provided you can describe the alley.) Be as creative as you like in finding examples, looking for patterns, and making conjectures, but try to prove your conjectures or at least supply some evidence or argument to support them.

For Further Reading

1. Budden, F. J. *The Fascination of Groups.* London: Cambridge University Press, 1972.

2. Davis, Philip J., and Reuben Hersh. *The Mathematical Experience.* Boston: Birkhäuser Boston, Inc., 1981, pp. 203–209.

3. Devlin, Keith. *Mathematics: The New Golden Age.* London: Penguin Books, 1988, Chapter 5.

4. Grossman, Israel, and Wilhelm Magnus. *Groups and Their Graphs.* New York: Random House, Inc., 1964.

---

[8]This is a defined term in the mathematical literature. There were several articles about loops in the *American Mathematical Monthly* during the mid-1960s.

5. Kemeny, John G. *Random Essays on Mathematics, Education and Computers*. Englewood Cliffs, NJ: Prentice-Hall, Inc., 1964, Chapter 7.

6. Newman, James R., ed. *The World of Mathematics*, Vol. 3. New York: Simon and Schuster, Inc., 1956, Part IX.

7. Schonland, David S. *Molecular Symmetry*. London: D. Van Nostrand Company, Ltd., 1965, Chapter 2.

CHAPTER

MATHEMATICS OF SPACE AND TIME: FOUR-DIMENSIONAL GEOMETRY

8.1 What Is Four-Dimensional Geometry?

One of the most profound and far-reaching insights of 20th-century science has been the realization that we live in a four-dimensional world. More precisely, it is the realization that a proper understanding of physical laws depends on treating time as a dimension just like length, width, and depth.

When Isaac Newton formulated his laws of motion in the late 17th century, he thought of the universe as existing in the stationary, uniform three-dimensional space described by Euclid's solid geometry. These laws of mo-

Sir Isaac Newton. His view of the universe as being described by Euclidean geometry dominated the physical sciences for two hundred years.

tion, which described the physical theory of gravitation, were based on the idea that time was an absolute concept, moving onward at a steady pace that is the same for all observers everywhere in the universe. Thus, for Newton, space was a fixed frame of reference within which the laws of gravity governed the relative motions of objects — from atoms to planets — as they lived out their allotted time span measured by the universal clock of the Creator. This view of the universe dominated physical science for two hundred years.

In 1881, however, a disturbing event occurred in Cleveland, Ohio. Two American physicists, A. A. Michelson and E. W. Morley, performed an experiment whose results shocked the scientific world. The specifics of that experiment need not concern us here. Suffice it to say that they developed an apparatus that could measure the speed of a light beam to within a fraction of a mile per second. Newton's laws of motion as applied to light rays flashed from a point on our moving planet Earth assert that the speed of those light rays (186,284 miles per second) must vary some 20 miles per second one way or the other, depending on whether the light is flashed with or against the direction of the earth's motion. The Michelson-Morley experiment showed that this variation does *not* occur! The results of this experiment, verified by later work of Morley and other scientists, called into question the theory of gravity and with it the entire structure of the Newtonian universe.

Albert Einstein was just a toddler when the Michelson-Morley experiment was performed; the problem it posed had to wait for a solution until he formulated his theory of relativity at the age of twenty-six. That theory scrapped not only Newton's laws of gravity, but also the very foundation on which they rested. Einstein rejected the notions of absolute space and time, claiming, instead, that length, width, depth, *and* duration could only be measured *in relation to* some observer, and that these dimensions varied from observer to observer.

We are used to dealing with some aspects of this relativity of space. For example, if I am looking out a restaurant window and notice that on the adjacent sidewalk a lamppost is to the left of a fire hydrant, I know that you, standing across the street, see the lamppost as being to the right of the fire hydrant. With Einstein's relativity theory, this reversibility of order, based on the perspective of the observer, is extended to the before/after relation of time. Two events that appear to occur in one time order to me may appear to occur in the opposite order to someone in a distant galaxy, and *both* of us could be correct! Just as there is no absolute sense of left and right, there is no absolute sense of before and after.

To locate an object or an event in our universe accurately, then, we must be able to say *where and when* it is, relative to some chosen frame of reference. This idea did not originate with Einstein. The view that the mathematical science of mechanics can be regarded as the geometry of four dimensions — three spatial dimensions and a time dimension — first appeared in the eighteenth century, in the writings of the French mathematicians d'Alembert (in 1754) and Lagrange (in 1788). However, it was not until the emergence of time-space questions in early 20th-century physics that this mathematical device was recognized as a valuable tool for describing reality. When Einstein solved the problem of motion by denying the absolute nature of time, scientists suddenly were faced with the need for four-dimensional coordinate systems to describe the where-when locations of objects in space-time.

The relative nature of time-and-space measurements with respect to these coordinate systems is a bit beyond the scope of this book. If you are interested in pursuing this topic further, there is a clear, simple explanation of relativity theory in Lincoln Barnett's book, *The Universe and Dr. Einstein*, [2] in the list *For Further Reading* at the end of this chapter; a more recent discussion of it appears in [8] of that list. We shall confine our attention to an explanation of the coordinate systems themselves, starting with the simple one-dimensional world of a single straight line and building step by step to the four dimensions of space and time. In the course of this development we shall see how a few elementary geometric and algebraic techniques and the insight provided by analogy combine to produce a mathematical device powerful enough to carry us past the three-dimension limit of our spatial intuition to the real and fictional worlds of four dimensions and beyond.

EXERCISES 8.1

WRITING EXERCISES

1. Using complete sentences, write a half-page summary that identifies the main theme of this section and outlines how that theme is carried out.

2. Do you think that Einstein's theory of relativity has any effect at all on your daily life? If so, give an example; if not, why not?

3. Using the library as a resource, write a short biographical sketch of Isaac Newton or Albert Einstein. Be sure to list the source(s) you use and be careful that you do not simply copy or paraphrase the material from your source(s).

4. From a political and/or social standpoint, what was going on in the United States around 1881 (the time of the Michelson-Morley experiment)? What about in 1905 (about the time that Einstein's theory of relativity emerged)? Feel free to consult a history book, but be sure to list your source(s).

8.2   One-Dimensional Space

The word *dimension* comes from the Latin word *dimensus*, which literally means "measured out" or "measured separately." This meaning expresses the underlying idea of dimension very well. For instance, a line segment is considered one dimensional because it has only one measure, its length; a rectangle, on the other hand, requires two separate measurements, length and width, so it is considered two-dimensional. A cardboard carton is three-dimensional because it requires three separate measurements — length, width, and depth (or height). Space-time objects are considered four-dimensional because four separate measurements are needed to describe them — length, width, depth, and duration. An object that requires five separate measurements (length, width, depth, duration, and temperature, for instance) is five dimensional, and so on. Thus, our study of dimensions must begin with the notion of measurement on a single straight line.

Note
Readers familiar with coordinates and distance on a plane can skim this material until the definition of "taxicab path" on page 363.

To measure length on a straight line, we must first choose a **unit length**; that is, we must decide what length will be assigned the number one. This choice can be made in any way we please. The measurement theory will be the same, regardless of our choice of unit, so long as we are consistent throughout the process. We could use feet or inches or meters or miles, or we could make up a unit, such as —————— .

Different units are convenient for different purposes — inches or centimeters for cabinet makers; kilometers, miles, or light years for astronomers; angstroms or microns for certain engineers; and so on. A unit length is even used to represent time — a second, a minute, an hour, a day, etc. To emphasize that our theoretical work is independent of the choice of unit length, we shall often just use the word *unit* to denote the chosen basis for measurement.

Example 8.1   The distance between points p and q on the line in Figure 8.1 can be described (approximately) as:

> 2, if the unit length is an inch;
> 5, if the unit length is a centimeter;
> $\frac{1}{6}$, if the unit length is a foot;
> 3, if the unit length is —————— .   □

Figure 8.1

Coordinate geometry is based on the fact that, once a unit length is chosen, every point on a straight line can be labeled with a real number in such a way that the line can be used as a sort of "infinite ruler" to measure distances from some specific fixed point. This fixed point is called the **origin** and is assigned the number 0. Once the point of origin and the unit length have been chosen, the line is "coordinatized" by choosing a positive direction and marking off the integer points in successive unit lengths. If the line is horizontal, we usually regard *right* as the positive direction; if the line is vertical, we usually think of *up* as positive. However, those choices are simply a matter of custom and may be varied if it is more convenient to pick a different direction as the positive one. All the other points on the line have numerical labels, too. Each real number can be matched with exactly one point on the line, and vice versa.

Specific methods for matching numbers with points need not concern us here; it is sufficient for us to know that each point on the line corresponds to a unique real number. However, we note in passing that the Greeks had a geometric-construction method capable of locating the point for any rational number (fraction), and that irrational-number locations can be approximated to any desired degree of accuracy by using decimal expansions.[1] For the diagrams of this chapter we shall simply estimate the approximate locations of numbers relative to the unit length and the origin.

A line labeled with real numbers by specifying an origin, a unit length, and a positive direction is called a **number line** or **real line**, and the points on that line are usually referred to by their numerical labels. A single straight line is a **one-dimensional space**, sometimes called a **1-space**.

If each point in that space has been assigned a real number, then the distance between two points is easy to find. Recall that the **absolute value** of a number x is defined by

$$|x| = \begin{cases} x & \text{if } x \text{ is positive or zero} \\ -x & \text{if } x \text{ is negative} \end{cases}$$

DEFINITION   The **distance** between two points p and q on a real line is $|p-q|$, the absolute value of their difference.

Example 8.2   The distance between 5 and 2 is 3 because $|5 - 2| = 3$. The absolute value is used so that the order in which the points are chosen does not affect the distance between them. In particular, the distance between 2 and 5 is the same as the distance between 5 and 2 because $|2 - 5| = |-3| = 3$.   □

---

[1]For a discussion of the decimal expansions of rational and irrational numbers, see Section 6.4.

Example 8.3   The distance between -4 and 7 is
$$|-4-7| = |-11| = 11$$
□

Example 8.4   The absolute value of any point is its distance from the origin:

the distance between 2 and 0 is $|2-0| = 2$

the distance between $\frac{-3}{8}$ and 0 is $|\frac{-3}{8} - 0| = |\frac{-3}{8}| = \frac{3}{8}$
□

If we were to travel from a point p to a point q on a line, our path would cover the part of that line lying between p and q. More formally, if p and q are points in a one-dimensional space, with $p < q$, then the **path** between p and q is the line segment consisting of p and q themselves and all points between them. We symbolize this by

$$[p,q] = \{x \mid p \le x \le q\}$$

Its **length** is the distance between p and q.

Note
Recall that $<$ means *less than* and $\le$ means *less than or equal to*. If you are unfamiliar with this way of writing sets, look up "set-builder notation" in Section 4.2. We shall call such an expression a **point-set** description of the line segment.

Example 8.5   The path between 1 and 4 is $[1,4] = \{x \mid 1 \le x \le 4\}$. It can be pictured as in Figure 8.2. The path between 4 and 1 is exactly the same set of points; we customarily write the smaller number first.
□

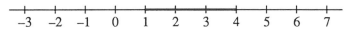

Figure 8.2   *The path between 1 and 4.*

Example 8.6   The path between -2 and 7 is $[-2,7] = \{x \mid -2 \le x \le 7\}$. Its length is $|-2-7| = |-9| = 9$.
□

The only basic geometric figures in a one-dimensional space are single points and line segments; all other figures are just collections of separate segments and points. Aesthetically speaking, this makes 1-space pretty dull. Moreover, travel in a one-dimensional world can be difficult. A single point provides a barrier that cannot be avoided. For instance, there is no way to travel from 4 to 6 without crossing the point 5. Two points completely enclose the segment between them; there is no way to get from a point in that segment to a point not in it without crossing one of the endpoints. Getting around a one-point barrier is like avoiding a waiting catcher on the way from third base to home plate — unless you go outside the base line, you're bound for a collision. Like it or not, there is no way to go outside the base line in one-dimensional space. That option requires a space of at least two dimensions.

EXERCISES 8.2

For Exercises 1–8, measure with a ruler the distance between the points p and q on the line in Figure 8.3, using each of the unit lengths given. (Make your answers accurate to the nearest whole number.)

1. an inch
2. a centimeter
3. a quarter-inch
4. a millimeter
5. a foot
6. a meter
7. ─────────────
8. ──────────────────

For Exercises 9–16, mark the approximate locations of the following points on the number line in Figure 8.4.

9. 3   10. −2   11. $\frac{1}{2}$ 12. $\frac{-4}{5}$
13. $\frac{-10}{3}$ 14. $\frac{39}{17}$ 15. $\sqrt{2}$ 16. $-\sqrt{3}$

For Exercises 17–26, find the distance between the given pair of points on a real (number) line.

17. 7 and 3
18. 15 and 2
19. −5 and 2
20. −5 and 5
21. −5 and −5
22. −5 and 0
23. $2\sqrt{2}$ and 0
24. $\sqrt{3}$ and $2\sqrt{3}$
25. 6 and x
26. x and $x+1$

For Exercises 27–32, write in set notation the path between the given pair of points, find the length of that path, and sketch the path on a number line.

27. 2 and 3
28. 0 and 6
29. −2 and 1
30. $\sqrt{2}$ and $\sqrt{3}$
31. −5 and $\frac{-1}{2}$
32. −2.5 and .75

If a centimeter represents 10 minutes, mark the time intervals in Exercises 33–38 on the number line in Figure 8.5, assuming that each one starts at 0.

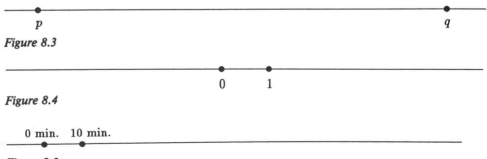

p q

Figure 8.3

        0   1

Figure 8.4

    0 min.   10 min.

Figure 8.5

33. 40 minutes   **34.** 5 minutes

35. one hour   **36.** half an hour

37. 37 minutes

38. 1 hour and 15 minutes

WRITING EXERCISES

1. Name two units of length measure besides the ones mentioned in this section and describe them *without referring to any other unit of measure*. (If you can't think of any, you might try looking up *fathom*, *furlong*, or *rod*, but there are many others. Try to find at least one on your own.)

2. Explain *absolute value*, as if you were talking to a ninth grader, without using any symbols at all.

8.3   Two-Dimensional Space

When we go from the world of a single line to the world of a plane, we need a second measurement to fix the location of a point. That is, if we have a single real line in a plane, then instructions for reaching any point in that plane from the origin can be specified by *two* numbers. We need one number to tell us which way and how far to go along the line, and a second number to tell us which way and how far to go above or below the line. Figure 8.6 shows two examples of locating points in this way. Starting at the origin of the line, we can reach the point P by traveling along the line 3 units in a positive direction, then moving up 2 units. Thus, assuming *up* is considered as the positive direction, we can give the "address" of P relative to the origin as the ordered pair of numbers 3, then 2. Similarly, the address of the point Q in Figure 8.6 can be given as 4, then -1. Because the location of any point in the plane can be determined by two numbers in this way, the plane is called a **two-dimensional space**, or simply, a **2-space**.

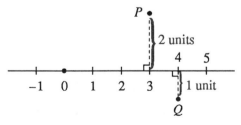

Figure 8.6   Locating points in two-dimensional space.

The second numbers in the location process just described indicate distance and direction, just as the first numbers do, so we can also consider the second numbers as describing locations on a single real line, a line perpendicular to the first one at a common origin point, as shown in Figure 8.7. This arrangement of perpendicular copies of the real line crossing at their common origin is called a **Cartesian coordinate system**,[2] and each of the two number lines is called a **coordinate axis**.

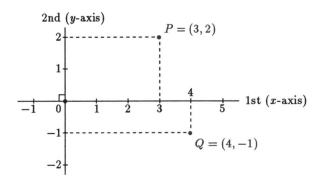

Figure 8.7   A Cartesian coordinate system.

The most common form of the Cartesian coordinate system uses a horizontal line as the first axis and a vertical line as the second. Then the location of any point in the plane is specified by an ordered pair (x, y) of real numbers, called the **coordinates** of the point. The first number denotes a position on the horizontal axis, which is usually called the **x-axis**; the second number denotes a position on the vertical axis, or **y-axis**. The point itself is the intersection of the two perpendicular lines through the axis points. Figure 8.7 shows this for the points $(3, 2)$ and $(4, -1)$. In this way the set of all points in the plane can be represented by the Cartesian product $\mathbf{R} \times \mathbf{R}$, the set of all ordered pairs of real numbers.

DEFINITION   In general, the **Cartesian product** of two sets A and B is the set of all ordered pairs with first elements from A and second elements from B. In symbols,

$$A \times B = \{(a, b) \mid a \in A, b \in B\}$$

The use of perpendicular coordinate axes makes it easy to compute the straight-line distance between two points in the plane by using the

---

[2]Named for 17th-century French mathematician René Descartes, who used this type of system as the key to his development of analytic geometry.

Pythagorean Theorem.[3] For example, the straight-line distance d between $(2, 1)$ and $(6, 4)$ is the length of the hypotenuse of a right triangle whose right-angle vertex is $(6, 1)$, as shown in Figure 8.8. The lengths of the other two sides of this triangle are 4 (the x-axis distance between the first coordinates) and 3 (the y-axis distance between the second coordinates). The Pythagorean Theorem then tells us that

$$d^2 = 4^2 + 3^2$$

and thus,

$$d = \sqrt{16 + 9} = \sqrt{25} = 5$$

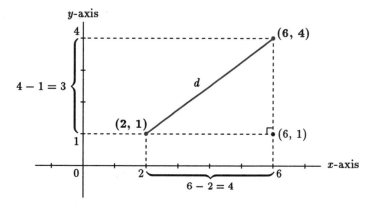

Figure 8.8   *The distance between (2,1) and (6,4).*

DEFINITION   The **distance** between two points (x_1, y_1) and (x_2, y_2) in the plane is

$$\sqrt{(x_2 - x_1)^2 + (y_2 - y_1)^2}$$

Example 8.7   The distance between $(3, 5)$ and $(8, -2)$ is

$$\sqrt{(8 - 3)^2 + (-2 - 5)^2} = \sqrt{25 + 49} = \sqrt{74}$$

It does not matter which point is taken first. For example, reversing the order of the points here, we get

$$\sqrt{(3 - 8)^2 + (5 - (-2))^2} = \sqrt{25 + 49} = \sqrt{74} \qquad \square$$

---

[3] "The square of the hypotenuse of a right triangle equals the sum of the squares of the other two sides." (See Figure B.1 of Appendix B.)

Example 8.8   The two-dimensional distance between two points on the same horizontal or vertical line (including the axes) is the same as the one-dimensional distance between them. For instance, the one-dimensional distance between -2 and 5 on the x-axis is $|-2 - 5| = 7$. In two-dimensional space, the distance between the same points, $(-2, 0)$ and $(5, 0)$, is

$$\sqrt{(5 - (-2))^2 + (0 - 0)^2} = \sqrt{49} = 7$$

Similarly, the distance between $(2, 3)$ and $(2, 7)$, which are on the same vertical line, is

$$\sqrt{(2 - 2)^2 + (7 - 3)^2} = \sqrt{4^2} = 4$$

which is the same as the one-dimensional distance between 3 and 7.   □

The shortest road between two points is not always the easiest one to travel or describe. For instance, a Manhattan taxicab seldom takes its passengers from pick-up to drop-off in a straight path because it has to follow the rectangular grid of the streets and avenues. We, too, shall find it convenient to describe paths "taxicab fashion" — that is, by using segments of lines parallel to the coordinate axes. Although they are not always the shortest paths between points, we can visualize them easily, describe them in a simple symbolic form, and generalize them to higher-dimensional spaces in an obvious way.

The coordinates of all the points on a line parallel to an axis are easy to describe. For instance, the x-axis itself is the set of all points whose second coordinate is 0 because all those points are at the 0 level relative to the y-axis. Thus, in point-set notation, the x-axis is the set of points given by

$$\{(x, 0) \mid x \in \mathbf{R}\}$$

The segment of the x-axis between 2 and 5 is the set of points

$$\{(x, 0) \mid 2 \leq x \leq 5\}$$

Similarly, a horizontal line is parallel to the x-axis. Thus, the horizontal line 3 units above the x-axis is the set of all points with second coordinate 3; that is,

$$\{(x, 3) \mid x \in \mathbf{R}\}$$

It is sometimes called "the line $y = 3$." The segment of this line between $x = 2$ and $x = 5$ is the set

$$\{(x, 3) \mid 2 \leq x \leq 5\}$$

Figure 8.9(a) illustrates these lines and segments. Lines and segments parallel to the y-axis are described similarly, as shown in the following examples.

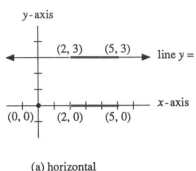

Figure 8.9 *Lines and segments.*

Example 8.9 (Figure 8.9(b) illustrates this example.) The y-axis is the set of all real-number pairs with first coordinate 0; that is,

$$y\text{-axis} = \{(0, y) \mid y \in \mathbf{R}\}$$

The segment on it between -1 and 3 is $\{(0, y) \mid -1 \leq y \leq 3\}$.   □

Example 8.10 (Figure 8.9(b) illustrates this example.) The vertical line at $x = 4$ is

$$\{(4, y) \mid y \in \mathbf{R}\}$$

and the segment on it between -1 and 3 is

$$\{(4, y) \mid -1 \leq y \leq 3\}$$

Similarly, the line $x = -2$ is

$$\{(-2, y) \mid y \in \mathbf{R}\}$$

and the segment on it between -1 and 3 is

$$\{(-2, y) \mid -1 \leq y \leq 3\}$$ □

Now it is easy to give numerical descriptions of "taxicab paths" in the plane. Suppose, for instance, we want to go from the origin, $(0,0)$, to the point $(5,3)$. We could travel along the x-axis from 0 to 5, then travel up along the line perpendicular to the x-axis at 5 from the 0 level to the 3 level. (See Figure 8.10.) The two line segments of this path are

$$\{(x,0) \mid 0 \leq x \leq 5\} \quad \text{and} \quad \{(5,y) \mid 0 \leq y \leq 3\}$$

(Note that the two segments have exactly one point in common — $(5,0)$, the place where they are joined.) Thus, a path from $(0,0)$ to $(5,3)$ is the union of these two line segments:

$$\{(x,0) \mid 0 \leq x \leq 5\} \cup \{(5,y) \mid 0 \leq y \leq 3\}$$

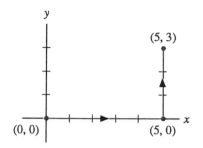

Figure 8.10   A taxicab path in the plane.

In general, we call two line segments **connected** if they have a common endpoint, and we describe a **taxicab path** as the union of connected line segments parallel to the coordinate axes. Because we shall use only taxicab paths for the remainder of this chapter, we shall use the term **path** to mean a taxicab path. Now, all but two of the endpoints of line segments in a (taxicab) path are common to two (or more) line segments. The remaining two points are called the **endpoints** of the path. If a path has endpoints P and Q, we call it a **path between** P and Q.

Example 8.11   A path between $(-1, 3)$ and $(4, -2)$ is

$$\{(x, 3) \mid -1 \le x \le 4\} \cup \{(4, y) \mid -2 \le y \le 3\}$$

The intersection point of the two segments is $(4, 3)$. Another path between $(-1, 3)$ and $(4, -2)$ is

$$\{(-1, y) \mid -2 \le y \le 3\} \cup \{(x, -2) \mid -1 \le x \le 4\}$$

The intersection of these two segments is the point $(-1, -2)$. A third path between $(-1, 3)$ and $(4, -2)$ is

$$\{(x, 3) \mid -1 \le x \le 1\} \cup \{(1, y) \mid -2 \le y \le 3\} \cup \{(x, -2) \mid 1 \le x \le 4\}$$

The segments in this path are connected at $(1, 3)$ and $(1, -2)$. All three of these paths are pictured in Figure 8.11.   □

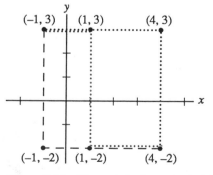

Figure 8.11   Three paths between $(-1, 3)$ and $(4, -2)$.

One of the most basic two-dimensional figures is the square. It can be built by using four copies of the basic one-dimensional figure, a line segment, in an obvious way: Place one copy of the segment horizontally in the plane and attach another one at each end so that they are pointing vertically upward from the first segment; then "cap" the two vertical segments with the fourth one by attaching it to their upper endpoints. This can also be done by attaching the four segments end-to-end on a single line, then folding them up, as in Figure 8.12. This diagram shows segment 2 held fixed while segments 1 and 3-4 are folded vertically upward. Then segment 4 is folded over until it meets segment 1, completing the square.

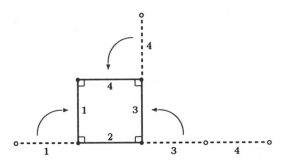

*Figure 8.12   Constructing a square
from line segments.*

The region enclosed by a square in the plane can be thought of in another way that has interesting implications for us. If a line segment is moved in a direction perpendicular to itself for a distance as long as itself, then it "sweeps out" a square region in the plane. Now, if this perpendicular direction is taken to represent time, then the square region represents the existence of the line segment during a period of time.

For example, consider a two-dimensional coordinate system in which the vertical axis represents height (in inches) and the horizontal axis represents time (in minutes). Suppose that you have a very thin 3-inch candle that burns at the constant rate of 1 inch every 2 minutes. Set up the candle (but don't light it) and start your stopwatch. At the end of exactly 3 minutes the unlit candle will have "swept out" a 3-by-3 square in the height-time plane. After 3 minutes light the candle. As it burns, its height decreases by 1 inch every 2 minutes, so that it will be completely gone 6 minutes after you light it. The candle, set up for 9 minutes and burning from the end of the third minute until the end of the ninth minute, can be represented by a single two-dimensional region, as shown in Figure 8.13. In this way,

regions in two-dimensional space-time can be used to represent
the existence in time of objects in one-dimensional space.

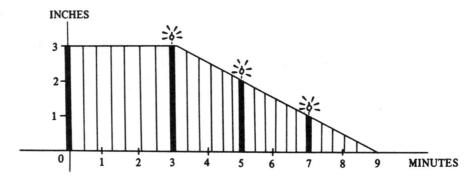

Figure 8.13   The "life" of a candle in the height-time plane.

Example 8.12   The outdoor temperature is represented by a column of mercury calibrated in degrees Celsius. During a 24-hour day the temperature rises from a midnight low of 10° to a high of 25° at 2 p.m., then drops back to 15° by the following midnight. All the changes in this column of mercury during that day are represented by the single height-time region in Figure 8.14. □

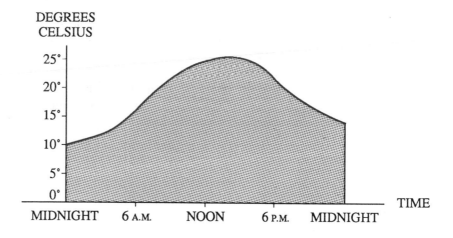

Figure 8.14   Temperature change represented by a height-time region.

Finally, let us consider the 1-space barrier problem described at the end of the previous section. It is impossible to find a path between the point 4 on the number line and a point inside the interval $[-2, 2]$ — say 1 — that does not pass through the endpoint 2. However, if we consider that number line as the x-axis in a two-dimensional coordinate system, then the problem becomes:

Find a path from $(4,0)$ to $(1,0)$ that does not pass through $(2,0)$.

That's easy; we just go up, over, and down. Figure 8.15 shows such a path; its point-set description is

$$\{(4,y) \mid 0 \le y \le 1\} \ \cup \ \{(x,1) \mid 1 \le x \le 4\} \ \cup \ \{(1,y) \mid 0 \le y \le 1\}$$

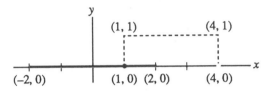

Figure 8.15   *A 2-space solution to the 1-space barrier problem.*

Even without a picture, it is clear from this description that $(2,0)$ is not a point of this path because the only points of the path that have second coordinate 0 are the endpoints $(4,0)$ and $(1,0)$. The rest of the path lies outside the original one-dimensional space.

There is an analogous barrier problem for 2-space:

> Between a point outside a square region of the plane and a point inside the region, find a path that does not pass through any side of the square.

There is no two-dimensional solution of this problem; but a third dimension makes it easy, as we shall see in the next section.

EXERCISES 8.3

For Exercises 1–8, set up a single Cartesian coordinate system for the plane, and mark on it the location of the given points.

1. $(2,4)$ **2.** $(3,3)$ **3.** $(-1,3)$

4. $(\frac{1}{2},4)$ **5.** $(-1,-1)$ **6.** $(1,-2)$

7. $(-5,0)$ **8.** $(\frac{3}{4},\frac{1}{4})$

For Exercises 9–14, compute the distance between the given pair of points.

9. $(2,1)$ and $(8,9)$ **10.** $(2,3)$ and $(4,5)$

11. $(0,0)$ and $(1,1)$ **12.** $(-2,7)$ and $(7,-5)$

13. $(9,-1)$ and $(9,4)$ **14.** $(-1,-3)$ and $(1,3)$

For Exercises 15–24, sketch the line, segment, or path in a Cartesian coordinate plane.

15. $\{(x,5) \mid x \in \mathbf{R}\}$

16. $\{(-1,y) \mid y \in \mathbf{R}\}$

17. $\{(x,0) \mid 0 \le x \le 3\}$

18. $\{(x,4) \mid -1 \le x \le 2\}$

19. $\{(1,y) \mid -1 \le y \le 1\}$

20. $\{(-3,y) \mid -3 \le y \le 0\}$

21. $\{(x, 2) \mid \frac{-1}{2} \leq x \leq \frac{1}{2}\}$

22. $\{(5, y) \mid \frac{1}{2} \leq y \leq \frac{3}{4}\}$

23. $\{(x, -1) \mid -2 \leq x \leq 3\}$
$$\cup \{(3, y) \mid -1 \leq y \leq 4\}$$

24. $\{(4, y) \mid 2 \leq y \leq 5\} \cup \{(x, 2) \mid 0 \leq x \leq 4\}$

For Exercises 25–32, describe in set-builder notation two taxicab paths between the two given points. Also, specify the intersection points that connect the segments in the paths. Sketch the paths.

25. $(0, 0)$ and $(3, 4)$ 26. $(2, 2)$ and $(0, 0)$
27. $(1, 2)$ and $(5, -1)$ 28. $(0, 1)$ and $(3, 0)$
29. $(2, 0)$ and $(-1, -1)$ 30. $(-1, 3)$ and $(4, 3)$
31. $(2, \frac{-5}{2})$ and $(2, \frac{7}{3})$ 32. $(\frac{1}{2}, 1)$ and $(5, \frac{2}{3})$

For Exercises 33–36, draw a two-dimensional space-time figure to represent (approximately) each situation.

33. A 2-inch candle that burns at the constant rate of one inch every 5 minutes is lighted and burns until nothing remains.

34. A 5-inch candle that burns at the constant rate of 1 inch every 2 minutes is lighted, burns for 4 minutes, is blown out for 3 minutes, then is relighted, and burns until nothing remains.

35. A stone is dropped from the top of a 140-foot cliff and hits the ground in 3 seconds.

36. A mercury thermometer measures the heat in a factory as it rises at a constant rate from 15° (Celsius) at 7 a.m. to 35° at 3 p.m.

37. Consider the point 3 and the interval $[0, 1]$ on the real line. By considering that line as part of 2-space, find a path between 3 and some point inside $[0, 1]$ that does not pass through the endpoint 1. Sketch that path and write its set-builder description.

38. (a) Write a set-builder description of a square (line figure) with corners $(0, 0)$, $(2, 0)$, $(0, 2)$, $(2, 2)$. Sketch it.

(b) Write a set-builder description of the planar region bounded by the square in Part (a).

(c) List the coordinates of three points that are inside the square region of Part (b) and of three points that are outside it.

WRITING EXERCISES

1. The construction of this section is analogous to that of Section 8.2. Write a comparative outline of the two sections that displays all points of the analogy.

2. Cartesian-product (ordered-pair) labeling is used to identify location in many situations outside of mathematics. For example, the location of a seat in a theater or stadium is often specified by an ordered pair consisting of a letter (or pair of letters) and a number, with the letter(s) specifying the row and the number specifying the seat within that row. Find and describe two more such examples.

3. Is the term "taxicab path" helpfully suggestive of its meaning, or do you find its imagery distracting? Explain. Can you think of a better term for this concept?

8.4   Three-Dimensional Space

Now that we have seen how to generalize basic spatial concepts from one dimension to two dimensions, the step from 2-space to 3-space should be easy. Let us begin with the floor beneath your feet, extending it to form an unbounded plane. Any point in our spatial universe that is not on the plane itself is directly above or directly below some point of this plane. Now, because the floor plane is a two-dimensional space, each of its points can be specified by an ordered pair of real numbers in some Cartesian coordinate system. We can extend this coordinate system in order to label any point in the universe by using a third number to indicate exactly how far directly above or below the plane that point is.

Figure 8.16 shows this for two specific points — P, which is 4 units above the point $(2, 3)$ of the plane, and Q, which is 2 units below the point $(5, -1)$ of the plane. If at the origin we attach a third coordinate axis perpendicular to the two for the plane, this third axis can be used to measure the height above or below the plane. By this means, every location in our spatial universe can be specified by three separate coordinate measurements; thus, we speak of the universe as a **three-dimensional space**, or simply, **3-space**.

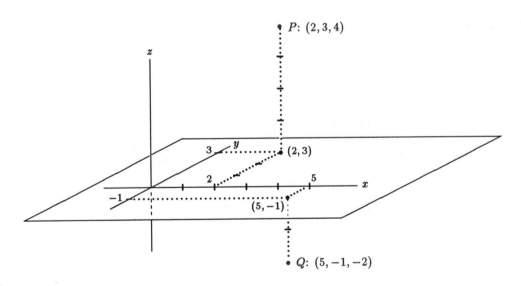

Figure 8.16 *Two points in 3-dimensional space.*

As in 2-space, this arrangement of perpendicular axes is called a *Cartesian coordinate system*; in fact, all the terminology carries over to higher-dimensional spaces in the obvious way. The third coordinate axis is usually called the **z-axis**. The plane determined by a pair of coordinate axes is named by the axis letters; we shall call such a plane a **basic plane**. (Sometimes a basic plane is also called a **coordinate plane**.) In 3-space there are three basic planes — the xy-plane, the xz-plane, and the yz-plane. The basic plane pictured in Figure 8.16 is the xy-plane.

Example 8.13   The ordered triple $(3, 1, 2)$ specifies locations on the x-, y-, and z-axes, respectively. As shown in Figure 8.17, the location of the point $(3, 1, 2)$ can be found by moving 3 units along the x-axis from the origin to $(3, 0, 0)$, then 1 unit parallel to the y-axis in the positive direction to $(3, 1, 0)$, then 2 units parallel to the z-axis in the positive direction to $(3, 1, 2)$.   □

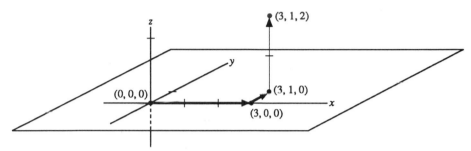

Figure 8.17   The point (3,1,2).

The formula for finding distances between any two points in 3-space is analogous to the distance formula for 2-space. As a typical example, let us compute the distance between $(1, 2, 3)$ and $(5, 7, 9)$. These points in space can be visualized as the diagonally opposite corners of a rectangular box with sides parallel to the basic planes, as shown in Figure 8.18.

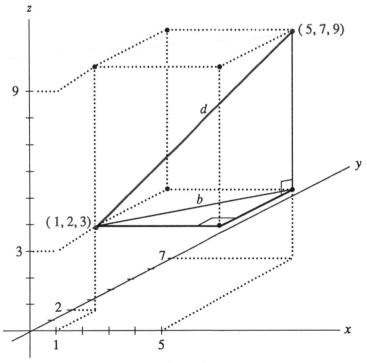

Figure 8.18   Points (1,2,3) and (5,7,9) at the diagonally opposite corners of a rectangular box.

The length d of the diagonal between those points can be found by two applications of the Pythagorean Theorem. First, the length b of the diagonal at the bottom of the box is found by using the x- and y-coordinates:

$$b = \sqrt{(5-1)^2 + (7-2)^2}$$

Then the length of d is found using b and the z-coordinates:

$$
\begin{aligned}
d &= \sqrt{b^2 + (9-3)^2} \\
&= \sqrt{(5-1)^2 + (7-2)^2 + (9-3)^2} \\
&= \sqrt{4^2 + 5^2 + 6^2} \\
&= \sqrt{77}
\end{aligned}
$$

DEFINITION   The **distance** between two points (x_1, y_1, z_1) and (x_2, y_2, z_2) in 3-space is given by

$$\sqrt{(x_2 - x_1)^2 + (y_2 - y_1)^2 + (z_2 - z_1)^2}$$

As in 2-space, lines and segments parallel to the coordinate axes are relatively easy to visualize and to describe in point-set notation. For example, Figure 8.19 shows the line parallel to the z-axis through the point $(4, 2, 0)$ on the xy-plane; it is the set of points

$$\{(4, 2, z) \mid z \in \mathbf{R}\}$$

The segment on it between $z = -1$ and $z = 3$ is the set

$$\{(4, 2, z) \mid -1 \le z \le 3\}$$

The general method for describing such lines and segments should be clear from this example and the ones that follow.

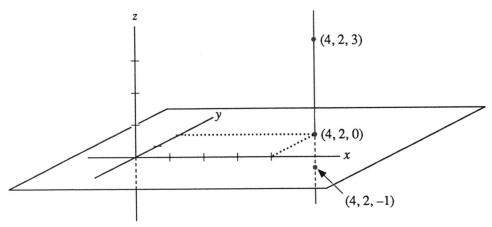

Figure 8.19   *The line through (4,2,0) parallel to the z-axis.*

Example 8.14   The segment between $x = -7$ and $x = 25$ on the line parallel to the x-axis through the point $(0, 9, -1)$ is

$$\{(x, 9, -1) \mid -7 \leq x \leq 25\} \qquad \square$$

Example 8.15   The following three lines all go through the point $(2, 8, .6)$:

$$\{(x, 8, .6) \mid x \in \mathbf{R}\} \text{ is parallel to the } x\text{-axis}$$
$$\{(2, y, .6) \mid y \in \mathbf{R}\} \text{ is parallel to the } y\text{-axis}$$
$$\{(2, 8, z) \mid z \in \mathbf{R}\} \text{ is parallel to the } z\text{-axis} \qquad \square$$

The descriptions for segments parallel to the coordinate axes make it convenient to use taxicab paths to connect points in 3-space. If we want to travel from one point to another, we can simply move in the axis directions, getting each coordinate of the "traveling point" to match those of our destination point, one by one. For instance, to go from $(1, 2, 3)$ to $(5, 7, 9)$, we can travel from $(1, 2, 3)$ along the segment

$$\{(x, 2, 3) \mid 1 \leq x \leq 5\}$$

to the point $(5, 2, 3)$, then along

$$\{(5, y, 3) \mid 2 \leq y \leq 7\}$$

to $(5, 7, 3)$, and finally along

$$\{(5, 7, z) \mid 3 \leq z \leq 9\}$$

to $(5, 7, 9)$. Thus, a path between $(1, 2, 3)$ and $(5, 7, 9)$ is

$$\{(x, 2, 3) \mid 1 \leq x \leq 5\} \cup \{(5, y, 3) \mid 2 \leq y \leq 7\} \cup \{(5, 7, z) \mid 3 \leq z \leq 9\}$$

The cube in 3-space is analogous to the square in 2-space. In Section 8.3 we saw how a square can be built from four copies of a line segment. A (hollow) cube can be built in a similar way, using six copies of a square region:

Consider one copy of the square in some plane to be a fixed "base" of the cube and attach another copy on each of the base edges. Attach a sixth square in the plane as shown in Figure 8.20; this sixth square acts as the "cap" of the cube. Fold them up in 3-space as shown in Figure 8.21. The edges in Figure 8.20 are numbered in such a way that the edges with the same number are matched in the folding process.

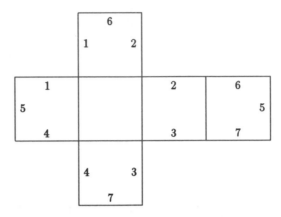

Figure 8.20 Six copies of a square, to be folded into a cube.

The three-dimensional solid cube also can be thought of as being made from a two-dimensional square region in a way like that of making the square region from a line segment. If a square region is moved in a direction perpendicular to its original plane for a distance as long as one of its sides, then it sweeps out a cubic region in 3-space, as in Figure 8.22. Now, if that perpendicular direction is taken to represent time, then the cubic solid represents the existence of the square region during a period of time. It is an infinite stack of square regions, one for each instant in the time interval.

For example, consider a square piece of paper, 6 centimeters on each side, lying on a desk during a 10-minute observation period. After 6 minutes, someone comes along with a razor, cuts the paper in half (top to bottom),

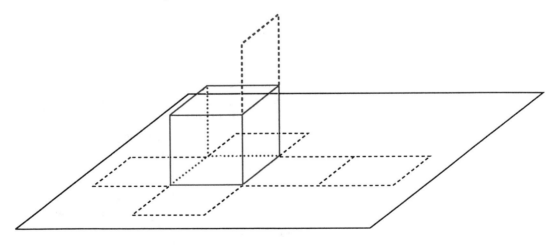

Figure 8.21   The construction of a cube by "folding up" six squares.

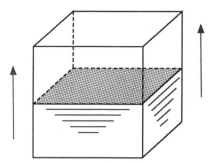

Figure 8.22 A cubic region as an infinite stack of square regions.

and throws one half away. The "life" of the paper on the desk during the 10-minute observation period can be represented by a single solid region in 3-space, in which two coordinates measure length (in centimeters) and one measures time (in minutes). The first 6 minutes of the paper's 10-minute existence on the desk are represented by a cube and the last 4 minutes by a rectangular box, as shown in Figure 8.23. In this way, regions in 3-

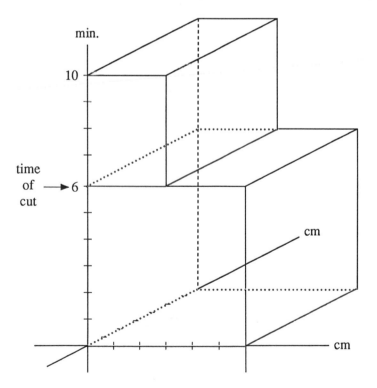

Figure 8.23 The existence in time of the cut piece of paper.

dimensional space-time can be used to represent the existence in time of objects in 2-dimensional space.

Example 8.16   A thin pancake is to be cooked on a hot griddle. The batter is poured slowly and steadily, so that by the end of 5 seconds a circular pancake 20 centimeters in diameter is formed on the griddle. After it has cooked for 40 seconds, the pancake is flipped over (in 2 seconds) and cooked for the rest of one minute, then is removed from the griddle. Figure 8.24 represents the

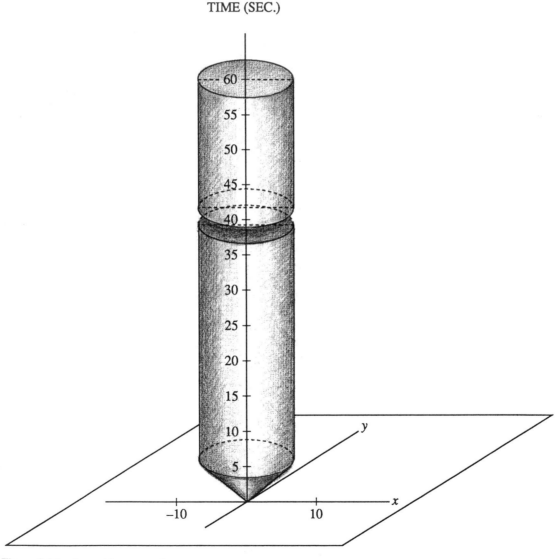

Figure 8.24   *A cooking pancake represented in space-time.*

cooking pancake. The xy-plane represents the griddle upon which the batter is about to be poured. The pancake, centered at the xy-origin, grows to a diameter of 20 centimeters in 5 seconds; this is represented by a space-time cone. It then cooks at that size until 40 seconds have elapsed, as is shown by a space-time cylinder 35 seconds high. It is removed from its cooking plane for seconds, then returned (upside down) to cook for the rest of the minute, as shown by another cylinder. Note that the gap between the cylinders at 40 seconds does not show the flipping of the pancake because that motion happens off the surface of the griddle, so it is outside the two-dimensional space being pictured in time.   □

Finally, let us look at the 2-space barrier problem described at the end of the preceding section. Consider, for example, a plane containing a square with vertices at $(0,0)$, $(2,0)$, $(0,2)$, and $(2,2)$. There is no way to make a 2-space path between the point $(1,1)$ at the center of this square and the point $(3,1)$ outside the square without intersecting the square itself. However, if we consider the plane as the xy-plane in a three-dimensional coordinate system, then the problem becomes:

Find a path between $(1,1,0)$ and $(3,1,0)$ that does not intersect the given 2-by-2 square in the xy-plane.

That's easy; we just go up, over, and down. Figure 8.25 shows such a path, whose point-set description is

$$\{(1,1,z) \mid 0 \le z \le 1\} \ \cup \ \{(x,1,1) \mid 1 \le x \le 3\} \ \cup \ \{(3,1,z) \mid 0 \le z \le 1\}$$

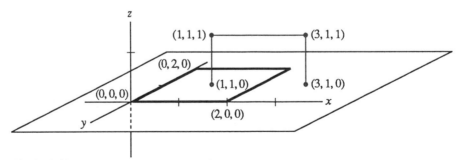

Figure 8.25 *A 3-space solution to the 2-space barrier problem.*

Even without a picture, it is clear from the description that this path does not intersect the square. All points of the square have third coordinate 0 (Why?), but the only points of the path set with third coordinate 0 are the endpoints $(3,1,0)$ and $(1,1,0)$. Thus, all the rest of the path lies outside the original two-dimensional space. (If we regard the third axis as representing time, this solution says: Move ahead one unit in time; then change your space location; then move back one time unit to the spatial world you started from!)

There is an analogous barrier problem for 3-space:

> Between a point outside a cubic region and a point inside this region, find a path that does not pass through any wall of the cube.

There is no three-dimensional solution of this problem, but a fourth dimension makes it easy. Before reading the next section, you might find it instructive to try constructing a solution analogous to our 3-space solution of the 2-space barrier problem.

EXERCISES 8.4

For Exercises 1–8, compute the distance between the two given points in 3-space.

1. $(2, 1, 1)$ and $(3, 8, 5)$

2. $(-4, 6, -1)$ and $(7, 7, -1)$

3. $(3, 2, 7)$ and $(0, 2, 3)$

4. $(0, 0, 0)$ and $(3, 3, -3)$

5. $(0, 0, 0)$ and $(1, 1, 1)$

6. $(7, 6, 4)$ and $(7, -5, 4)$

7. $(\sqrt{2}, \sqrt{3}, \sqrt{4})$ and $(0, 0, 0)$

8. $(\frac{1}{2}, \frac{1}{3}, \frac{1}{4})$ and $(1, 1, 1)$

For Exercises 9–11, describe each line and segment in set-builder notation.

9. The line through $(-8, 6, 1)$ parallel to the x-axis, and the segment of this line between $x = 4$ and $x = 7$.

10. The line through $(7, 0, -3)$ parallel to the y-axis, and the segment of this line between $y = -1$ and $y = 1$.

11. The line through $(-1, 2, -3)$ parallel to the z-axis, and the segment of this line between the xy-plane and $z = 2$.

For Exercises 12–16, write the set-builder description of a taxicab path between the two given points. Find the intersection points that connect the segments in your paths.

12. $(1, 2, 3)$ and $(4, 5, 6)$

13. $(0, 0, 0)$ and $(2, -5, -1)$

14. $(2, 1, 1)$ and $(2, 2, 1)$

15. $(-4, 3, 5)$ and $(7, -1, -9)$

16. $(\sqrt{2}, \sqrt{3}, k)$ and $(0, 0, 0)$

17. On a piece of cardboard, draw six copies of the same square, attached as in Figure 8.20. Then cut out the entire region in one piece and fold it up to form a cube, as shown in Figure 8.21.

18. Find the coordinates for each vertex of the 3-space solid in Figure 8.23.

19. A 30-centimeter square pane of glass is frosted with a thin layer of ice. The bottom of the glass is heated so that the ice melts evenly upward at a constant rate, and at the end of 5 minutes all the ice is gone.

    (a) Represent the ice in this process by a three-dimensional space-time solid.

    (b) Determine the coordinates of each vertex of the solid figure of Part (a).

20. A kitchen table, whose top is a square, 1 meter on a side, is standing with one edge against a wall. The table is constructed so that a 40-centimeter (by 1-meter) center leaf can be inserted with its longer sides parallel to the wall. Ten seconds into a 30-second observation period, the table is opened by pulling the edge farthest from the wall directly away from the wall without disturbing the opposite edge. The table is pulled open at a constant rate for 4 seconds, so that the gap is just wide enough to accept the leaf. Six seconds later the leaf is put in place.

    (a) Represent the 30-second observation period for this table top by a three-dimensional space-time solid. (Draw a picture.)

    (b) Find the coordinates of each vertex of the solid figure of Part (a). (*Hint*: There are 20 vertices in all.)

For Exercises 21–24 describe a path between the given point and the point $(1, 1)$ at the center of a 2-by-2 square, constructed so that the path does not intersect any side of the square. Write a set-builder description of the path, and draw a sketch.

21. $(1, 3)$

22. $(5, 4)$

23. $(0, 6)$

24. $(-3, 1)$

WRITING EXERCISES

1. The construction of this section is analogous to that of Section 8.3. Write a comparative outline of the two sections that displays all points of the analogy.

2. The parallel construction of Sections 8.2, 8.3, and 8.4 suggests that an analogous discussion of 4-dimensional space will appear in Section 8.5. Without looking ahead, try to anticipate this discussion by writing analogous descriptions for 4 dimensions of each of these items:

    (a) a point;
    (b) the distance between two points;
    (c) the construction of a hollow cube-like figure;
    (d) the construction of a solid cube-like figure;
    (e) a solution to the 3-space barrier problem.

3. Choose a homework problem that is giving you difficulty and write three simpler questions that you think might provide helpful steppingstones for solving it. Be as specific as possible.

4. Describe how at least three of the dozen problem-solving tactics of Section 1.2 can be used to help in the solution of Exercise 20. (Even if you can't complete Exercise 20, you should be able to do this writing exercise.)

8.5 Four-Dimensional Space

In the light of Sections 8.3 and 8.4, we can now view the transition from three dimensions to four as a natural extension of the step from 2-space to 3-space. At this step the guidance of visual imagination must be replaced by a sense of analogy. **Four-dimensional space** (or just **4-space**) is a world where the address of every location requires four separate measurements. In other words, every point in 4-space is specified by four real numbers, so four copies of the real line are needed as the axes for a Cartesian coordinate system.

In lower dimensions we have visualized the coordinate axes as perpendicular to each other, but the requirement of the word *dimension* is just that they be separate in an essential way. (The technical term is *independent.*) To replace our visual imagination, we might imagine 4-space as infinitely many copies of 3-space, each one corresponding to a specific real number that is the fourth coordinate of every point in that copy of 3-space. This is analogous to thinking of 3-space as an infinite collection of planar "layers," each layer corresponding to a specific number on the z-axis, which is the third coordinate of every point at that level. (Compare this with the view of a cube as an infinite stack of squares, illustrated by Figure 8.22 on page 373.)

If the fourth coordinate is considered to be time, then 4-space is the whole succession of copies of our spatial universe, one copy for each instant of its existence — much as motion in a cartoon world is drawn by an animator, one still frame at a time, for viewing in rapid succession. This is the space-time view of our physical space; we shall emphasize this view throughout the rest of the chapter by calling the fourth coordinate axis the **t-axis**.

Example 8.17   If we choose three spatial coordinate axes (measured in meters, say) and a starting time for a fourth coordinate (measured in hours), then the space-time point $(3, 5, 7, 2)$ represents the 3-space point $(3, 5, 7)$ two hours after the starting time. The space-time point $(3, 5, 7, 6)$ represents the same spatial point 4 hours later, and $(3, 9, 7, 2)$ represents a point 4 meters away from the first point, but simultaneous with it.   □

The formula for finding straight-line distances in 4-space is exactly analogous to the 3-space distance formula.

DEFINITION   The **distance** between two points (x_1, y_1, z_1, t_1) and (x_2, y_2, z_2, t_2) in 4-space is given by

$$\sqrt{(x_2 - x_1)^2 + (y_2 - y_1)^2 + (z_2 - z_1)^2 + (t_2 - t_1)^2}$$

Example 8.18   The distance between $(1, 2, 3, 4)$ and $(5, 7, 9, 11)$ is

$$\sqrt{(5 - 1)^2 + (7 - 2)^2 + (9 - 3)^2 + (11 - 4)^2} \;=\; \sqrt{16 + 25 + 36 + 49} \;=\; \sqrt{126}$$

□

As in 3-space, lines and segments parallel to the coordinate axes are easy to describe. For instance,

$$\{(2, y, -1, 4) \mid y \in \mathbf{R}\}$$

is a line through the point $(2, 0, -1, 4)$ parallel to the y-axis. In space-time this is a line parallel to the (spatial) y-axis at time 4. Similarly,

$$\{(3, -5, 8, t) \mid t \in \mathbf{R}\}$$

is a line through $(3, -5, 8, 0)$ parallel to the t-axis. In space-time this line represents the single spatial point $(3, -5, 8)$ existing for all eternity; the segment

$$\{(3, -5, 8, t) \mid 1 \le t \le 7\}$$

represents six time-units in the life of that point. This ease of description means that, once again, "taxicab paths" provide a convenient way to connect points in 4-space.

Example 8.19   A path between $(1, 2, 3, 4)$ and $(5, 7, 9, 11)$ is given by

$$\{(x, 2, 3, 4) \mid 1 \le x \le 5\} \cup \{(5, y, 3, 4) \mid 2 \le y \le 7\}$$
$$\cup \{(5, 7, z, 4) \mid 3 \le z \le 9\} \cup \{(5, 7, 9, t) \mid 4 \le t \le 11\}$$

The intersection points connecting the segments of this path are $(5, 2, 3, 4)$, $(5, 7, 3, 4)$, and $(5, 7, 9, 4)$. Notice that, without drawing a picture, a path such as this is easy to construct. We just change one coordinate at a time, connecting the points by line segments parallel to the axis of the coordinate being changed. In this case, the pattern of changes is:

$$
\begin{array}{cccc}
(\quad 1, & 2, & 3, & 4 \quad) \\
\downarrow & & & \\
(\quad 5, & 2, & 3, & 4 \quad) \\
& \downarrow & & \\
(\quad 5, & 7, & 3, & 4 \quad) \\
& & \downarrow & \\
(\quad 5, & 7, & 9, & 4 \quad) \\
& & & \downarrow \\
(\quad 5, & 7, & 9, & 11 \quad)
\end{array}
$$

This path can be visualized in space-time measured, say, in meters and hours: From $(1, 2, 3, 4)$ the path is 4 meters parallel to the x-axis in the positive direction, followed by 5 meters parallel to the y-axis (positive direction), followed by 6 meters parallel to the z-axis (positive direction), followed by a 7-hour wait.   □

The four-dimensional analogue of a cube is called a **hypercube**, or a **tesseract**. We have seen (in Figures 8.20 and 8.21 on page 372) how a cube can be constructed by connecting six square regions in a plane and then "folding them up" and attaching their edges in 3-space. A tesseract is made in much the same way, using eight copies of a cube: Consider one copy as the "base" cube and attach a copy to each of its six faces. Then attach the eighth cube to one of the outer six, as in Figure 8.26; this last cube is the "cap" that will close up the tesseract when it is folded up in 4-space.

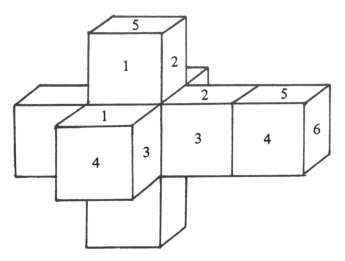

Figure 8.26   Eight copies of a cube to be "folded up" in 4-space.

The folding process itself is impossible to visualize because it requires more than three spatial dimensions. Nevertheless, the analogy with the cube construction gives us some helpful information. Recall that all the exposed edges of the unfolded cube in the plane must be sealed together in pairs when the cube is folded up in 3-space. In the same way, all the exposed square faces of the unfolded tesseract must be sealed together in pairs when it is folded up in 4-space. Some of the pairings are shown by the numbers in Figure 8.26; the rest are left for you in Exercise 21.

If we allow some distortion of the cubes in the folding process, we can get a rough idea of what must happen. Try to visualize the faces marked 1, 2, and 3 in Figure 8.26 coming together in pairs. Of course, the cubes will have to be distorted to do this in 3-space; Figure 8.27 illustrates this distortion for two of the cubes, with faces numbered as in Figure 8.26. Notice that the two faces numbered "2" have been joined here. The partial cube in this figure is the "cap" cube, which would have to be cut open, turned inside out, and snapped around the rest of the cubes to close the tesseract in 3-space. None of these distortions are needed in 4-space; with the extra dimension to allow "folding," all of these matchings can be done without stretching, cutting, etc.

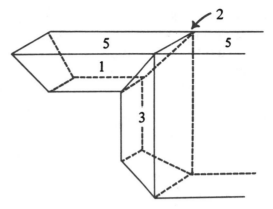

Figure 8.27   Faces of two cubes sealed together in the folding process.

The analogous "folding up" of an unfolded cube if the process were restricted to 2-space would require similar types of distortions, as you can see from Figure 8.28. The final figure (c) does not look very much like a cube if you look at it as a planar figure. However, if you pretend to be looking

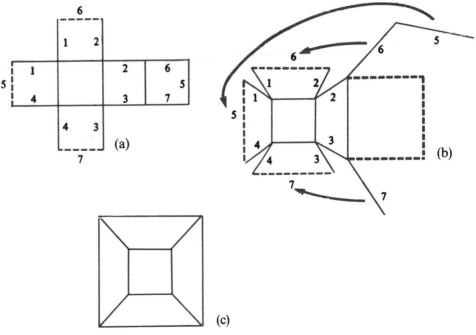

Figure 8.28   An analogy: "Folding up" a cube from six squares without leaving the plane requires distortion of angles and line segments.

down a mine shaft and think of all the angles you see as right angles, you may get a feeling for the 3-space object it represents. A similar reinterpretation of what we see is required to understand the folding up of the tesseract. Unfortunately, our visual intuition is restricted from making the leap into 4-space, so we must be content with reasoning from the cube analogy.

The *solid* hypercube can be thought of as the 4-space region swept out by a cube moving in the direction of the fourth axis for a distance as long as one of its sides. This is analogous to the construction of square areas and cubic solids. If the fourth axis is taken to represent time, then a tesseract is a cube whose "life span" in time units equals the length (in distance units) of one of its edges.

Example 8.20   In the space-time of meters and hours, a tesseract 2 units on a side is a 2-meter-by-2-meter-by-2-meter[4] cube that exists for 2 hours. It has 16 corners. (Why?) If one corner of this tesseract is at the origin and its edges are parallel to the coordinate axes (in their positive directions), then the coordinates of the corners are

$(0,0,0,0)$	$(0,0,0,2)$
$(2,0,0,0)$	$(2,0,0,2)$
$(0,2,0,0)$	$(0,2,0,2)$
$(0,0,2,0)$	$(0,0,2,2)$
$(2,2,0,0)$	$(2,2,0,2)$
$(2,0,2,0)$	$(2,0,2,2)$
$(0,2,2,0)$	$(0,2,2,2)$
$(2,2,2,0)$	$(2,2,2,2)$

In space-time, the eight corners listed in the left column are the corners of the cube at time 0, and the eight corners listed in the right column are the corresponding corners of the cube two hours later.   □

We close this section by solving the 3-space barrier problem posed at the end of Section 8.4. Consider a 2-by-2-by-2 cube in 3-space, situated as in Figure 8.29. The point $(1,1,1)$ is at the center of the cube and the point $(3,1,1)$ is outside it. Clearly, every path in 3-space between $(1,1,1)$ and $(3,1,1)$ intersects a wall of the cube. For instance, the straight line segment

$$\{(x,1,1) \mid 1 \leq x \leq 3\}$$

between the points intersects the cube wall at $(2,1,1)$. However, if we consider this situation as part of a three-dimensional "slice" of a four-

---

[4]We abbreviate this as "2-by-2-by-2-meter cube." Descriptions of 3- and 4-dimensional figures measured by coordinate units that are clear from the context will be abbreviated similarly from here on.

dimensional space, it is easy to find a path between the two points $(1,1,1,0)$ and $(3,1,1,0)$ that does not intersect the cube walls: Just go "out" in the fourth coordinate direction, move over, then come back "in" to this copy of 3-space. For instance, the path

$$\{(1,1,1,t) \mid 0 \le t \le 1\} \cup \{(x,1,1,1) \mid 1 \le x \le 3\} \cup \{(3,1,1,t) \mid 0 \le t \le 1\}$$

connects the point outside the cube with the point inside it. Moreover, this path does not intersect the cube walls anywhere — every point of the cube has fourth coordinate 0, but none of the path points (except the endpoints) have fourth coordinate 0.

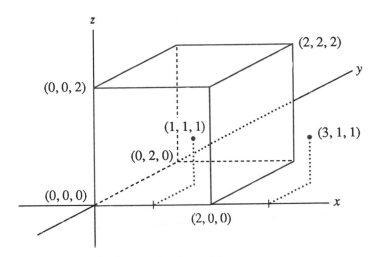

Figure 8.29   *The 3-space barrier problem.*

EXERCISES 8.5

For Exercises 1–4, compute the distance between the two given points in 4-space.

1. $(0,0,0,0)$ and $(1,1,1,1)$

2. $(2,-1,5,8)$ and $(-2,3,0,5)$

3. $(3,3,3,3)$ and $(5,5,5,5)$

4. $(6,1,5,-7)$ and $(-2,-4,3,-1)$

For Exercises 5–8, describe each line and segment in set-builder notation.

5. The line through $(8,6,4,2)$ parallel to the x-axis, and the segment of this line between $x = -1$ and $x = 5$.

6. The line through $(1,0,1,0)$ parallel to the y-axis, and the segment of this line between $y = 0$ and $y = 1$.

7. The line through $(-3,7,-4,2)$ parallel to the z-axis, and the segment of this line between $z = -4$ and $z = -8$.

8. The line through $(2,3,9,\frac{1}{2})$ parallel to the t-axis, and the segment of this line between $t = -.5$ and $t = 2.6$.

For Exercises 9–14, write the set-builder description of a taxicab path between the two given 4-space points, and specify the intersection points that connect the segments in the path.

9. $(0,0,0,0)$ and $(1,1,1,1)$

10. $(2,3,4,5)$ and $(6,7,8,9)$

11. $(7,-3,4,-1)$ and $(2,5,4,0)$

12. $(1,0,-1,0)$ and $(0,1,0,-1)$

13. $(2,7,6,-1)$ and $(5,7,6,3)$

14. $(8,9,-6,-7)$ and $(4,-2,1,0)$

15–20. Propose a space-time interpretation of each of the paths in Exercises 9–14.

21. Find or construct eight cubes of the same size and glue them together to form an unfolded tesseract, as shown in Figure 8.26. Label each of the 34 exposed square faces with the numbers 1 through 17 to indicate how these faces must be paired together when the tesseract is folded up in 4-space.

22. The tesseract described in Example 8.20 has 32 edges.

(a) Write set-builder descriptions of five of those edges. For each edge you describe, specify the two corners it connects.

(b) Write a taxicab path that connects $(0,0,0,0)$ and $(2,2,2,2)$.

(c) The points $(0,0,0,0)$ and $(2,2,2,2)$ are diagonally opposite corners of the tesseract. Match the other fourteen corners in diagonally opposite pairs.

(d) Describe a taxicab path that connects $(2,0,0,2)$ and its diagonally opposite corner.

23. This exercise refers to the cube of Figure

8.29, considered as existing in the 3-space "slice" of 4-space, for which $t = 0$.

(a) List the coordinates of the eight corners of the cube.

(b) Describe a path between $(3,1,1,0)$ and $(1,1,1,0)$ that intersects the cube wall in the xy-plane.

(c) Describe a path between $(1,1,1,0)$ and $(3,3,3,0)$ that does not intersect any of the cube walls.

24. Prove: If each edge of a tesseract is n units long, then the length of its diagonal is $2n$.

WRITING EXERCISES

1. The construction of this section is analogous to that of Section 8.4. Write a comparative outline of the two sections that displays all points of the analogy.

2. What is the etymology of the word "tesseract"? That is, what are its linguistic origin and root meaning?

3. As we have seen, one interpretation of 4-space treats time as if it were a spatial dimension, like length, width, or height. In what ways are time and these three spatial dimensions similar? In what ways are they different?

4. Write a brief description of a tesseract as you would explain it to a curious high-school sophomore. Do not use any pictures.

5. The "folding-up" construction of a tesseract from eight cubes (described in this section by Figures 8.26 and 8.27) yields a *hollow* tesseract, even though the eight cubes are solid. Write a convincing justification of this statement.

8.6   Figures in 4-Space

Now we examine some elementary 4-dimensional figures. Let us begin by taking a second look at the space-time tesseract described in Example 8.20 on page 382 — a 2-by-2-by-2-meter cube that exists for two hours. The eight points listed in the left column are the corners of the "beginning" cubic face (at 0 hours) and the points in the right column are the corners of the "ending" cubic face (at 2 hours). If we choose any particular instant of time and consider all points of the tesseract whose time coordinate is that instant, we get a 3-dimensional "cross section" of the tesseract and this cross section is itself a cube. For example, if we fix the time at exactly one hour, the cross section of the tesseract we get is the cube exactly halfway through its life span; its eight corners are:

$$(0,0,0,1) \qquad (2,2,0,1)$$
$$(2,0,0,1) \qquad (2,0,2,1)$$
$$(0,2,0,1) \qquad (0,2,2,1)$$
$$(0,0,2,1) \qquad (2,2,2,1)$$

We define a **cross section** of a figure as the set of all points of the figure that have one coordinate fixed at a particular value.[5] Then, as we have just seen, any cross section of the space-time tesseract obtained by specifying the time coordinate is a cube. But we have said that time in 4-space behaves just like a spatial coordinate. This suggests that a tesseract cross section obtained by fixing *any one* of the coordinate values should be a cube, and it is, as the following examples show.

Example 8.21   If we take all the points of the 2-by-2-by-2-meter cube of Example 8.20 with x-coordinate 1, then we get a square cross section of that cube; we have "sliced it in half" parallel to the yz-plane. Now, if we consider that 2-by-2 square slice for its entire 2-hour life span, we get a 2-by-2-by-2 cube, which is the cross section of the tesseract at $x = 1$. □

Example 8.22   The six faces of the cube in the previous example are just the square cross sections obtained by fixing the x-, y-, or z-coordinate at either 0 meters or 2 meters. The corresponding six cross sections of the tesseract are the cubes formed by these six squares "living out" the two-hour time span. □

The fact that all the tesseract cross sections are cubes might tempt you to make some bad guesses about the behavior of 4-space, so let us consider

---

[5] Actually, these cross sections are all of a particularly simple kind, in that each one is "parallel" to a lower-dimensional subspace. However, since this is the only kind we shall work with, we use the term *cross section* without any qualifying adjectives.

another example, a little more complicated, but perhaps more enlightening. To keep the mathematics as simple as possible, we shall describe our example in a somewhat artificial way; but this should only serve to make the point of the illustration clearer.

Consider a 2-by-2-by-2-meter block of ice enclosed in a porous box of the same size, and suppose the ice melts so that each edge of the block shrinks at the constant rate of 1 meter per hour. Tilt the box slightly toward one corner, so that as the ice shrinks, it always touches that corner of the box. Regard that corner as the origin of the spatial coordinate system. (See Figure 8.30.) The ice block takes two hours to melt; its shrinking existence during those two hours can be represented by a single four-dimensional space-time figure. Again, our imagination is not able to visualize that 4-space figure all at once; however, we can get an idea of it by looking at some of its cross sections, just as a builder might learn about a structure by looking at its blueprints.

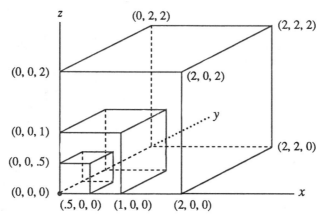

Figure 8.30   *Four cross sections of the melting ice block.*

If we specify a particular time t, the cross section we get is a cube because the ice block has shrunk uniformly during the time from 0 hours to t hours. In fact, because of our simplifying assumption about the rate of shrinkage, the cross section at a fixed time t is a cube $2 - t$ meters on a side.

Example 8.23   Figure 8.30 shows four cross sections of the melting ice block for fixed times:

When $t = 0$, the ice block is 2 meters on a side;

when $t = 1$, the ice block is 1 meter on a side;

when $t = \frac{3}{2}$, the ice block is $\frac{1}{2}$ meter on a side;

when $t = 2$, the ice block is 0 meters on a side, and hence it is just the point $(0, 0, 0, 2)$.   □

On the other hand, if we specify a particular *spatial* coordinate value, the resulting cross section is not a cube at all. For instance, if we choose $y = 0$, the cross section of the cube in Figure 8.30 is the front face, on the xz-plane. Now, as time progresses from 0 hours to 2 hours, this ice face shrinks uniformly from a 2-by-2 square to a point. Thus, the three-dimensional space-time cross section representing that melting face over time is the off-center pyramid shown in Figure 8.31(a).

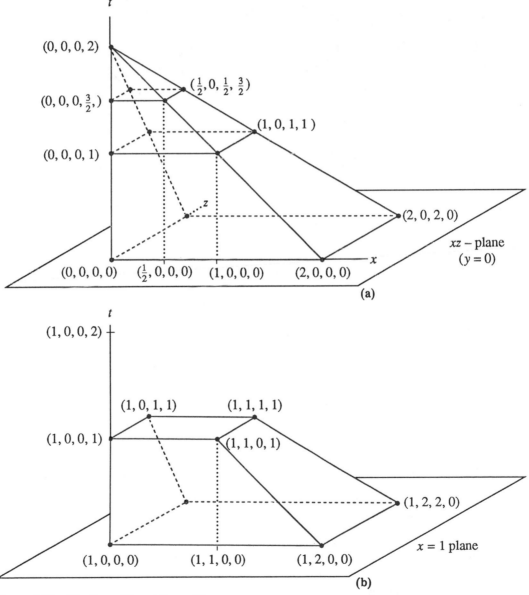

Figure 8.31   *The "y = 0" and "x = 1"
cross sections of the melt-
ing ice block.*

If we choose $x = 1$, the cross section of the cube is again a 2-by-2 square. Think of cutting the box that holds the ice block in half at $x = 1$. As time passes from 0 hours to 1 hour, the ice square at that cut in the box shrinks from 2 meters by 2 meters to 1 meter by 1 meter. But after 1 hour has elapsed, there is no ice at all at that cut in the box; hence, the space-time cross section at $x = 1$ is a truncated pyramid, as shown in Figure 8.31(b).

Another example of an elementary 4-space figure is the 4-dimensional analogue of the circle and the sphere. A circle is the set of all 2-space points that are a given distance (radius) from a particular point (center), and a sphere is the set of all 3-space points that are a given distance from a particular point. The extension to four dimensions is natural:

A **hypersphere** is the set of all 4-space points that are a given distance (**radius**) from a particular point (**center**).

To simplify our calculations, when we describe these figures using coordinates, *we shall always assume they are centered at the origin.* Thus, a circle of radius 5, for example, can be described as the set of all points (x, y) whose distance from $(0, 0)$ is 5. Using the distance formula of Section 8.3, this becomes

$$\{(x, y) \mid \sqrt{(x - 0)^2 + (y - 0)^2} = 5\}$$

which simplifies to

$$\{(x, y) \mid x^2 + y^2 = 25\}$$

In general, then, a circle of radius r is given by:

$$\{(x, y) \mid x^2 + y^2 = r^2\}$$

In 3-space, the distance formula of Section 8.4 provides an analogous description for a sphere of radius r centered at $(0, 0, 0)$:

$$\{(x, y, z) \mid x^2 + y^2 + z^2 = r^2\}$$

The analogous description for a hypersphere of radius r centered at $(0, 0, 0, 0)$ should be clear. Using the 4-space distance formula of Section 8.5, we obtain

$$\{(x, y, z, t) \mid x^2 + y^2 + z^2 + t^2 = r^2\}$$

Example 8.24   To find out whether a point in the plane is inside, outside, or on a circle, just apply the distance formula to that point and the center, then compare the result to the radius. Thus, the point $(3, 4)$ is on the circle of radius 5 centered at the origin because

$$\sqrt{(3 - 0)^2 + (4 - 0)^2} = \sqrt{9 + 16} = \sqrt{25} = 5$$

Similarly, using the easier version of the distance formula given in the definition of this circle:

- $(2, 4)$ is inside the circle because $2^2 + 4^2$ is less than 5^2;
- $(-1, 5)$ is outside the circle because $(-1)^2 + 5^2$ is greater than 5^2.   □

Example 8.25   Consider the sphere of radius 3 centered at the origin.

- $(2, 2, 1)$ is on the sphere because $2^2 + 2^2 + 1^2 = 3^2$;
- $(1, 0, 2)$ is inside the sphere because $1^2 + 0^2 + 2^2 < 3^2$;
- $(2, 2, 2)$ is outside the sphere because $2^2 + 2^2 + 2^2 > 3^2$.   □

Example 8.26   Consider the hypersphere of radius 4 centered at the origin.

- $(2, 2, 2, 2)$ is on the hypersphere because $2^2 + 2^2 + 2^2 + 2^2 = 4^2$;
- $(1, 2, 3, 1)$ is inside the hypersphere because $1^2 + 2^2 + 3^2 + 1^2 = 15$, which is less than 4^2;
- $(1, 2, 3, 2)$ is outside the hypersphere because $1^2 + 2^2 + 3^2 + 2^2 = 18$, which is greater than 4^2.   □

Although the entire hypersphere defies our visual imagination, its cross sections are easy to describe. Moreover, they provide us with another confirmation of the close analogy between 4-space and 3-space. Consider first the cross sections of a sphere in 3-space. If we slice it directly through the center, the resulting cross section is a circle of the same radius. If we slice it elsewhere, the cross section is still a circle but of smaller radius; the farther from the center we slice it, the smaller the radius of the circular cross section is. Now let us look at a hypersphere of radius 5 centered at the origin:

- The cross section at $t = 0$ is

$$\{(x, y, z, 0) \mid x^2 + y^2 + z^2 + 0^2 = 5^2\}$$

which is just a sphere of radius 5 in the 3-space formed by the x-, y-, and z-coordinate axes.

- The cross section at $t = 2$ is

$$\{(x, y, z, 2) \mid x^2 + y^2 + z^2 + 2^2 = 5^2\}$$

which is just a copy of the set

$$\{(x, y, z) \mid x^2 + y^2 + z^2 = 5^2 - 2^2\}$$

a sphere of radius $\sqrt{5^2 - 2^2} = \sqrt{21}$.

- Similarly, the cross section at $t = 4$ is

$$\{(x, y, z, 4) \mid x^2 + y^2 + z^2 = 5^2 - 4^2\}$$

a copy of the sphere

$$\{(x, y, z) \mid x^2 + y^2 + z^2 = 9\}$$

whose radius is 3.

Thus, the cross sections of the hypersphere are spheres, and the farther from the center the cross section is taken, the smaller the radius of the spherical cross section is.

This last observation about cross sections provides a way of visualizing a hypersphere in 4-dimensional space-time. A hypersphere of radius 5 in the space-time of meters and hours, for instance, is a sphere that starts its life as a point, grows like a balloon for 5 hours until it reaches its maximum radius of 5 meters, then shrinks back to a point during the next 5 hours. Moreover, the growing and shrinking of the radius is determined by

$$5^2 - t^2$$

where the time t varies from -5 hours to 5 hours. Thus, a hypersphere in space-time is a growing and shrinking sphere whose life span in hours is equal to its maximum diameter in meters!

The extension of the concept of space need not stop with four dimensions. As we have seen, the key to higher-dimensional spaces is the concept of a coordinate system. We can go beyond four dimensions to five or six or however many we want simply by using more coordinate axes. Thus,

5-space is the set of all ordered quintuples of real numbers

6-space is the set of all ordered sextuples of real numbers

and so forth. The formulas and figures we have developed in 4-space have analogues in these higher-dimensional spaces.

Spaces of large dimension have become quite useful in many areas of science. For example, because the location of a point in space can be specified by three numbers, the location of two points can be specified by six numbers. Thus, just as the movement over time of a particle in 3-space can be represented by a single figure in 4-space, so the movement of a two-particle system can be represented by a single figure in 7-space, and so forth. The geometric properties of these figures often provide insights about the dynamic properties of the systems they represent. In this way, higher-dimensional geometry is a basis for the mathematical treatment of dynamical systems in physics.

EXERCISES 8.6

1. Specify the coordinates of all corners of each of the first three cross sections given in Example 8.23.

For Exercises 2–9, describe the cross section of the melting ice block of Figure 8.30 for each given coordinate value. In your description be sure to specify the coordinates of all corners of the cross section.

2. $t = \frac{1}{2}$ hour   3. $t = 45$ minutes
4. $x = 0$ meters   5. $z = 0$ meters
6. $x = \frac{2}{3}$ meter   7. $y = 1$ meter
8. $z = 1.5$ meters   9. $y = 2$ meters

For Exercises 10–15, indicate whether the given point is *inside*, *outside*, or *on* the circle of radius 5 centered at the origin. Justify your answers.

10. $(2, 3)$ 11. $(-5, 0)$ 12. $(4, -3)$
13. $(7, -2)$ 14. $(2, \sqrt{21})$ 15. $(4, 4)$

For Exercises 16–21, indicate whether the given point is *inside*, *outside*, or *on* the sphere of radius 7 centered at the origin. Justify your answers.

16. $(5, 1, 4)$ 17. $(2, 6, 3)$
18. $(4, 4, 4)$ 19. $(5, 0, -5)$
20. $(\sqrt{13}, 6, 1)$ 21. $(-4, 5, \sqrt{8})$

For Exercises 22–27, indicate whether the given point is *inside*, *outside*, or *on* the hypersphere of radius 6 centered at the origin. Justify your answers.

22. $(1, 2, 3, 4)$ 23. $(2, 3, 4, 5)$
24. $(3, 3, 3, 3)$ 25. $(6, 0, -6, 0)$
26. $(\sqrt{10}, 4, 2, 2)$ 27. $(5, 0, 3, \sqrt{2})$

For Exercises 28–31, consider a hypersphere of radius 5 centered at the origin. Describe the cross section determined by the given coordinate value, and specify a point on it.

28. $t = 3$ 29. $x = 2$
30. $y = 1$ 31. $z = 0$

For Exercises 32–35, describe, as explicitly as you can, the 5-space analogue of the given 4-space concept.

32. the distance between two points

33. a taxicab path

34. a tesseract

35. a hypersphere

WRITING EXERCISES

1. What part of the ice-block example do you find most confusing or difficult to understand? Write three specific questions that you think might help to clarify it for you.

2. What is the difference between a solid 4-dimensional figure and a "hollow" one? Is the 4-dimensional ice-block figure described in this section solid or hollow? What about the hypersphere? When we construct a 4-space figure from a 3-space figure, what determines whether the result is solid or hollow?

3. Picking up on the idea explained in the last paragraph of this section, explain how two balls on a billiard table can be used to represent 4-space. Test your explanation by applying it to the following questions.

   (a) What is a *point* in this example?

   (b) What are the four coordinate axes?

   (c) Can you describe a line segment? A taxicab path? How?

   (d) How can you describe a tesseract or a hypersphere in this setting? (These may be difficult. You might try just "thinking with your pen" about this question, just writing down what comes to mind without worrying about paragraph structure or organization.)

8.7   Cylinders and Cones [optional]

In this section we investigate the 4-dimensional analogues of two other 3-dimensional shapes — the cylinder and the cone. As in the case of the sphere–hypersphere discussion in Section 8.6, we shall see how the interplay of geometry and algebra lets us use algebraic methods to "see" the 3-dimensional cross sections of a figure, thereby getting some sense of the shape of the 4-dimensional figure as a whole. That is, for each of these shapes, we shall do the following:

- describe the 3-space figure geometrically, and look at its 2-dimensional cross sections;
- translate these geometric descriptions into algebraic ones;
- generalize the algebraic description of the figure to 4-space by the appropriately analogous addition of a fourth coordinate;
- look at the algebraic descriptions of 3-dimensional cross sections of the 4-space figure by fixing points on different axes;
- interpret these algebraic descriptions geometrically to get a visual image of the cross sections in 3-space.

Don't let this apparently complicated menu discourage you. It is stated here merely as a guide to help you sort out the main steps in the material to come. If you refer to it as you work through this section, you should find that it useful in breaking the discussion into manageable pieces.

Cylinders

Let us begin with the **cylinder**, a hollow tube with a (very thin!) wall perpendicular to a circular base. If we place coordinate axes so that the cylinder is sitting on the xy-plane with the center of its base at the origin, as in Figure 8.32(a), then it is easy to give an algebraic description of this figure. Suppose the cylinder is h units high and the radius of the base is r units. Then the base, which is a circle of radius r centered at $(0,0,0)$, is just "copied" all the way up the cylinder; that is, we can think of the cylinder as an infinite stack of circles, from height 0 to height h (in the positive z direction). Algebraically, we have

$$(8.1) \qquad \{(x,y,z) \mid x^2 + y^2 = r^2 \text{ and } 0 \le z \le h\}$$

If the cylinder is cut horizontally (parallel to the xy-plane) at any height k between 0 and h, its cross section on the plane $z = k$, then, is just a circle

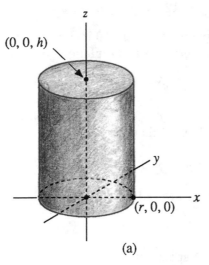

Figure 8.32 *A cylinder and some of its cross sections.*

of radius r, as in Figure 8.32(b); that is,

$$\{(x, y, k) \mid x^2 + y^2 = r^2\}$$

If, on the other hand, the cylinder is cut vertically, then its cross section is just a pair of parallel lines. Figure 8.32(b) shows two such cross sections — one on the yz-plane and one parallel to the yz-plane but over a little in the x-direction. Notice that the parallel lines get closer as the cut moves further out. Example 8.27 shows how the algebra reflects what we see in the figure.

Example 8.27 Consider a cylinder of radius 5 inches and height 12 inches, situated as in Figure 8.32. Its algebraic description is

$$\{(x, y, z) \mid x^2 + y^2 = 25 \text{ and } 0 \leq z \leq 12\}$$

At height $z = 6$, its horizontal cross section is the circle

$$\{(x, y, 6) \mid x^2 + y^2 = 25\}$$

Since the yz-plane is the plane where $x = 0$, its (vertical) cross section on this plane is

$$\{(0, y, z) \mid 0^2 + y^2 = 25 \text{ and } 0 \leq z \leq 12\}$$

But $y^2 = 25$ means that y must equal either 5 or -5, so this cross section is just

$$\{(0, y, z) \mid y = 5 \text{ and } 0 \leq z \leq 12\} \cup \{(0, y, z) \mid y = -5 \text{ and } 0 \leq z \leq 12\}$$

that is, two 12-inch line segments that are 10 inches apart on opposite sides of the z-axis. If we take the vertical on the plane $x = 3$, we get

$$\{(3, y, z) \mid 3^2 + y^2 = 25 \text{ and } 0 \leq z \leq 12\}$$

Now, $3^2 + y^2 = 25$ simplifies to $y^2 = 25 - 9 = 16$, implying that y must equal 4 or -4 in this case. Thus, this cross section is

$$\{(3, y, z) \mid y = 4 \text{ and } 0 \leq z \leq 12\} \cup \{(3, y, z) \mid y = -4 \text{ and } 0 \leq z \leq 12\}$$

two 12-inch line segments that are only 8 units apart on the plane $x = 3$.

$\square$

Now, what is a natural 4-space analogue of the cylinder? One way to proceed is to start with the 3-dimensional analogue of the circular base, which is a sphere of radius r, and extend that for some fixed "height" h in the t direction.[6] We shall call this 4-dimensional object a **hypercylinder**. If we again agree to put the origin of our coordinate system at the center of the base, this hypercylinder has an easy algebraic description:

(8.2) $$\{(x, y, z, t) \mid x^2 + y^2 + z^2 = r^2 \text{ and } 0 \leq t \leq h\}$$

To examine 3-dimensional cross sections of this object, we can:

- fix one of the four coordinates at 0 or at some other number;
- work out the resulting algebraic expression; then
- interpret this expression as a geometric figure in 3-space.

Rather than carrying out these steps for the general case of a hypercylinder of base radius r and height h, let us look at the following specific, but typical, example, which is the direct analogue of Example 8.27 — a hypercylinder of base radius 5 inches and height 12 inches.

Example 8.28   Consider the hypercylinder

$$\{(x, y, z, t) \mid x^2 + y^2 + z^2 = 25 \text{ and } 0 \leq t \leq 12\}$$

---

[6]This is not the only way to proceed. Another, slightly more cumbersome approach is to consider a figure with a *cylindrical* base that extends for some fixed length in the t direction. In time-space this would represent a 3-dimensional cylinder existing for some length of time. We shall not explore this alternative; you might like to consider it on your own.

As you might expect from the analogy used in constructing the hypercylinder, its cross section for any fixed t — 6, for instance — is just a sphere of radius 5 inches in the appropriate — $t = 6$ — hyperplane:

$$\{(x, y, z, 6) \mid x^2 + y^2 + z^2 = 25\}$$

The analogous cross section obtained from fixing one of the other variables is not so obvious geometrically, but the algebra is easy. If we let $x = 0$, we get

(8.3) $$\{(0, y, z, t) \mid 0^2 + y^2 + z^2 = 25 \text{ and } 0 \leq t \leq 12\}$$

The equation in this set description, which reduces immediately to $y^2 + z^2 = 25$, is a circle of radius 5 inches in the yz-plane. Therefore, the cross section $x = 0$, described by (8.3), is a (3-dimensional) cylinder of base radius 5 inches and height 12 inches in yzt-space! Continuing the analogy with Example 8.27, let us see what happens when we fix x at some other number.

For $x = 3$, we get

$$\{(3, y, z, t) \mid 3^2 + y^2 + z^2 = 25 \text{ and } 0 \leq t \leq 12\}$$

But $3^2 + y^2 + z^2 = 25$ reduces to

$$y^2 + z^2 = 25 - 9 = 16$$

which is a circle of radius 4 in the yz-plane. Therefore, the cross section $x = 3$ is a cylinder of base radius 4 inches and height 12 inches. Similarly, if we choose $x = 4$, we get

$$\{(4, y, z, t) \mid 4^2 + y^2 + z^2 = 25 \text{ and } 0 \leq t \leq 12\}$$

and, simplifying the equation to $y^2 + z^2 = 25 - 16 = 9$, we can see that the cross section $x = 4$ is again a cylinder of height 12 inches; but this one has a base radius of only 3 inches. Thus, the cross sections for values of x chosen farther and farther away from 0 are cylinders that get thinner and thinner.

Here is an illustration of how the algebraic formulas can be used to get more precise information than our visual intuition provides. Observing that the thinness of a cylindrical cross section depends on its distance from the origin, we can formulate the exact radius of the cylindrical cross section in terms of its distance from the origin. For instance, if we think of x as being some fixed number k, then the distance of the cross section $x = k$ from the origin is just $|k|$, and the formula for the radius of this cylinder is

$$k^2 + y^2 + z^2 = 25$$

or

$$y^2 + z^2 = 25 - k^2$$

Thus, the radius of the circular base of this cylinder is $\sqrt{25 - k^2}$, a number that shrinks to 0 as k grows from 0 to 5.   □

The coordinates x, y, and z in the algebraic description of the hyper-cylinder are interchangeable in some sense; that is, each of these three co-ordinates is squared and they are all added together in the same equation. Thus, cross sections obtained by picking a particular value for y or z will look like those we got by picking a value for x, as in Example 8.28. Thus, a hypercylinder "looks" something like a bunch of spheres stacked up in the t direction, and cross sections parallel to the t-axis in any of the other direc-tions are cylinders that shrink in diameter as we get farther away from the origin! In case you think that you are beginning to get a picture of what a hypercylinder "really" looks like in 4-space, let us offer one more observation about cylinders and hypercylinders to test your intuition.

Unlike a sphere, a cylinder does *not* separate 3-space into two separate pieces, an inside and an outside. Its hollow "interior" is actually just part of all the rest of 3-space. As you can see from Figure 8.33, it is possible to put a taxicab path right through the middle of the cylinder from any point P not on the cylinder, regardless of where P is, without intersecting the cylinder itself. That's not surprising; a piece of pipe wouldn't be very useful if water couldn't get through it! The surprise is that, despite the fact that a hypercylinder has a spherical base, it is hollow in exactly the same way in 4-space as a cylinder is in 3-space! That is, *a hypercylinder does* not *separate 4-space into two separate pieces*. It is possible to put a path right through the middle of the hypercylinder from any point P not on the hypercylinder, regardless of where P is, without intersecting the hypercylinder itself. To verify this (perhaps surprising) claim, let us look at some specific taxicab paths, using the cylinder and the hypercylinder of Examples 8.27 and 8.28 as typical figures.

Referring first to Example 8.27, consider a point P somewhere fairly far away from the cylinder — at $(10, 0, 14)$, for instance. Since the "core" of the cylinder centers around the z-axis, a taxicab path passing through the cylinder is easy to describe. Here is one possible path:

> Go down in the z direction to a level below the base of the cylinder $(z = -1)$; move over to the z-axis; travel up the z-axis to the level of P, beyond the top of the cylinder $(z = 14)$; move back out to point P.

The point-set description of this path is

$$\{(10, 0, z) \mid -1 \le z \le 14\} \ \cup \ \{(x, 0, -1) \mid 0 \le x \le 10\}$$
$$\cup \ \{(0, 0, z) \mid -1 \le z \le 14\} \ \cup \ \{(x, 0, 14) \mid 0 \le x \le 10\}$$

This path is pictured in Figure 8.33. Of course, if P were less conveniently placed, the path would be a little more complicated, but the main idea would be exactly the same: go down to a level below the base of the cylinder, travel up the z-axis to a point above the top, then go back to your starting point. Notice that the point-set descriptions of this path and the cylinder make it quite clear that the path doesn't intersect the cylinder anywhere:

- All points of the cylinder have $0 \leq z \leq 12$;
- No point of the cylinder has both x and y equal to 0 (because $x^2 + y^2 = 25$);
- No point of the cylinder has $x = 10$ (because $x^2 + y^2 = 25$, so $-5 \leq x \leq 5$);
- The only points of the path for which $0 \leq z \leq 12$ either have both x and y equal to 0 or have $x = 10$.

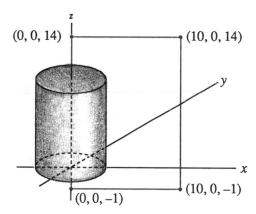

Figure 8.33 *A "taxicab loop" through a cylinder.*

Now consider the point $Q = (10, 0, 0, 14)$ in 4-space, which is fairly far away from the hypercylinder in Example 8.28. We can construct a taxicab "loop" from Q through the hypercylinder and back to Q, just as we did for the cylinder in 3-space:

Go "down" in the t direction to a level below the base of the cylinder ($t = -1$); move over to the t-axis; travel "up" the t-axis to the level of Q, beyond the "top" of the cylinder ($t = 14$); move back out to point Q.

The point-set description of this path is

$$\{(10,0,0,t) \mid -1 \leq t \leq 14\} \cup \{(x,0,0,-1) \mid 0 \leq x \leq 10\}$$
$$\cup \{(0,0,0,t) \mid -1 \leq t \leq 14\} \cup \{(x,0,0,14) \mid 0 \leq x \leq 10\}$$

Again, the point-set descriptions of this path and the hypercylinder make it quite clear that the path doesn't intersect the hypercylinder anywhere:

- All points of the cylinder have $0 \leq t \leq 12$;
- No point of the cylinder has its x-, y-, and z-coordinates *all* equal to 0 (because $x^2 + y^2 + z^2 = 25$);
- No point of the cylinder has $x = 10$ (because $x^2 + y^2 + z^2 = 25$, so $-5 \leq x \leq 5$);
- The only points of the path for which $0 \leq t \leq 12$ either have x, y, and z *all* equal to 0 or have $x = 10$.

Thus, a hypercylinder does not enclose any part of 4-space. ("Hyperwater" can flow through a hypercylindrical pipe!)

Cones[7]

In order to give a simple coordinate description of a cone in 3-space, we shall think of it as "upside down" (relative to its use for holding ice cream), so that it has a circular base and tapers upward to a point. (See Figure 8.34(a).) We can think of a **cone**, in fact, as an infinite stack of circles whose radii shrink uniformly from the radius of the base down to 0, the radius of the "circle" that is the top point of the cone. That is, the cross section of this cone parallel to its base at any height from bottom to top is just a circle, and the closer to the top it is, the smaller the circle is.

The algebra in the coordinate description of such a cone is a little more complicated that that for a cylinder, but this stack-of-circles viewpoint makes it a little easier to understand. Let r stand for the base radius of the cone and h stand for its height. To get the radii of the circles in the stack to shrink from r to 0 as z goes from 0 to h, we ought to subtract more and more from the initial radius r as the height z increases — but how much more? Let us look first at a typical example:

Example 8.29
(a)   Consider a cone with base radius 5 inches and height 10 inches, situated as in Figure 8.34. To find the radius of its circular cross section at a height

---

[7]Parts of this discussion presuppose some familiarity with slopes and equations of straight lines.

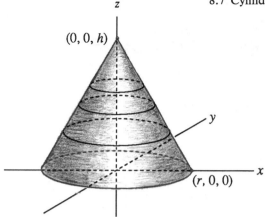

Figure 8.34   A cone in 3-space.

z between 0 and 10, first observe that the cone is exactly twice as high as the radius of its base. Thus, we would expect to shrink the base radius by half as much as the height of the cross section. This means that the circular cross section at height z should have radius $5 - \frac{1}{2}z$, so its equation is

$$x^2 + y^2 = (5 - \tfrac{1}{2}z)^2$$

Thus, the point-set description of this cone is

$$\{(x, y, z) \mid x^2 + y^2 = (5 - \tfrac{1}{2}z)^2 \text{ and } 0 \leq z \leq 10\}$$

The cross section at $z = 4$ is

$$\{(x, y, 4) \mid x^2 + y^2 = (5 - \tfrac{1}{2} \cdot 4)^2\}$$

a circle in the plane $z = 4$ with radius 3 inches. Similarly, the cross section $z = 7$ is

$$\{(x, y, 7) \mid x^2 + y^2 = (5 - \tfrac{1}{2} \cdot 7)^2\}$$

a circle with radius $1\frac{1}{2}$ inches in the plane $z = 7$.   □

As Example 8.29a illustrates, the algebraic description of a cone depends on the "shrinkage ratio" for the circles. The general form of that ratio, in terms of base radius r and height h, is probably not obvious from this one example, but it is not too hard to figure out with some experimentation and/or a little algebra. For every unit we go up from the base, we want to subtract $\frac{r}{h}$ units; that is, we want the radius at height z to be $r - \frac{r}{h}z$.[8] At the base, $z = 0$, so that

$$r - \tfrac{r}{h}z = r - \tfrac{r}{h}0 = r - 0 = r$$

while, at the top, $z = h$, so that

$$r - \tfrac{r}{h}z = r - \tfrac{r}{h}h = r - r = 0$$

as required. The general algebraic description of such a cone, then, is

(8.4) $\{(x, y, z) \mid x^2 + y^2 = (r - \tfrac{r}{h}z)^2 \text{ and } 0 \le z \le h\}$

Cross sections of the cone by planes other than horizontal ones have been of particular interest ever since early Greek times. The so-called **conic sections** were first thoroughly studied by Apollonius in the 3rd Century B.C. They are the four essentially different kinds of curves that result from the intersection of a plane and a cone — the *circle*, the *ellipse*, the *parabola*, and the *hyperbola*. Each of these figures can be represented by a quadratic equation[9] in two variables! Since we have been considering only cross sections that are parallel to coordinate axes, our treatment just covers two of the four kinds of conic sections. One, the circle, we have already seen; the other results from cross sections perpendicular to the base of the cone, as Example 8.29b shows.

Example 8.29(b)

The cross section $x = 0$ of the cone described in Example 8.29a is an inverted V-shaped figure on the yz-plane, as shown in Figure 8.35. Algebraically, it is found by setting $x = 0$ in the defining equation for the cone:

$$0^2 + y^2 = (5 - \tfrac{1}{2}z)^2$$
$$y = \pm(5 - \tfrac{1}{2}z)$$

The "$\pm$" sign indicates that there are really two equations represented here:

$$y = -\tfrac{1}{2}z + 5 \quad \text{and} \quad y = \tfrac{1}{2}z + 5$$

Each of these equations represents a straight line in the yz-plane, and they cross when $y = 0$ and $z = 10$ (at the tip of the cone). The vertical cross

---

[8]This can also be found by observing that slope of the line through $(0, 0, h)$ and $(r, 0, 0)$ in the xz-plane is $\frac{-h}{r}$ and solving the equation $z = \frac{-h}{r}x + h$ for x. The radius at height z is the x-value of the point $(x, 0, z)$ on the cone.

[9]A *quadratic equation* is an equation of degree 2.

sections away from the z-axis are somewhat different, however. At $x = 3$, for example, we get

$$
\begin{aligned}
3^2 + y^2 &= \left(5 - \tfrac{1}{2}z\right)^2 \\
y^2 &= \left(25 - 5z + \tfrac{1}{4}z^2\right) - 9 \\
y^2 - \tfrac{1}{4}z^2 &= -5z + 16
\end{aligned}
$$

the equation of the hyperbola that is partly shown in Figure 8.35.   □

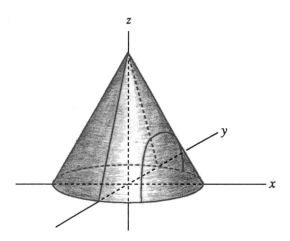

Figure 8.35   *Vertical cross sections of a cone.*

Generalizing the idea of *cone* from 3-space to 4-space using the same approach as we did for the cylinder, we think of a **hypercone** as an infinite stack of spheres whose radii shrink uniformly from the radius of the base down to 0.[10] If we again put the origin of our coordinate system at the center of the base, the algebraic description of the cone, given by (8.4), generalizes easily to 4-space:

$$(8.5) \qquad \{(x, y, z, t) \mid x^2 + y^2 + z^2 = (r - \tfrac{r}{h}t)^2 \text{ and } 0 \le t \le h\}$$

For each value of t between 0 and h, the 3-dimensional cross section is a sphere of radius $r - \tfrac{r}{h}t$; this radius value decreases from r to 0 as t increases from 0 to h.

Example 8.30   The point-set description of a hypercone with base radius 5 inches and height

[10] Again, this is not the only way to proceed. Another approach is to consider a figure with a *conical* base that extends for some fixed length in the t direction. In time-space, this would represent a 3-dimensional cone existing for some length of time. We shall not explore this alternative, but you might like to do so.

10 inches is

$$\{(x, y, z, t) \mid x^2 + y^2 + z^2 = (5 - \tfrac{1}{2}t)^2 \text{ and } 0 \leq t \leq 10\}$$

Its cross section at $t = 4$ is

$$\{(x, y, z, 4) \mid x^2 + y^2 + z^2 = (5 - \tfrac{1}{2} \cdot 4)^2\}$$

a sphere with radius 3 inches in the space $t = 4$. Similarly, the cross section at $z = 7$ is

$$\{(x, y, z, 7) \mid x^2 + y^2 + z^2 = (5 - \tfrac{1}{2} \cdot 7)^2\}$$

a sphere with radius $\tfrac{3}{2}$ inches in the space $t = 7$. As in the 3-dimensional case, cross sections perpendicular to the base are less obvious. For $x = 0$, we get

$$\{0, y, z, t) \mid y^2 + z^2 = (5 - \tfrac{1}{2}t)^2 \text{ and } 0 \leq t \leq 10\}$$

Comparing this with (8.4), it is not hard to see that this figure is a 3-dimensional cone in yzt-space! For $x = 3$, however, we get

$$\{(3, y, z, t) \mid 3^2 + y^2 + z^2 = (5 - \tfrac{1}{2}t)^2 \text{ and } 0 \leq t \leq 10\}$$

The defining equation here is not something we have seen before, but in some ways it is exactly the analogous result you might expect. Just as the vertical cross sections of the (3-dimensional) cone change from an inverted "V" at $x = 0$ to a hyperbola at $x = 3$, the corresponding cross sections of the hypercone change from a cone at $x = 0$ to an inverted cup with a hyperbolically curved top at $x = 3$!   □

EXERCISES 8.7

1. Give the coordinates of two points that are on the cylinder of Example 8.27 and have no coordinate equal to zero.

2. Give the coordinates of two points that are on the hypercylinder of Example 8.28 and have no coordinate equal to zero.

3. Give the coordinates of two points that are on the cone of Example 8.29 and have no coordinate equal to zero.

4. Give the coordinates of two points that are on the hypercone of Example 8.30 and have no coordinate equal to zero.

For Exercises 5–8, write a point-set description of each figure.

5. A cylinder with base radius 3 and height 7.

6. A hypercylinder with base radius 12 and height 6.

7. A cone with base radius 4 and height 20.

8. A hypercone with base radius 7 and height 15.

In Exercises 9–20, fill in the coordinate blank so

that the resulting point is on the figure described in the example noted.

9. $(3, \underline{\quad}, 2)$ on the cylinder of Example 8.27.

10. $(\underline{\quad}, 1, 9)$ on the cylinder of Example 8.27.

11. $(4, \underline{\quad}, 0, 3)$ on the hypercylinder of Example 8.28.

12. $(1, 2, \underline{\quad}, 4)$ on the hypercylinder of Example 8.28.

13. $(1, \underline{\quad}, 8)$ on the cone of Example 8.29.

14. $(\underline{\quad}, 1, 6)$ on the cone of Example 8.29.

15. $(3, 4, \underline{\quad})$ on the cone of Example 8.29.

16. $(0, 4, \underline{\quad})$ on the cone of Example 8.29.

17. $(0, 0, 0, \underline{\quad})$ on the hypercone of Example 8.30.

18. $(2, 1, 2, \underline{\quad})$ on the hypercone of Example 8.30.

19. $(3, \underline{\quad}, \sqrt{3}, 2)$ on the hypercone of Example 8.30.

20. $(\underline{\quad}, 1, 1, 6)$ on the hypercone of Example 8.30.

Exercises 21–24 refer to Example 8.27. Describe the cross section of this cylinder for each value.

21. $z = 1$ **22.** $y = 2$

23. $x = -4$ **24.** $y = 5$

Exercises 25–32 refer to Example 8.28. Describe the cross section of this hypercylinder for each value.

25. $t = 3$ **26.** $t = 12$

27. $y = 0$ **28.** $z = 0$

29. $x = 1$ **30.** $x = 2$

31. $x = 5$ **32.** $x = 6$

Exercises 33–36 refer to Example 8.29a. Write an equation for the circular cross section of this cone determined by each value.

33. $z = 0$ **34.** $z = 5$

35. $z = 8$ **36.** $z = 10$

For Exercises 37–40, write the point-set descriptions of *two* cross sections of the given figure that contain the specified point.

37. $(-3, 4, 9)$ on the cylinder of Example 8.27.

38. $(4, \sqrt{5}, 2, 7)$ on the hypercylinder of Example 8.28.

39. $(3, \sqrt{7}, 2)$ on the cone of Example 8.29.

40. $(1, 2, -2, 4)$ on the hypercone of Example 8.30.

41. This exercise refers Example 8.27. Write the point-set description of a taxicab path that starts at $(7, 8, 9)$, passes through the "inside" of the cylinder without touching the cylinder itself or the z-axis, and ends up back at $(7, 8, 9)$.

42. This exercise refers to Example 8.28. Write the point-set description of a taxicab path that starts at $(6, 7, 8, 9)$, passes through the "inside" of the hypercylinder without touching the cylinder itself or the t-axis, and ends up back at $(6, 7, 8, 9)$.

WRITING EXERCISES

1. Explain how a cylinder in 3-space is the appropriate analogue of two parallel line segments in a plane. (*Hint*: What is a "circle" in 1-space?)

2. Extending the analogy used in this section by one more dimension, describe a "hyperhypercylinder" and a "hyperhypercone."

3. Use the point-set descriptions of a cone and a hypercone to describe how a cone-shaped cup in 4-space can hold "hyperwater" without leaking.

LINK: 8.8

4-Space in Fiction and in Art

Cubism. An awareness of the notion of four-dimensional space in the first decade of the 20th century in Paris led to this important movement in art.

About 100 years ago, a Shakespearian scholar and theologian named Edwin Abbott Abbott (whose initials might be abbreviated EA^2) wrote a short novel of mathematical fiction entitled *Flatland*. Its purported narrator, A. Square, is an inhabitant of a 2-dimensional world visited by a 3-dimensional being, a Sphere. The Sphere tries to explain the existence of a third independent direction, "Upward, not Northward," with little initial success. The Sphere can only show himself to the Square as a circle, his cross section formed by intersecting with the planar world of Flatland. A. Square can only observe the Sphere's third dimension as growth of the circle in time, as the Sphere slowly descends through the plane. The Sphere finally convinces A. Square of the reality of the third dimension by "lifting" him into it. As the Square ponders this new world taught to him first by analogy, he is led to a startling conclusion:

> ...just as there was close at hand, and touching my frame, the land of Three Dimensions, though I, blind senseless wretch, had no power to touch it, no eye in my interior to discern it, so of a surety there is a Fourth Dimension....[11]

Later in the narrative, the skepticism of the Sphere about a space of more dimensions than his own is at once an insightful comment on human nature and an invitation to the further exploration of the analogies established earlier in the book.

As fiction, *Flatland* is light and enjoyable; as science, it was prophetic in that it preceded Einstein's relativity theory and the resulting interest in space-time by some thirty years or so; as pedagogy, *Flatland* remains to this day one of the clearest, simplest explanations of the idea of 4-space that has been written. But the very fact that *Flatland* was written deliberately to explain 4-space makes it a somewhat artificial example of 4-space in fiction. Our next example is more in keeping with the spirit of this section. It is a novel in which 4-space occurs as a natural part of the story, not as a vehicle for teaching the reader about mathematics.

Robert A. Heinlein is one of the acknowledged masters of 20th-century science fiction, and one of his best-known novels is *Stranger in a Strange Land*, first published in 1961. The stranger in the story is Valentine Michael Smith, a human raised as a Martian. Among his many unusual abilities, Smith had a peculiar way of disposing of "wrong things":

> He stepped toward Berquist; the gun swung to cover him. He reached out — and Berquist was no longer there.[12]

---

[11] Abbott, Edwin A.: *Flatland*, page 94 of [1] in the list *For Further Reading* at the end of this chapter.

[12] Heinlein, Robert A. *Stranger in a Strange Land*. New York: G. P. Putnam's Sons, 1961. (New York: Berkley Books, 1981, p. 67.) ©1961 Robert A. Heinlein; used by permission.

In an attempt to understand this strange power, his friends had him demonstrate it while they photographed the process from two different angles (on solid-sight-sound 4 mm. film, projected in a 3-dimensional tank.) An empty brandy case was thrown into the air, and, in the words of one witness:

> "The box did not simply vanish. The process lasted some fraction of a second. From where I am sitting it appeared to shrink, as if it were disappearing into the distance. But it did not go outside the room; I could see it up to the instant it disappeared."[13]

> In slow motion, the film from the first camera showed the box shrinking, smaller and smaller, until it was no longer there.... When they ran the second film, Duke cursed.

> "Something fouled the second camera, too.... It was shooting from the side so the box should have gone out of the frame to one side. Instead it went straight away from us again. You saw it."

> "Yes," agreed Jubal. "Straight away from us."

> "But it can't — not from both angles."

> "What do you mean, 'it can't'? It did.... Duke, the cameras are okay. What is ninety degrees from everything else?"

> "I'm no good at riddles."

> "It's not a riddle. I could refer you to Mr. A. Square from Flatland, but I'll answer it. What is perpendicular to everything else? Answer: two bodies, one pistol, and an empty case."[14]

As the excerpts quoted above suggest, both *Flatland* and *Stranger in a Strange Land* treat the fourth dimension primarily as another spatial direction. Time as the fourth dimension has also been a popular theme in the fiction of the last hundred years. One of the best-known early examples of this is H. G. Wells' *The Time Machine*, published in 1895.[15] Early in the book, Wells' "Time Traveller" makes the case for the credibility of time travel by explicitly citing the similarity between time and the spatial dimensions, saying (in part):

> There are really four dimensions, three which we call the three planes of Space, and a fourth, Time. There is, however, a tendency to draw an unreal distinction between the former three dimensions and the latter, because it happens that our conscious-

---

[13]ibid., p. 111.

[14]ibid., pp. 124-125.

[15]Wells, H. G. *The Time Machine: An Invention.* London: W. Heinemann, 1895.

ness moves intermittently in one direction along the latter from the beginning to the end of our lives.

The fourth dimension became a popular topic in European and American fiction late in the 19th century. A comprehensive catalog of 19th- and 20th-century fiction in which it appears as a theme would run to many pages and would include the names of many well-known authors. We list here just a few of the most prominent early writers and their works to provide some sense of how pervasive this topic has been. The fourth dimension is a major theme in H. G. Wells' *The Wonderful Visit* (1895) and *The Invisible Man* (1897), in Oscar Wilde's *The Canterville Ghost* (1891), in George Macdonald's *Lilith* (1895), and in Joseph Conrad and Ford Madox Hueffer's *The Inheritors* (1901). It is mentioned in Fyodor Dostoevsky's *The Brothers Karamazov* (1879), in Marcel Proust's *Duc Sté de chez Swann* (1913), in P. G. Wodehouse's *The Amazing Hat Mystery* (1922), and (at least implicitly) in Gertrude Stein's writings in the 1910s.

The late 19th-century surge of popular interest in the fourth dimension can be traced back to several major mathematical events that had occurred in the middle of the century. In the 1840s, Arthur Cayley in England and Hermann Grassmann in Germany published major works describing the theory of n-dimensional space. At the same time, Bernhard Riemann was developing a type of non-Euclidean geometry[16]; a decade later (in 1854), he gave a landmark speech describing this geometry for a space of n dimensions. From the next several decades of mathematical research and the continuing philosophical debate engendered by the conflict between the theory of non-Euclidean geometry and the philosophy of Immanuel Kant, there gradually evolved increasingly concrete representations of the idea of dimensions higher than three. These representations caught the popular interest, aided in part by the deliberate efforts of mathematicians such as Henri Poincaré in France and James Joseph Sylvester in England, who sought to make the mathematical theories more intelligible to interested nonmathematicians.

By the early 1900s, "the fourth dimension" was discussed widely in the popular press. In the United States, for example, there were frequent articles in magazines such as *Science*, *Harper's Weekly*, and *Current Literature*. In 1909, *Scientific American* offered a $500 prize for "the best popular explanation of the Fourth Dimension." It is worth noting that all of the twenty or so essays published in this contest described the fourth dimension in *spatial* terms. Within the following decade, the announcement of Einstein's Theory of Relativity abruptly changed the interpretation of the fourth dimension to *time* (and led to another *Scientific American* essay contest).

The last quarter of the 19th century also saw the emergence of a popular

---

[16]The non-Euclidean geometries are described in Section 3.6.

philosophy based on the acceptance of the reality of four spatial dimensions. The most influential early figures in this movement were J. C. F. Zöllner in England and Charles Howard Hinton in England and then in America. In the early 20th century, P. D. Ouspensky in Russia and Claude Bragdon in the United States were widely read proponents of this philosophy. Zöllner was also an active supporter of Henry Slade, an American medium who was a controversial figure in London in the late 1870s. The publicity surrounding this link between spiritualism and the fourth dimension added a mysterious, mystical flavor to the mathematics of 4-space.

In France, the popular writings of Henri Poincaré, a highly respected mathematician and philosopher, brought an awareness of four-dimensional space to Paris in the first decade of the 20th century. It was here that Cubist painters became intrigued by "new measures of space which, in the language of the modern studios, are designated by the term *fourth dimension*."[17] They were working at a time when H. G. Wells' novels were enjoying widespread popularity, as were the 4-space science-fiction writings of French authors (especially Gaston de Pawlowski and Alfred Jarry). Several of the Cubist artists in Paris studied the writings of Poincaré and other mathematicians on the geometry of four-dimensional space. Maurice Princet, a friend of the artist Jean Metzinger and an actuary with an interest in modern painting, often joined in their discussions and helped work through the connections between geometry and Cubist art.

Perhaps the most famous of the Cubist artists involved in those Paris discussions in the 1910s was Marcel Duchamp. During the years 1912-1914, Duchamp began working on *The Bride Stripped Bare By Her Bachelors, Even* (*The Large Glass*). The notes he made explicitly identify aspects of the mathematical writings of Poincaré and others in his design of "The Bride...as if it were the projection of a four-dimensional object."[18] Other Duchamp works depicting aspects of four-dimensionality are *Portrait of Chess Players* (1911) and *Nude Descending a Staircase, No. 2* (1912). The latter painting is noteworthy because it is one of the few Cubist works that views the fourth dimension as time. (All three of these works may be seen at the Philadelphia Museum of Art.)

Cubist art arrived in the United States with the opening of the 1913 Armory Show in New York City. To say it got mixed reviews would be to err on the charitable side of accuracy. Typical of the reaction in the popular press was a page-one story in the March 20, 1913, *Chicago Record-Herald*:

---

[17]Apollinaire, Guillaume. "La Peinture nouvelle: Notes d'art." *Les Soirées de Paris*, no. 3 (Apr. 1912), p. 90.

[18]As quoted in Henderson, *The Fourth Dimension...*, [5], from Cabanne, Pierre. *Dialogues with Marcel Duchamp* (1967) (trans. Ron Padgett). New York: Viking Press, 1971.

CUBIST ART IS HERE/AS CLEAR AS MUD...

CUBIST ART — ...a 'woosy' attempt to express the fourth dimension.

FUTURIST ART — Same as the former, only more so, with primeval instincts thrown in; Cubism carried to the extreme or fifth dimension.

Negative "popular" reaction notwithstanding, the Cubist art in the show was clearly recognized as related somehow to four-dimensional space. This relationship covered a wide range of work, however, including not only Duchamp's analytical *Nude Descending A Staircase*, but also the completely abstract work of Francis Picabia, who used the Cubists' appeal to 4-space to justify the total rejection of perspective. In this regard, Picabia is more representative than Duchamp of the modern artist's view of the fourth dimension. In the words of one art historian:

> ...the fourth dimension was primarily a symbol of liberation for artists....Specifically, belief in a fourth dimension encouraged artists to depart from visual reality and to reject completely the one-point perspective system that for centuries had portrayed the world as three-dimensional.[19]

The 1913 Armory Show provided the impetus for American artists to explore the theory of four-dimensional space. Both Picabia and Duchamp were key figures in the development of American Cubism. However, the popularization of Einstein's theory of relativity a few years later shifted the focus of interest in the fourth dimension away from the spatial interpretation of the early Cubists to a time-space interpretation. That, along with an increasing tendency to reject perspective of any kind, led, by about 1930, to a fading of interest in 4-space among most artistic movements.

The one notable exception to this trend was the Surrealist movement. The Surrealists adopted the spatial interpretation of four-dimensional space, along with many of its mystical overtones. One of the most striking examples of 4-space imagery in Surrealist art hangs in the Metropolitan Museum of Art in New York City. It is a painting titled *The Crucifixion*, by Salvador Dali. It was painted in 1954, a period during which Dali was deeply involved with both mathematical and religious themes, and he originally titled it *Corpus Hypercubus (Hypercubic Body)*. The "cross" in the scene is, in fact, a hypercube, or tesseract, unfolded in three dimensions. Its "shadow" on the planar surface below is an unfolded cube, and a checkerboard pattern of squares stretches out toward the horizon.

---

[19]Henderson, *The Fourth Dimension...*, [5], p. 340.

As we approach the present, there are signs of recurring artistic interest in the visual representation of 4-space. Some of this interest has been sparked by the development of powerful techniques in computer graphics, techniques that represent accurately on a screen the various cross sections and projections of four-dimensional objects. Other interest stems from a more speculative urge, similar to the kind of curiosity that drives a research scientist or mathematician. In the words of one modern artist:

> Artists who are interested in four dimensional space... are motivated by a desire to complete our subjective experience by inventing new æsthetic and conceptual capabilities.... Our reading of the history of culture has shown us that in the development of new metaphors for space artists, physicists, and mathematicians are usually in step.[20]

The foregoing examples by no means exhaust the occurrences of 4-space in fiction and art. In fact, they are intended merely as illustrations to whet your appetite for finding more examples on your own. Now that you have had a glimpse of the world of four dimensions, you should have little trouble finding other instances of it in the literature you read and the art you see.

Acknowledgement
Much of the material for this section was drawn from Linda Dalrymple Henderson's superb book, *The Fourth Dimension and Non-Euclidean Geometry in Modern Art*, [5] in the list *For Further Reading* at the end of the chapter. If you are interested in learning more about the connection between modern geometry and modern art, this is a richly detailed, very readable source.

Topics for Papers – Chapter 8

1. In this chapter we have explored some 4-dimensional figures that are natural analogues of 3-dimensional figures — the tesseract, hypersphere, hypercylinder, and hypercone (analogues of the cube, sphere, cylinder, and cone, respectively). These figures have been described in several different ways, starting with an extension of their geometric construction in 3-space, then by considering the fourth dimension as time, then by looking at cross sections, etc.

---

[20]Robbin, Tony. "The New Art of 4-Dimensional Space: Spatial Complexity in Recent New York Work." *Artscribe* (London), no. 9: 20, 1977.

This paper is to be a thorough description of a 4-dimensional figure of your choosing. You may choose to describe the 4-dimensional analogue of any one of the following figures —

rectangular box   cylinder   cone   half-cylinder ("quonset hut")

    hemisphere   prism   pyramid   torus ("doughnut")

— or you may choose something else (at your own risk. See your instructor if you have any doubts about your choice). Your paper should include the following ingredients:

- A preliminary note indicating the typical readers you have in mind and the mathematical background they would need.
- A general geometric description of your 3-dimensional figure and also a coordinate description of a particular example of it;
- A description of the 4-dimensional analogue of your figure, both in general geometric terms (how to build or visualize it) and by using specific coordinates in 4-space;
- A time-space interpretation of your 4-dimensional figure;
- Some discussion of its 3-dimensional cross sections;
- Anything else you find interesting about your 4-dimensional object;
- Any problem-solving strategies you used in attacking this project.

A word of advice: As in many exhibition sports, originality and degree of difficulty undoubtedly will be factors considered in grading your performance. You might take this into account in choosing your figure and in deciding on your overall approach to the topic. (In particular, if you choose a figure that is described in detail in the chapter, you must do more with it than just repeat what is in the book.) If you are whimsically inclined, an allegorical or fictional approach to this topic would be appropriate, *provided that it contains substantial mathematical ingredients.*

2. Any 2-dimensional figure can be generalized to 3 dimensions in two standard ways: it can be "pushed" perpendicularly in the third coordinate direction or it can be shrunk to a point located somewhere off the original plane. A circular region that is pushed perpendicularly is a cylinder; one that is shrunk to a point (directly "above" or "below" its center) is a cone. A square that is pushed perpendicularly is a cube or a rectangular box (depending on how far it is pushed); one that is shrunk to a point (directly above its center) is a pyramid. In addition, the way some planar figures are constructed can be generalized to 3-space, as we have seen in this chapter. The circle and square constructions generalize to yield a sphere and a cube, respectively.

These three ways of generalizing figures can be extended from 3-space to 4-space. A solid cube "pushed" perpendicularly in the fourth coordinate direction yields a solid tesseract; one that is shrunk to a point becomes a figure somewhat like the ice block in Section 8.6; the figure obtained by generalizing the construction of the cube (as in Section 8.5) is a hollow tesseract. A sphere pushed perpendicularly in the fourth coordinate direction yields a hypercylinder; one that is shrunk to a point becomes a hypercone; the generalization of the construction of the sphere yields a hypersphere.

Describe the three 3-space figures that are obtained when each of these three generalization methods is applied to *one* of the following planar figures:

- a right triangular region;
- an equilateral triangular region;
- a non-square rectangular region;
- a half-disk.

Then generalize from 3-space to 4-space using these three methods on *each* of the three 3-dimensional figures you obtained in the first part. (At this point, you should have nine 4-dimensional figures, but they may not all be different.) Compare and contrast these nine figures using whatever techniques from this chapter you find useful for clarifying your ideas — algebraic and/or geometric descriptions, time-space, cross sections, etc. Consider the reader of your paper to be an anonymous student in your class — someone who knows the material of Chapter 8 fairly well, but does not know anything about your particular topic. Introduce your topic clearly and take your reader step by step through an orderly development of your ideas to your desired conclusion.

For Further Reading

1. Abbott, Edwin A. *Flatland*, 1884. Reprint, 5th Ed., Rev. New York: Barnes & Noble, Inc., 1969.

2. Barnett, Lincoln. *The Universe and Dr. Einstein*, Rev. Ed. New York: William Sloane Associates, 1957.

3. Dirac, P. A. M. "The Evolution of the Physicist's Picture of Nature," *Mathematics in the Modern World*. San Francisco: W. H. Freeman and Company, 1968.

4. Hawking, Stephen. *A Brief History of Time.* New York: Bantam Books, 1988.

5. Henderson, Linda Dalrymple. *The Fourth Dimension and Non-Euclidean Geometry in Modern Art.* Princeton: Princeton University Press, 1983.

6. Manning, Henry P. *Geometry of Four Dimensions.* New York: Dover Publications, Inc., 1956.

7. Marr, Richard F. *4-Dimensional Geometry.* Boston: Houghton Mifflin Company, 1970.

8. Rucker, Rudolf v. B. *Geometry, Relativity and the Fourth Dimension.* New York: Dover Publications, Inc., 1977.

9. Weeks, Jeffrey R. *The Shape of Space.* New York: Marcel Dekker, Inc., 1985.

CHAPTER

MATHEMATICS OF CONNECTION: GRAPH THEORY

9.1 What Is Graph Theory?

In this chapter we look at a relatively new branch of mathematics called *graph theory*. This topic has virtually no connection with the notion of graphs as you saw them in high school algebra, where a graph is a geometric representation of relationships between numbers. Instead, a *graph* as referred to here is an abstract "map" of some situation, often geometric, which allows us to understand certain features of that situation by concentrating on its essence and ignoring its accidental features.

The origins of graph theory are traditionally traced to a paper presented by Leonhard Euler[1] in 1736. The paper was a mathematical treatment of a famous puzzle, known as the Königsburg Bridge Problem. Throughout the following years, the few problems that were addressed by Euler's methods generally tended to fall into the category of mathematical games, and they attracted little attention among most mathematicians. More recently, the techniques of graph theory have proved to be of considerable practical value in analyzing a number of applied problems. We shall touch on some of these later, but let us ease ourselves into the topic gently (and historically) by considering several puzzles. Many are quite old, so you may well recognize an old friend (or nemesis).

Example 9.1 **Problem:** Can Figure 9.1 be traced in one continuous path without lifting one's pencil from the paper and without tracing any line segment more than once? □

---

[1]See Appendix Section B.7.

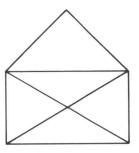

Figure 9.1

Example 9.2 **Problem: (The Königsburg Bridge Problem)** Königsburg[2] was a major city of Prussia. It was situated on both sides of a river and on two islands in the river, interconnected by seven bridges, as indicated in Figure 9.2. The townspeople took great pleasure in strolling through the various parts of the city, and posed themselves the problem of finding a route that would cross each bridge exactly once. Can this be done? □

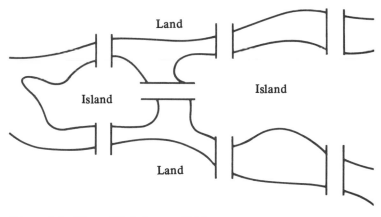

Figure 9.2 *The Königsburg Bridge Problem.*

Example 9.3 **Problem:** A city street crew wishes to install new street signs in a section of the city indicated in Figure 9.3. Signs should be posted at every intersection. What route should be taken by the crew so that each intersection is visited exactly once? □

---

[2]Königsburg was renamed Kaliningrad after World War II, when it became part of the Soviet Union.

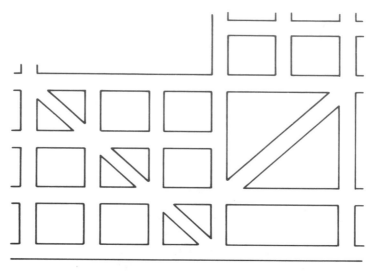

Figure 9.3

Example 9.4   **Problem**: Another city street crew is sent to the same section of the city (Figure 9.3) to repair potholes. Since every street must be inspected, this crew would like to find a route that travels each street exactly once. Can it be done?   □

Example 9.5   **Problem**: Referring to Figure 9.4, can you draw a continuous curve in the plane of this page that crosses each segment exactly once?   □

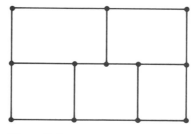

Figure 9.4

Before moving on to the next section, try your hand at the puzzles described in these problems. For each one you solve successfully, give yourself a well-deserved pat on the back. For those you cannot solve, consider the possibility that there may be *no* solution of the type called for. (Can you *prove* that there is no solution? If so, this would also "solve" the problem, although in a different sense.)

Exercises 9.1

WRITING EXERCISES

1. Choose one of the problems presented in Examples 9.1 – 9.5 and describe how at least four of the dozen problem-solving techniques of Section 1.2 can be used to help solve it. (Even if you can't completely solve the prob-

lem, you should be able to do this writing exercise.)

2. Examine the dozen problem-solving tactics of Section 1.2 and decide which two of them are likely to be *least* useful in solving the problems described in Examples 9.1 – 9.5. Justify your choices.

9.2   Some Basic Terms

We begin our study of graph theory by introducing some of the basic terminology. We shall see that all the problems posed in Section 9.1 can be rephrased in simple terms using the language of graph theory. (Note that graph theory, being a mere two centuries old, is still in a developmental stage, and its terminology is not yet universally agreed upon. Thus, there may be some discrepancies between terms we introduce here and those you find in the references. In some instances the difference is deliberate, to avoid the confusion of dissimilar terms for similar concepts, a confusion caused by historical accident.)

DEFINITION   A **graph** is a nonempty, finite set of points together with a finite set of line segments or arcs. The points are called **vertices** (singular, **vertex**) and the lines are called **edges**. Each edge must have a vertex at each of its two ends.

We shall label the vertices of a graph with capital letters. Edges can normally be identified by the labels of their endpoints. Thus, in Figure 9.5, the vertices are A, B, C, and D; the edges are AB, AC, BC, and CD (or BA, CA, and so on).

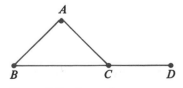

Figure 9.5   A graph.

The definition of a graph allows the possibility of more than one edge joining a given pair of vertices, or of a single vertex being "both" endpoints of an edge. A graph with these features is shown in Figure 9.6. We have assigned numbers to the multiple edges there in order to distinguish, for example, E_1F and E_2F (or F_1E and F_2E). In many instances, the only important issue is whether or not there is at least one edge joining a given pair of vertices; in such cases, multiple edges need not be numbered. The edge HH in Figure 9.6 is an example of a **loop**, an edge whose endpoints coincide at a single vertex. It is important that you not think of a loop as an edge with no end or with only one end, but rather as one with two ends that coincide. Every edge *must* have two endpoints, both of which are vertices; these vertices need not be distinct, though.

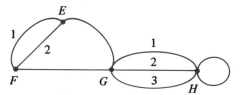

Figure 9.6 *A graph with multiple edges between vertices and with a loop.*

Example 9.6   Figure 9.7 as labeled is *not* an example of a graph; the line segment with J at one end has no vertex at its other end, and thus does not represent an edge. Figure 9.8, on the other hand, *is* a graph because every edge connects two vertices. Note that vertex M is not an endpoint of any edge. Also, this graph is in several "pieces," but is nevertheless a single graph.   □

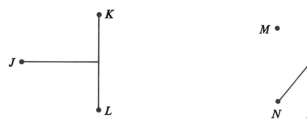

Figure 9.7 *Not a graph as labeled.*   **Figure 9.8** *A disconnected graph.*

To help describe a graph such as the one in Figure 9.8, we introduce the following terms.

DEFINITION A **path** (**from** A **to** B) in a graph is a finite sequence of (one or more) successive edges that form a route from one vertex (A) to another (B). An **open path** is one that begins at one vertex and ends at a different vertex; a **closed path** is one that begins and ends at the same vertex.

Example 9.7 In Figure 9.5, the edges AB, BC, CD form a path from A to D. For brevity, we shall denote this path by $ABCD$. (Notice that with this notation, a three-edge path is denoted by four letters.) Another path from A to D is given by ACD. $ABCBA$ is a closed path from A to A. (A path may use some edges more than once.) Note that ABD does not designate a path from A to D because there is no edge from B to D. □

Where there are multiple edges between vertices, we can distinguish their use in paths by numbering, as indicated previously. Thus, in Figure 9.6, a

A graph pattern on an urban road.

path from E to H is given by E_2FG_1H. If we are not concerned with which edge is used, we may omit the numerical subscripts.

Note
You may have observed that our notation seems to allow some ambiguity: Does XY name the edge joining X and Y or a one-edge path from X to Y? The answer is "Yes" to both. In other words, every edge is a path. We do not really have two *different* ideas represented by an ambiguous label; rather, one idea incorporates the other, and this is represented by the notation.

Turning to the graph in Figure 9.8, we can distinguish its "piecemeal" character from those in Figures 9.5 and 9.6 by the existence or nonexistence of paths among vertices.

DEFINITION   If every pair of distinct vertices of a graph is joined by a path, it is a **connected** graph. Otherwise, it is **disconnected**.

One more clarification is in order here. In Figure 9.9, notice that the two edges SU and TV have no vertex in common, although they do intersect. Nevertheless, Figure 9.9 is a graph because every edge does, in fact, connect two vertices. It is a *disconnected* graph! Connectedness depends on the existence of paths, not on the geometric configuration. The graph of Figure 9.9 is distinguished from earlier graphs by the following terminology:

DEFINITION   A **planar graph** is one in which two edges, or two parts of the same edge, have points in common only at a vertex.

Thus, Figures 9.5 and 9.6 show connected planar graphs; Figure 9.8 is planar and disconnected; Figure 9.9 is nonplanar and disconnected; and Figure 9.10 is nonplanar but connected.

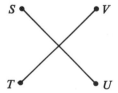

Figure 9.9   This graph is disconnected.

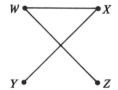

Figure 9.10   A nonplanar, connected graph.

Let us now return to the puzzles of Section 9.1 and rephrase them using the language of graph theory. The figure to be traced in Example 9.1 (Figure

9.1) is essentially a graph already; we need only label the vertices. This is done in Figure 9.11. The problem of tracing translates into one of finding a path with each edge used exactly once. For future reference, we define such a path as an **edge path**.

Turning to the Königsburg Bridge Problem, we can translate Figure 9.2 into a graph containing all the significant features of the original picture. Observing that, since walking around on one side or the other of the river or on one of the islands has no effect on the problem (the challenge is finding a way to cross the bridges appropriately), we may represent each of these land areas as a vertex and each bridge as an edge. This is done in Figure 9.12. The problem of crossing each bridge exactly once then translates into finding an edge path in this graph — the same kind of problem as the tracing puzzle!

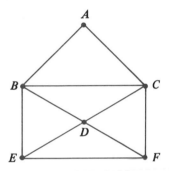

Figure 9.11   Figure 9.1 with vertex labels.

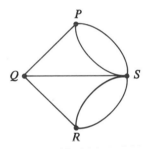

Figure 9.12   The Seven Bridges of Königsburg.

The street map of Examples 9.3 and 9.4 translates in an obvious way into a graph, with each intersection becoming a vertex and each street an edge. This is left as an exercise. The problem for the pothole-repair crew, to travel each street exactly once, is *again* the edge-path problem. The task for the street-sign crew is different, but analogous. Visiting each intersection exactly once translates to finding a path that uses each vertex exactly once. Quite naturally, we define such a path as a **vertex path**.

As is the case with the first puzzle, Figure 9.4 is already a graph, except for the labeling of the vertices. The problem, though, does not translate into finding a path of any sort. Here we are challenged to find some continuous curve that *crosses* edges and *avoids* vertices, instead of following them as in a path. It is useful to have a name for such curves, too: a **crossing curve** in a graph is a continuous curve that crosses each edge exactly once and passes through no vertex. Notice that a crossing curve may have none, one, or both of its ends inside one of the regions (if any) enclosed by edges of the graph. Notice also that a crossing curve may cross itself one or more times; however, it must cross each edge of the graph once and only once.

EXERCISES 9.2

1. Which of the drawings in Figure 9.13 represent graphs?

2. Which of the graphs in Figure 9.13 are connected?

3. Which of the graphs in Figure 9.13 are planar?

4. Which of the graphs in Figure 9.13 contain a loop?

5. Referring to Figure 9.13(g):

   (a) Identify a path from Z to X that uses no edge more than once.

   (b) Identify a path from W to X that uses

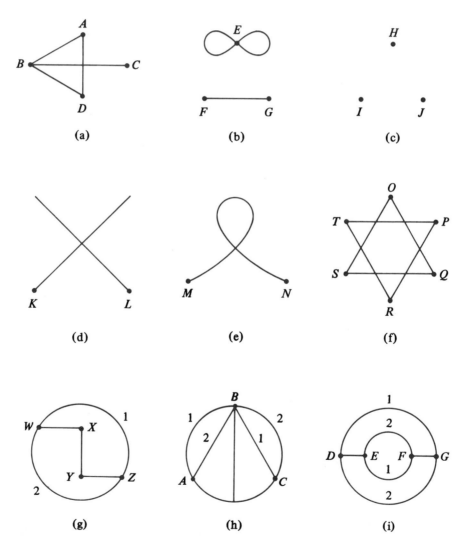

Figure 9.13

every edge at least once.

(c) Identify a path from W to Z that is an edge path.

(d) How many different edge paths are there from W to Z?

(e) How many different edge paths are there from X to Y?

(f) How many different paths are there from X to Y?

6. In a certain graph, $AFEBBCEAGCB$ is a path. Determine whether the following statements are *true, false,* or *cannot be determined* without additional information.

(a) This is an open path.

(b) This path contains 11 edges.

(c) The graph contains at least 10 edges.

(d) The graph contains at least 8 edges.

(e) The graph contains at least 7 vertices.

(f) The graph contains a loop.

(g) There is no edge joining A and B.

(h) There is a two-edge path joining A and B.

(i) There is a closed path that uses only vertices A, B, and C.

(j) The graph contains multiple edges joining some pairs of vertices.

(k) The graph is connected.

(l) The graph is planar.

7. Answer Exercise 6 with the additional assumption that the given path is an edge path.

8. Draw a graph that represents the street map given in Figure 9.3.

9. The Town of Mud Flats is built on the banks of a river, much like Königsburg. (See Figure 9.14.) Many years ago the river followed a different channel (indicated by dashed lines) and one of the old bridges was left standing for its scenic beauty. Draw a graph representing the current land masses as vertices and the bridges as edges. Can you find an edge path for this graph?

10. For each of the following sets of conditions, give an example of a graph with four vertices that satisfies all of the given conditions. If this is not possible, explain why not.

(a) Disconnected, nonplanar.

(b) Disconnected, with an edge path.

(c) Disconnected, with a vertex path.

(d) Connected, with no vertex path.

(e) With an edge path, but no vertex path.

(f) With a vertex path, but no edge path.

(g) With an edge path and a vertex path.

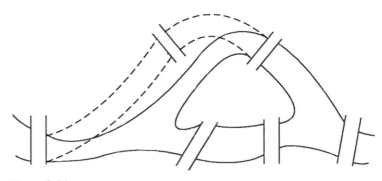

Figure 9.14

11. Repeat Exercise 10 by giving examples of graphs with three vertices, rather than with four.

WRITING EXERCISES

1. As an alternative to the graph of Figure 9.11, the puzzle of Example 9.1 could also be represented by a graph that *looks* the same, but has no vertex *D*. Write a one-page paper explaining the difference between the two graphs and *either* supporting or contradicting the position that one graph is as good as the other as a representation of the puzzle.

2. Consider a graph that represents the Königsburg Bridge Problem by treating each bridge as a vertex and streets connecting the bridges as edges. Write a one-page paper on whether or not finding a vertex path in this graph is essentially the same problem as finding an edge path in the graph of Figure 9.12. If they are different problems in graph theory, which one represents the original puzzle? If they are essentially the same, is there any reason to choose one over the other?

3. Select one part of Exercise 10 whose answer is different from the corresponding part of Exercise 11. Write a paragraph explaining why the number of vertices makes a difference.

4. Two different closed paths in a given graph might contain the same edges. (For example, in Figure 9.13(a), consider *ABDA* and *BDAB*.) How could the definition of path be rephrased so that this distinction would be eliminated?

9.3   Edge Paths

Let us examine thoroughly the first type of puzzle identified in the preceding sections — that of tracing each edge (or each bridge, or each street) exactly once; that is, of finding an edge path. By trial and error, you may have found an edge path in some of the puzzles of Section 9.1. Maybe you were lucky and found an edge path with little effort; maybe you found one only after many trials. But you surely did not find an edge path in all of them, for none exists in some of the puzzles posed.

Unfortunately, failure to find an edge path does not solve the puzzle because, for even moderately complex graphs, trial and error alone will not guarantee that there is *no* edge path; we may have merely overlooked one. In principle, we could consider each pair of vertices in the graph, record every path joining them that uses no edge more than once, stopping if an edge path is found; if no edge path is found between any two vertices, we

The front of a building in the form of a graph.

could conclude that no edge path exists. For any but the simplest graphs (where the existence or nonexistence of edge paths is already obvious), the huge number of possibilities makes this approach thoroughly impractical. For example, in the relatively simple "envelope" graph of Figure 9.11 there are twenty-one pairs of vertices (including such pairs as C, C) and over fifty different paths that use no edge more than once between each pair. We must find a more efficient strategy!

Even when an edge path is found and it solves the particular puzzle at hand, there are still unsettled questions. Have we found the only edge path or are there others? Is there an edge path between any two vertices or only between the pair we happened upon? Why? Why does one graph have an edge path and another not have one? Is there some property or combination of properties that will allow us to distinguish between these two classes of graphs without the immense effort required by exhaustive trial and error?

The distinction between graphs that have edge paths and graphs that do not can be made by using a common mathematical technique. Without constructing a particular example, we assume that we have a "typical" graph that has an edge path, and try to deduce whatever we can about this graph. Effectively, then, we are discovering properties possessed by *every* graph that has an edge path. With luck, we may find that some of these properties will apply *only* to graphs with edge paths, and thus we can identify such graphs just by finding these critical properties.

Suppose, then, that A is a vertex of our hypothetical graph-with-edge-path; and suppose further that A is neither the beginning nor the end of the path, but is some vertex encountered along the way. Clearly, there must be at least two edges at A, one by which the path first comes to A and one by which it leaves. If the path does not return to A, then there are *exactly* two edges at A (Figure 9.15). On the other hand, if the path does return to A, it cannot do so by one of the two edges previously mentioned because it is an edge path; thus, there must be a third edge to return by *and a fourth edge* to leave by (Figure 9.16). If the path does not return to A again, then there are exactly four edges at A. If the path does return a second time, there must be a fifth and a sixth edge at A, to allow the path to come and go. In general, although we cannot say exactly how many edges there are at A, we can conclude that *there must be an even number* because the edge path neither begins nor ends at A.

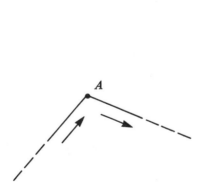

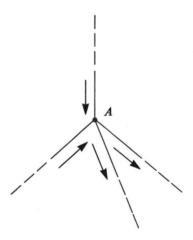

Figure 9.15   A path that does not return to A.

Figure 9.16   A path that returns to A.

There is one slight flaw in this analysis. We have assumed that when the path leaves A, it goes on to some *other* vertex (and comes to A from some other vertex). If there is a loop at A, the path leaves A and returns on the same edge, so that only one edge is involved in this leaving and returning.[3]

---

[3] For this reason, among others, some authors exclude loops in defining a graph.

From the localized viewpoint, however, we can incorporate this possibility in our earlier analysis by noting that there are still an even number of ends of edges at A. This discussion motivates the next definition.

DEFINITION   The **degree** of a vertex is the number of ends of edges incident with it. An **even vertex** is one of even degree; an **odd vertex** is one of odd degree. An **isolated vertex** is one of degree zero.[4]

Example 9.8   In Figure 9.8, vertices P, Q, and R each have degree 2 and thus are even vertices; N and O each have degree 1 and are odd vertices; and M is an isolated vertex, having degree zero. (Of course, M is also an even vertex because zero is an even number.)   □

As we have seen, in a graph with an edge path, a vertex that neither begins nor ends the path must be an even vertex. By a similar analysis, a vertex that is an endpoint (beginning or ending) of an open edge path must be an odd vertex. If the edge path is closed, then any vertex can be used as the beginning and ending point and it must be even. Thus, we have proved the following general property:

(9.1)

> If a graph has an edge path, then either all the vertices are even and the path is closed, or there are exactly two odd vertices and the edge path is open, beginning at one odd vertex and ending at the other.

Example 9.9   In Figure 9.17, $PRTSRQ_1S_2QP$ is a closed edge path beginning and ending at vertex P. Observe that each vertex in the graph is even. In Figure 9.18, $ACDAB$ is an open edge path beginning at one odd vertex, A, and ending at a second, B; all the other vertices are even.   □

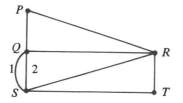

Figure 9.17   Each vertex of this graph is even.

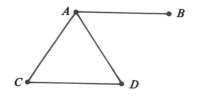

Figure 9.18   This graph has two odd vertices, A and B.

---

[4]Some authors exclude isolated vertices in defining a graph.

Because a graph with an edge path can have at most two odd vertices, we have a relatively easy way to rule out a large number of graphs as having no possible edge paths. We need merely look at each vertex to see if it is odd; if we find three such vertices, there is no need to proceed any further. This is formalized in the following statement.

(9.2) | If a graph has more than two odd vertices, then it has no edge path.

Does the converse of this statement hold? In other words, does it follow that a graph with at most two odd vertices has an edge path? Clearly not, as Figure 9.19 shows. The problem in this graph, of course, is its disconnectedness.

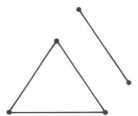

Figure 9.19 This graph has only two odd vertices, but no edge path.

Returning to our examination of our hypothetical graph-with-edge-path, we can see that the existence of an edge path guarantees our ability to get from one vertex to another — with one exception. That exception is the possible presence of isolated vertices in a graph with an edge path. Since no edges are incident with an isolated vertex, the presence of such a vertex has no effect on the existence of an edge path, although it does cause the graph to be disconnected. To coin a phrase for the sort of graph we are now considering, we will describe a graph that is connected, or connected except for one or more isolated vertices, as a **nearly connected** graph.

Example 9.10   Figure 9.17 is connected, and thus is nearly connected. Neither Figure 9.18 nor Figure 9.19 is connected. Figure 9.18 is nearly connected, but Figure 9.19 is not.   □

With this terminology, the following statement becomes obvious.

(9.3) | A graph that has an edge path is nearly connected.

Now, if we combine the two properties we have identified, we obtain a characteristic way to distinguish graphs with edge paths:

(9.4) | If a graph is nearly connected and has at most two odd vertices, then it has an edge path.

Since near-connectedness is obvious at a glance, this theorem effectively reduces the problem of determining the existence of an edge path to that of counting the degrees of the vertices. (Actually, since only the odd or even parity need be determined, it is literally as simple as performing the numerical analog of "She loves me, she loves me not," by "counting": "odd, even, odd, even, ...")

Although our analysis of edge-path graphs should make Statement 9.4 plausible, *we have not yet proved that it is true.* The proof is somewhat lengthy, but it is not difficult. In studying it, proceed slowly, making sure you understand each step before going on to the next one.

Proof of (9.4)

We shall prove the case of two odd vertices; the remainder is left to the exercises.

Step 1. Let us assume that the two odd vertices are labeled P and Q. We start at one, say P, and *randomly* construct a path, using no edge more than once, continuing as long as there are unused edges available when the path comes to a vertex.

Each time this path comes to a vertex other than P or Q, the path uses one edge of an even number of available edges, so there must be another edge at this vertex that is not yet used, by which the path can continue. Coming to and leaving such a vertex uses two of an even number of edges (or ends of edges), leaving an even number yet to be used. Thus, the path cannot end at such a vertex.

One edge of an odd number of edges at P is used (the very first one of the path), leaving an even number. Thus, if the path happens to return to P the situation is exactly like coming to an even vertex; that is, the path cannot end at P. Eventually, the path must end, since there are only a finite number of edges. Our analysis shows that the only place it can end is at Q. By the time the random path ends, all the edges at Q must have been used up — otherwise, the path could continue.

At this stage, an odd number of edges at P have been used, all the edges at Q have been used, and an even number of edges (possibly zero) at every other vertex have been used. There still may be edges in the graph not used in this path. If there are any, there are an even number remaining at any vertex where they exist.

Step 2. If the path constructed in *Step 1* has used all the edges, then it is an edge path and we are finished. If there are unused edges, then there must be some vertex having some edges that were used in the path and some that were not. (Otherwise, the graph would not be nearly connected.) We choose one such vertex, labeling it X for reference.

Step 3. Starting at X, we again randomly construct a path, using no edge that has previously been used (in *Step 1* or in this step), and continuing as long as there are available edges. As was the case in *Step 1*, this path must end, but it cannot stop where an even number of edges are available. Thus, it must stop at X, where the use of one edge to start left an odd number of edges unused.

Step 4. We now "cut and paste" the two paths constructed in *Steps 1* and *3*, at the place where we know they share a common vertex, X. That is, we follow the original (*Step 1*) path from P until we come to X; we then follow the path constructed in *Step 3* to its end at X, where we resume the original path, following it to its end at Q. The result is an expanded path from P to Q, using no edge more than once.

Step 5. If all edges have now been used, the expanded path is an edge path and we are finished. If not, the analysis of *Step 2* applies; we choose an appropriate vertex — one with both "used" and "unused" edges — and repeat *Steps 3* and *4*. Each time we do this, we use up more edges. Eventually, all edges in the graph must be included in the expanded path, and we have an edge path.

Statements 9.1, 9.3, and 9.4 completely characterize graphs that have edge paths. Note that there are effectively four slight variations possible among graphs with edge paths: the graph may be connected or not (but it must at least be nearly connected), and the edge path may be open or closed. We will not create special terminology for each of these cases, but we would be remiss not to identify the classic label for the "nicest" of these possibilities: An **Euler graph** is a connected graph with a closed edge path.

EXERCISES 9.3

1. For each of the seven graphs in Figure 9.20, determine which vertices are even and which are odd.

2. In Figure 9.20(g), find the degree of each vertex.

3. For each of the seven graphs in Figure 9.20, how many odd vertices are there?

4. (a) Which of the seven graphs in Figure 9.20 have no edge path?

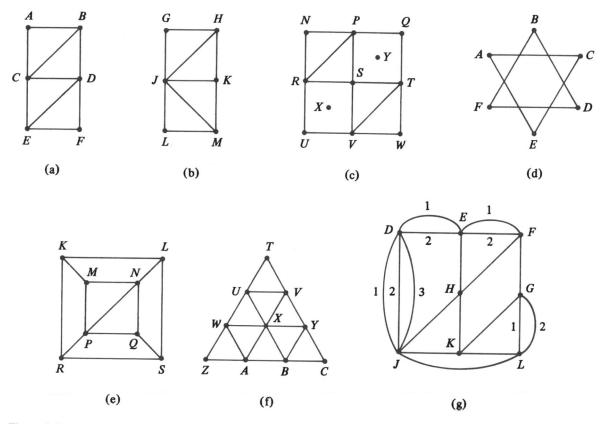

Figure 9.20

(b) Which have an open edge path?

(c) Which have a closed edge path?

(d) Which are Euler graphs?

5. In the text only one case of Statement 9.4 was proved—that of two odd vertices. Modify this proof to cover the case of no odd vertices.

6. Can you construct a graph with exactly one odd vertex? (*Hint*: Study the reasoning of *Step 1* in the proof of Statement 9.4.)

7. Generalize the reasoning of *Step 1* in the proof of Statement 9.4 to show that every graph (connected or not) has an even number of odd vertices.

8. Apply the methods of this section to Examples 9.1, 9.2, and 9.3 to answer the questions posed there.

9. Show that if the people of Königsburg built one additional bridge from any land mass to another, they could find a path that crosses each of the bridges exactly once. (See Example 9.2 and Figures 9.2 and 9.12.) Show that the same would be true if they tore down one bridge.

10. How many new bridges should be built in Königsburg if the townspeople wish to be able to cross every bridge exactly once and return to their starting place? (See Exercise 9.) Where should the new bridge(s) be built?

11. If a connected graph has exactly six odd vertices, how many new edges must be added to change it into an Euler graph? (See Exercises 9 and 10.) Where should these edges be added? Apply this idea to the graph in Figure 9.20(e); which vertices can be joined by edges to convert the graph to an Euler graph?

12. If a connected graph has exactly n odd vertices (n is some natural number), how many new edges must be added to change it into an Euler graph? (See Exercise 11.)

13. Show that an Euler graph must have at least as many edges as it has vertices.

14. Show that by deleting certain edges from the graph of Figure 9.20(e) you can convert it into an Euler graph. (See Exercise 11.) What is the smallest number of edges that can be deleted to accomplish this? What is the largest number? (See Exercise 13.)

15. Give an example of a connected graph that is not an Euler graph and that cannot be converted into one by the deletion of any number of edges. (See Exercise 14.)

16. A **null graph** is the name given to a graph with no edges (that is, to a nonempty finite set of vertices). A certain graph is not a null graph and is not an Euler graph. What is the smallest number of vertices this graph can have?

17. A certain graph is not a null graph and has no edge path. What is the smallest number of vertices this graph can have? (See Exercise 16.)

WRITING EXERCISES

1. Define the term *graph* so as to exclude loops and isolated vertices.

2. Write a one-page paper arguing the merits of the definition described in Exercise 1 when one is studying the question of which graphs have edge paths.

3. The text uses visual aids (Figures 9.15, 9.16) to support the argument that a vertex that neither starts nor ends an edge path must be even. Write a paper making the same argument, but based on a consideration of the "verbal" representation of a path as a string of letters. (Don't forget to consider the possibility of $\ldots AA \ldots$ appearing somewhere in the string.)

4. Carefully consider the appropriate definitions (See Exercise 16) to write a short paper that answers the following questions. Explain the reasoning behind each of your answers.

   (a) Is a null graph connected?
   (b) Is a null graph nearly connected?
   (c) Is there a path in a null graph?
   (d) Is a null graph an Euler graph?

9.4 Vertex Paths

We now turn to the problem suggested by Example 9.4 of Section 9.1 — finding in a graph a path that passes exactly once through each vertex. Noting the close parallel in form with the problem of Section 9.3, we shall approach it with the same method.

Specifically, let us imagine that we have some arbitrary graph that has a vertex path. What can we logically assert about such a graph? Perhaps the most obvious fact is that the graph must be connected; because the vertex path passes (exactly once) through every vertex of the graph, some part of the vertex path must be a path between any two vertices we choose. Thus, we have quickly and easily proved our first result:

(9.5)

> If a graph has a vertex path, it is connected.

Connectedness is not sufficient to guarantee that a graph has a vertex path, as Figure 9.21 shows. Although this graph is connected, it cannot have a vertex path. Vertex A has degree 1, and thus any vertex path must either begin or end at A. But the same is true of vertices C and F, and a path cannot have three ends. For brevity, we will call a vertex of degree 1 a **terminal vertex**, and the one edge at such a vertex, a **terminal edge**. With this terminology, we have the following statement:

(9.6)

> If a graph has a vertex path, it has at most two terminal vertices.

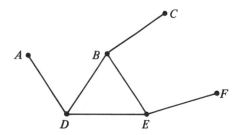

Figure 9.21 *A connected graph with no vertex path.*

A little thought should make it clear that a vertex path in a graph with one or two terminal vertices must be an open vertex path. In other words:

(9.7)

> If a graph has a closed vertex path, it has no terminal vertices.

A church window frame forming a graph.

We could continue to pursue conditions pertaining to open versus closed vertex paths, stating appropriate pairs of theorems, but this would quickly become tiresome. Instead, let us restrict ourselves to considering one type of vertex path, leaving the similar observations about the other for the exercises. For simplicity and historical interest, we shall consider only closed vertex paths for the rest of the section. This analogy to an Euler graph, namely, a graph with a closed vertex path, is called a **Hamilton graph.**[5]

Suppose we have a connected graph with no terminal vertices. Is it a Hamilton graph? Figure 9.22 shows us that the answer is "No." Here we see a graph that has no terminal edges and is connected, but is not a Hamilton graph. The problem is that it has two "halves," with only one vertex, T, connecting them. A closed vertex path must pass through vertices on either side of T and form a closed path. No matter how we try to do this, it requires passing through T twice, contradicting the definition of a vertex path; therefore, no closed vertex path exists. In order to characterize Hamilton graphs, we must describe the feature exemplified by vertices such as T. This we do in Statement (9.8), using the following definition.

---

[5]Named after William Rowan Hamilton; see Appendix Section B.8.

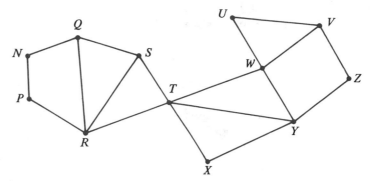

Figure 9.22 *This is not a Hamilton graph.*

DEFINITION   A vertex in a connected graph is a **critical vertex** if its removal, together with all edges incident with it, would leave a disconnected graph.

(9.8)

> A Hamilton graph contains no critical vertices.

Do we now have a characteristic set of properties for a Hamilton graph? If a graph is connected and has neither terminal vertices nor critical vertices, is it a Hamilton graph? Again, the answer is "No," as is shown in Figure 9.23.

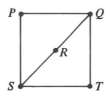

Figure 9.23 *This is not a Hamilton graph, either.*

In the graph of Figure 9.23, which meets all the conditions specified in the preceding paragraph, a closed vertex path must go through edges PQ and SP because this is the only way to reach (and leave) vertex P. But the same is true of edges SR and RQ, relative to vertex R, and of edges ST and TQ, relative to vertex T. Thus, three edges at Q (and also at S) must be included in any possible closed vertex path in order to reach every vertex; but this implies passing through Q (and S) more than once. Again, we have a new necessary condition that uses a new term:

DEFINITION   An edge incident with a vertex of degree two is called a **Hamilton edge**. (We abbreviate this as H-**edge**.)

(9.9) | A closed vertex path of a graph must contain every H-edge.

(9.10) | In a Hamilton graph, no vertex is incident with more than two H-edges.

We still do not have a characterization of Hamilton graphs, however. Figure 9.24 suggests two easy ways to satisfy the necessary condition of Statement (9.10) without having a Hamilton graph. In Part (a), multiple edges join the vertices; in Part (b), loops have been added. The net effect in either case is to increase the degree of the vertices, thus technically eliminating H-edges, but the graphs still have the same basic drawback as Figure 9.23 in failing to be Hamilton graphs. As this example suggests, multiple edges and loops do not provide useful additional edges to be used in a closed vertex path. They only serve to obscure our search for a characteristic of Hamilton graphs. So we shall ban them for the remainder of this section, restricting ourselves to what we shall call **simple graphs**, graphs with no loops and no multiple edges.

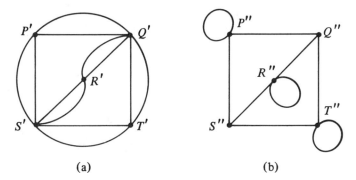

(a)   (b)

Figure 9.24   *Figure 9.23 with (a) added edges and (b) added loops.*

Collecting all our observations so far, we are ready to pose the characterization question again. If a simple, connected graph has no terminal vertices, no critical vertices, and no vertex incident with more than two H-edges, is it a Hamilton graph? You guessed it. The answer is still "No." Figure 9.25 shows a graph with all the conditions called for; each H-edge is marked with a short cross-mark. We can identify our new problem in terms of these edges, but it is not that there are too many at one vertex. Rather, the H-edges at A, B, C, and D form a closed path. By Statement (9.9), each of these edges must be used in a closed vertex path; by definition, the other vertices (E, F, G, H) must also be used. But there is no way to do both of these things. Therefore, we have another exclusion to add to our growing list.

(9.11) | If the H-edges of a graph form a closed path among some, but not all, vertices, then the graph is not a Hamilton graph.

Are we finished? If a simple graph passes all the tests implied in Statements (9.5), (9.7), (9.8), (9.10), and (9.11), is it a Hamilton graph? Unfortunately, no. Figure 9.26 is a counterexample. As before, H-edges are marked with a single cross-mark. Thus, any closed vertex path must contain $QRSTU$. Clearly, this cannot be accommodated in a vertex path that must reach both P and V and also be closed. One way to describe and classify this problem is to note that since RS and ST are H-edges, PS and SV cannot be used. Effectively, then, PQ, QV, VU, and UP are "secondary" H-edges because they are the only edges actually available to reach P and V after the H-edges have been identified. (These are marked with double cross-marks.) Then the H-edges and "H_2-edges" together rule out a closed vertex path by either of the statements (9.10) or (9.11).

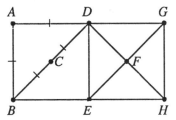

Figure 9.25   *Still not a Hamilton graph.*

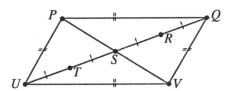

Figure 9.26   *Still not a Hamilton graph.*

We could formally define "H_2-edges," "H_3-edges," and so on, and generalize (9.10) and (9.11) accordingly. But even then we would fail to cover every property that characterizes Hamilton graphs. Figure 9.27 shows a simple, connected graph with no isolated or critical vertices and no H-edges Yet it takes very little study to conclude that there is no closed vertex path for this graph.

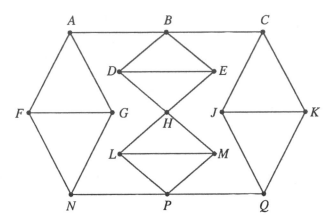

*Figure 9.27   Even this is not a Hamil-
ton graph.*

What we have experienced in this section is a taste of the frustration met by every individual who has tried to characterize Hamilton graphs. Although the problem is over a century old, no one has yet found any combination of properties whose presence in a graph will guarantee the existence of a closed vertex path. This is not to say that our efforts have been wasted. We have identified a number of properties that are necessary for a Hamilton graph. Their absence, then, allows us easily to identify when a given graph is *not* Hamilton. And our observations about *H*-edges can help us find a closed vertex path if one exists, even though we have no theorems to *guarantee* the existence of the desired path. Similarly, although we cannot characterize whether or not a graph has an open edge path, we can modify the statements about closed edge paths to provide useful information.

In all, determining whether or not a graph has a vertex path remains largely a matter of trial and error, but the necessary conditions we have stated can greatly reduce the number of errors and provide an organized guide as to what to try.

EXERCISES 9.4

Exercises 1–4 refer to Figure 9.28.

1. Which graphs have terminal vertices?

2. Identify the *H*-edges in Graphs (b), (e), and (h).

3. Which graphs are Hamilton graphs?

4. Which graphs have open vertex paths?

5. Can the street-sign crew of Example 9.3 do their job by visiting each intersection exactly once? (See Figure 9.3.)

6. Show that every Hamilton graph has an open

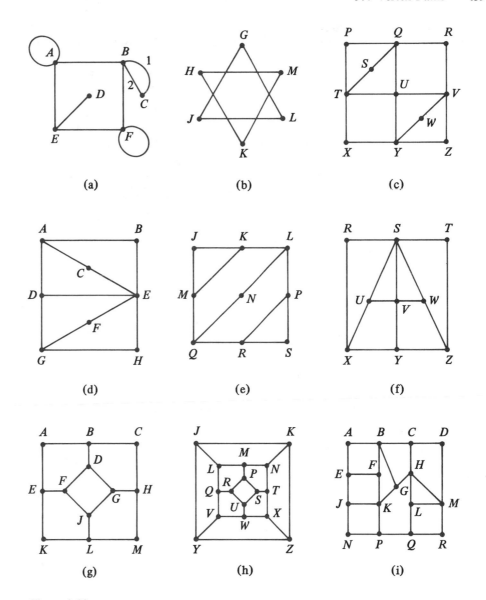

Figure 9.28

vertex path. Give an example to show that the converse is not true.

7. Prove that an open vertex path may fail to contain at most two *H*-edges in a graph. (*Hint*: See Statement (9.9).)

8. Prove that if a graph has a vertex incident

with five or more *H*-edges, it contains no vertex path. (*Hint*: See Statement (9.10).)

9. Prove that a connected graph with more than two vertices and with a terminal vertex must have a critical vertex.

10. Show that if a closed path is both a vertex

path and an edge path for a graph, the number of vertices must be equal to the number of edges.

WRITING EXERCISES

1. Give a formal definition of an H_2-**edge**.

2. Using the definitions of an H-edge and an H_2-edge (see Writing Exercise 1), formulate the generalizations of Statements (9.10) and (9.11), and present some argument for their validity.

3. Write a paper identifying the reason that Figure 9.27 fails to be a Hamilton graph. If possible, formulate that identification into a definition, and state the appropriate theorem linking that concept to Hamilton graphs.

4. Identify as many distinctions as you can between edges and vertices, and discuss how these distinctions might explain the difference between the ease of characterizing Euler graphs and the continuing failure to do so for Hamilton graphs.

9.5   Crossing Curves

Now we consider the type of problem posed in the puzzle of Example 9.5 — finding a continuous curve crossing each line segment in a figure exactly once. In that puzzle, we asserted that the figure could be labeled in an obvious and unambiguous way to become a graph. This was a correct assertion for that graph and many others, but it is not always the case; and the choice of how to represent a figure as a graph may dramatically alter the answer to the crossing-curve question.

Figure 9.29(a) shows a simple figure which, with appropriate identification of vertices, is a graph. Two different ways of making this identification are given in Figure 9.29 (b) and (c). In (b), the resulting graph is nonplanar, has six edges, and has a crossing curve, as indicated. But this technically correct crossing curve *for the graph* seems inadequate to represent the problem posed by the original picture, almost as if we were cheating somehow. In (c), the resulting graph is planar (and has eight edges) and appears to be a more honest graph representation of (a), but no crossing curve is possible. For this reason, we shall restrict ourselves to planar graphs in our consideration of crossing curves. We will also limit ourselves to connected graphs because for the purpose of finding a crossing curve, a disconnected graph can be treated as several different connected graphs.

DEFINITION   A **polygonal graph** is a connected planar graph. Each region of the plane determined by the edges of a polygonal graph is called a **face** of the graph.

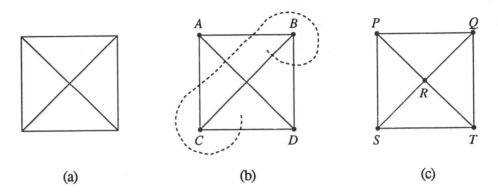

Figure 9.29 *Two different graphs for a figure.*

Note

The choice of the phrase "determined by," rather than "contained within," is deliberate, since the portion of the plane "outside" a polygonal graph is also considered to be a face. For example, in Figure 9.29(c), there are *five* faces — the four triangular regions that share vertex R, and the fifth portion of the plane, outside of the square $PQST$.

A crossing curve, then, is one that crosses from one face to another, intersecting each edge exactly once. An exception occurs if a graph has a terminal edge. In such an example, the crossing curve remains within a given face when crossing such an edge. Terminal edges pose no barrier to the existence of a crossing curve; as long as we are able to reach every edge, a crossing curve exists. A curve will be impossible if all the edges surrounding some face are used up (crossed once already) and the curve is "trapped" within that face while some edges, elsewhere, are not yet crossed. Thus, the existence of a crossing curve is determined by the number of times a possible curve must enter or leave a face. We define a term to identify this number.

DEFINITION The **order** of a face is the number of edges adjacent to it. A terminal edge is counted twice.

The peculiarity of this definition with regard to terminal edges can be explained by observing that a crossing curve, in crossing a terminal edge, leaves and enters the face simultaneously. Alternatively, we could think of each edge as having two sides, and the process of determining the order of a face as that of counting these; for most edges, one side is in one face and the other is in a neighboring face, but both sides of a terminal edge lie in the same face.[6]

Example 9.11   **Problem**: Determine the order of each face in the graphs of Figure 9.30.

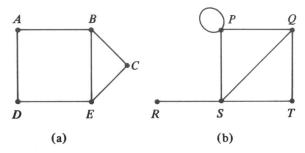

(a)   (b)

*Figure 9.30   One of these has a crossing
curve; the other does not.*

Solution: In Figure 9.30(a), there are three faces; one of order three, one of order four, and one (the "outside" face) of order five. In Figure 9.30(b), there are four faces: two of order three, one (contained by the loop) of order one, and one of order seven.   □

Let us look for a crossing curve for Figure 9.30(a). Since the face enclosed by the path *BCEB* has order three, a crossing curve can enter this face twice and leave once, or leave twice and enter once. In other words, the start or finish of the curve must lie in this face. The same is true of the "outside" face of order five. Fortunately, the face of order four must be entered twice and exited twice, so that the curve does not have to begin or end in this face. Therefore, a crossing curve is possible — one such curve is indicated in Figure 9.31.

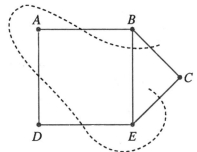

*Figure 9.31   A crossing curve for Fig-
ure 9.30(a).*

---

[6]This peculiarity could be avoided by excluding terminal edges in the definition of polygonal graphs. However, that option also has its problems.

In Figure 9.30(b), each of the four faces has an odd order, thereby requiring one more entry than exit, or vice versa. Thus, a crossing curve must have one of its ends in each of four faces — an impossibility. There is no crossing curve for this graph.

If this "even versus odd" analysis is reminiscent of the discussion of edge paths, there is good reason. The problem of finding a crossing curve from face to face across edges is quite similar to finding an edge path from vertex to vertex along edges, not in the given graph, of course, but in an "analogous" graph constructed for the purpose. We make the connection formally by using the following definition and Statement (9.12).

DEFINITION The **dual** of a polygonal graph is a graph such that each face of the given (original) graph has a corresponding vertex in the dual, and each edge in the given graph has a corresponding edge in the dual. The face-vertex and edge-edge correspondences must be such that an edge in the given graph that separates two faces corresponds to an edge in the dual that connects the vertices corresponding to those faces.

Example 9.12 The graph of Figure 9.32(a), with faces labeled by lowercase letters, has as its dual the graph of Figure 9.32(b). The face-vertex and edge-edge correspondences are given in Table 9.1. For example, because in Graph (a) (the "given graph") face q is separated from face r by edges BC and CE, in Graph (b) (the dual graph) vertex Q is joined to vertex R by edges $Q_1 R$ and $Q_2 R$.   □

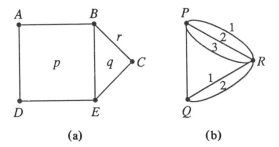

(a)   (b)

Figure 9.32 *A graph and its dual.*

(9.12) | A polygonal graph has a crossing curve if and only if its dual graph has an edge path.

Graph (a) Face	Dual (b) Vertex	Graph (a) Edge	Dual (b) Edge
p	P	AB	P_1R
q	Q	AD	P_2R
r	R	BC	Q_1R
		BE	PQ
		CE	Q_2R
		DE	P_3R

Table 9.1

The proof of Statement (9.12) is essentially contained in the previous discussion. We need only observe that in the correspondence from a polygonal graph to its dual, the order of a face corresponds to the degree of the corresponding vertex, so that the discussion and proofs in Section 9.3 apply exactly to crossing curves by means of the dual-graph correspondence. For example, the crossing curve shown in Figure 9.31 corresponds to the edge path $Q_2R_3P_2R_1PQ_1R$ in the dual graph (Figure 9.32(b)). Conversely, any edge path in the dual can be translated to a crossing curve in the original graph.

In order to determine whether or not a crossing curve exists, then, we can construct the dual graph and apply the results of Section 9.3. More simply, we can just translate the statements of Section 9.3 to corresponding statements about crossing curves by replacing *vertex* with *face*, *degree of vertex* with *order of face*, and so on.

Example 9.13 **Problem:** Is it possible to draw a crossing curve in the graph of Figure 9.33?

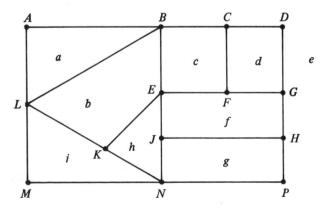

Figure 9.33

Solution: The order of each face is given in Table 9.2. Because there are only two odd faces (a and f), there is a crossing curve, with one end in face a and the other in face f. The drawing of a crossing curve is left as an exercise.   □

Face	Order	Face	Order
a	3	f	5
b	4	g	4
c	4	h	4
d	4	j	4
e	10		

Table 9.2

EXERCISES 9.5

1. Find a crossing curve for the graph of Figure 9.33.

2. Does the graph in Example 9.5 have a crossing curve? If so, draw one; if not, why not?

3. Which of the graphs in Figure 9.34 have crossing curves? For each one that does, draw a crossing curve.

4. Construct a graph that is the dual of Figure 9.34(c).

5. In the graph of Figure 9.32(b), label the faces with the letters a, b, c, d, and e to show that Figure 9.32(a) is the dual graph, with its vertices corresponding to these faces.

6. A **self-dual** graph is one that, with the appropriate correspondence, is dual to itself. Give a correspondence of faces and edges in Figure 9.34(a) to show that it is self-dual.

7. Show that Figure 9.34(b) is self-dual. (See Exercise 6.)

8. If possible, construct a self-dual graph with exactly three vertices. (See Exercise 6.) If impossible, explain why.

9. If possible, construct a self-dual graph with exactly four vertices, which is different from the graph of Figure 9.34(b) (that is, not with four faces each of order three). (See Exercise 6.) If impossible, explain why.

10. In the process of forming dual graphs, the concepts *face* and *vertex* are duals of each other; *crossing curve* and *edge path* are duals, and so on. Identify the dual of each of the following:

    (a) Terminal edge.
    (b) Terminal vertex.
    (c) Isolated vertex.

11. Construct a graph with exactly one odd face. If impossible, explain why.

WRITING EXERCISES

1. The definition of the order of a face somewhat arbitrarily counts a terminal edge twice. Discuss the pros and cons of counting a terminal edge once.

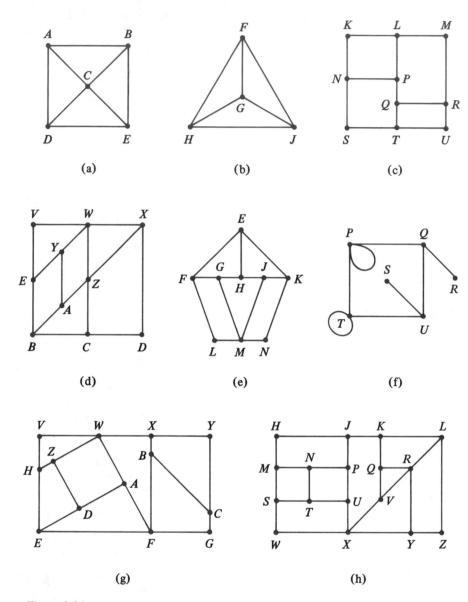

Figure 9.34

2. Discuss the pros and cons of ignoring terminal edges in determining the order of a face.

3. Give definitions of "open crossing curve" and "closed crossing curve" that are dual to

"open edge path" and "closed edge path." (See Exercise 10.)

4. Explain why the dual of the dual of a graph is the original graph. (See Exercise 5.)

9.6   Euler's Formula

We have already noted some strong connections between graph theory and numbers in searching for edge paths and crossing curves. In both instances, the even or odd degree of vertices (or order of faces) largely determines the existence of the path (or curve). There are many other numerical properties of graphs. For example, if we add up the degrees of all of the vertices in a graph, the result must be even because we have effectively counted all the ends of all the edges, which must come in pairs. This fact allows us to prove that there must be an even number of odd vertices (if any), and it provides an alternative method of proof from that suggested in Exercise 7 of Section 9.3.

A more significant, and useful, relation exists among the numbers of vertices, edges, and faces in a polygonal graph. For a source of data, let us turn to the graphs of Figure 9.34 in the exercises at the end of Section 9.5. In Table 9.3 we have recorded the numbers of vertices, edges, and faces in each of these graphs; they are denoted by V, E, and F, respectively.

Graph	V	E	F
(a)	5	8	5
(b)	4	6	4
(c)	10	13	5
(d)	10	15	7
(e)	9	13	6
(f)	6	8	4
(g)	12	18	8
(h)	17	25	10

Table 9.3   The numbers of vertices, edges, and faces in the graphs of Figure 9.34.

At first glance, there may not seem to be much of a pattern. For instance, graphs with the same number of vertices do not have the same number of edges, and vice versa. Those with more vertices tend to have more edges and faces, but in apparently random fashion. However, we can see that edges tend to be more abundant than either vertices or faces, and if we look at the excess of edges over vertices, a striking pattern emerges. (See Table 9.4.)

Comparing the last two columns of Table 9.4, we see that $E - V$ is exactly 2 less than F in every instance. Let us state this observation in the form of an algebraic equation:

(9.13)
$$E - V = F - 2$$

Graph	V	E	F	E − V
(a)	5	8	5	3
(b)	4	6	4	2
(c)	10	13	5	3
(d)	10	15	7	5
(e)	9	13	6	4
(f)	6	8	4	2
(g)	12	18	8	6
(h)	17	25	10	8

Table 9.4   *A pattern in the data of Table 9.3.*

This equation can be rearranged algebraically in many ways; regardless of the form, the relation is called *Euler's Formula*.[7]

(9.14)

> **Euler's Formula**: In any polygonal graph, $V - E + F = 2$, where V, E, and F are the number of vertices, edges, and faces, respectively.

Of course, the fact that this formula is correct for eight examples is not proof that it is always correct. The general argument is not too difficult, as we shall see.

Proof of Euler's Formula

Any polygonal graph can be constructed by starting with one vertex and expanding step by step with one of two types of procedures:

Type 1 Add a new vertex and join it by an edge to some vertex already present.

Type 2 Add a new edge (possibly a loop) joining two (not necessarily distinct) vertices already present.

For example, we can construct the polygonal graph of Figure 9.35(g) by starting with a single vertex in Figure 9.35(a) and adding vertices and edges according to these two types of steps. Each intermediate step is shown in Figure 9.35, labeled "Type 1" or "Type 2." This is not the only sequence of steps you might follow, but this, or any, polygonal graph can be constructed using only these two types.

Now, let us keep track of how the quantity represented by Euler's Formula, $V - E + F$, changes at each stage of construction. Starting with $V = 1$,

---

[7]This formula reportedly was known to earlier mathematicians, but the earliest recorded proof was in a brief paper by Euler.

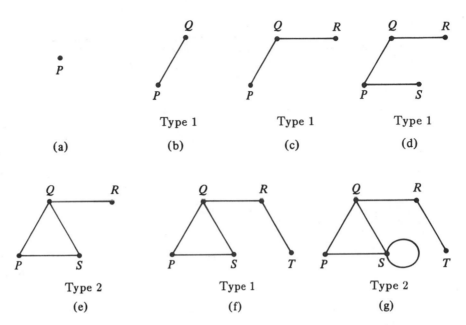

Figure 9.35 *Proving Euler's Formula.*

$E = 0$, and $F = 1$ (the plane is the one face), Euler's Formula is satisfied. Each time a step of Type 1 is used, we add a vertex and an edge, so V and E each increase by 1. Since the graph we are building is polygonal, it is planar. Thus, the new edge lies entirely within a face that already exists, and no new faces are created by this construction (that is, F remains unchanged). Thus, $V - E + F$ does not change from its value before the step. Each time a step of Type 2 is used, E increases by 1 and V remains fixed. Since the added edge lies within some face (planar graph) and the graph is connected at each step (new vertices are always connected by a new edge by Type-1 steps), this added edge must "cut off" part of the "old" face, creating a new, additional face; that is, F increases by 1. Again, the value of $V - E + F$ does not change. In this way, starting with $V - E + F = 2$, we build up the graph by a series of construction steps that leave the quantity $V - E + F$ unchanged at each step. Therefore, we must end with a graph satisfying Euler's Formula.

There are a number of applications and generalizations of Euler's Formula that are of importance in graph theory, geometry, and topology. One classical case involves solid geometric figures, such as cubes, prisms, and pyramids. Specifically, let us look at any solid whose edges are straight-line segments, whose faces are polygons (triangles, quadrilaterals, and so on), and which have no "holes" (no square doughnuts allowed, for instance). If we think of the solid as reduced to an open-latticework figure (such as a toothpick model), with elastic edges, we can flatten the solid into a plane,

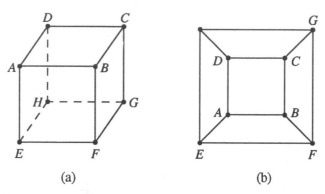

(a)   (b)

Figure 9.36   *A polygonal graph.*

forming a polygonal graph.[8]   For example, in Figure 9.36(b), we have a
"graphic" representation of the cube indicated in Figure 9.36(a).

Notice that, although the graph distorts the geometric shape of most of
the faces (from square to trapezoidal), it accurately represents the number of
vertices, edges, and faces of the solid, as well as the number of edges meeting
at each vertex (the degree) and the number of edges surrounding each face
(the order).   Note that one face of the cube (*EFGH*) corresponds to the
"outside" face of the graph. Since the numbers of vertices, edges, and faces
translate unchanged from solid to graph and vice versa, Euler's Formula
must hold for such solids.   This allows us to establish several important
facts about geometric solids.

As one example, we focus on **regular solids** — that is, solids with each
face an equilateral, equiangular polygon, and with the same number of faces
at each vertex. The cube is probably the most familiar example of a regular
solid. Are there others? If so, how many kinds are there, and what do they
look like? Euler's Formula and a little numerical reasoning can answer these
questions.

Consider a typical regular solid. If we let D be the degree of each vertex
(the degree must be the same for each one if the solid is regular), then $D \cdot V$
counts the sum of all the degrees of all the vertices. This must equal twice
the number of edges, as noted earlier. Similarly, letting R be the order of
each face, $R \cdot F$ must also equal $2 \cdot E$; this follows either by duality or by
noting that each edge has two faces adjacent to it. Thus, we know that:

(9.15)
$$D \cdot V = 2 \cdot E \quad \text{or} \quad V = \frac{2 \cdot E}{D}$$

and

---

[8]It is precisely this connection that lends the name "polygonal" to polygonal graphs.

(9.16)
$$R \cdot F = 2 \cdot E \quad \text{or} \quad F = \frac{2 \cdot E}{R}$$

Substituting these into Euler's Formula, we obtain

(9.17)
$$\frac{2 \cdot E}{D} - E + \frac{2 \cdot E}{R} = 2$$

and dividing this expression by E produces

(9.18)
$$\frac{2}{D} - 1 + \frac{2}{R} = \frac{2}{E}$$

Since we made no special assumptions about the regular solid under discussion, this formula *must hold for every regular solid*.

We can now try some particular values of D and R, and determine the corresponding values of E from Statement (9.18). The order of each face, R, must be at least 3 (a triangle is the simplest regular polygon) and D must be at least 3 (each vertex of a solid must join at least three faces). If we let $D = R = 3$, the left side of Statement (9.18) becomes $\frac{2}{3} - 1 + \frac{2}{3}$, which equals $\frac{1}{3}$. Then E must equal 6 to balance the equation. Statements (9.15) and (9.16) tell us that V and F are each 4 in this case, and we have described a solid with four vertices and four triangular faces, three meeting at each vertex. This figure is called a (regular) **tetrahedron**.

As a consequence of Statement (9.18), the number of allowable choices for D and R turns out to be quite limited. Because E, and thus $\frac{2}{E}$, must be positive, the left side of the equation must also be positive. If either D or R is too large, $\frac{2}{D}$ and $\frac{2}{R}$ become so small that their sum is less than 1, and the left side is negative or zero. Moreover, as noted previously, D and R must each be at least 3. By trial, there are only five possible pairs of values for D and R. These values are given in Table 9.5, together with the resulting values of E, V, and F required by Statements (9.18), (9.15), and (9.16).

D	R	E	V	F	Name of Solid
3	3	6	4	4	tetrahedron
3	4	12	8	6	hexahedron (cube)
3	5	30	20	12	dodecahedron
4	3	12	6	8	octahedron
5	3	30	12	20	icosohedron

Table 9.5 *The five regular solids.*

These five regular solids were known to the ancient Greek mathematicians. They could find no others. But they could not *prove* that there were no others. It was not until the advent of graph theory that the relatively simple numerical relationships, which were "hidden" in the original geometric forms of these figures, became clear from their graph-theoretic forms and then could easily be stated and proved.

EXERCISES 9.6

1. The graph in Figure 9.37 is planar, but not connected. If we stretch the definition of *face* a little, we can identify three faces. Count the vertices and edges and compute

$$V - E + F$$

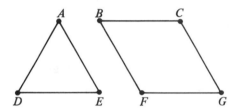

Figure 9.37

2. Construct several planar graphs that are disconnected into two parts. (See Exercise 1.) For each graph, compute $V - E + F$. Conjecture a formula for such graphs. Prove your conjecture if you can.

3. Repeat Exercise 2 for planar graphs disconnected into three parts.

4. Let P be the number of parts in a planar graph. (For a connected graph, $P = 1$; the graph of Figure 9.37 has $P = 2$; etc.) Determine a formula for planar graphs that relates V, E, F, and P. (See Exercises 2 and 3.)

5. In a polygonal graph, does the degree of a vertex always equal the number of faces meeting at that vertex? If not, why not?

6. In Statement (9.18), if $R = 3$ and $D = 6$, the left side of the equation becomes zero. The right side, $\frac{2}{E}$, can be thought of as equaling zero if E is infinite. As a polygonal graph, this corresponds to a figure with six $(D = 6)$ equilateral triangles $(R = 3)$ at each vertex,

extending infinitely throughout the plane (so that there are an infinite number of edges, vertices, and faces). This is called a **regular tesselation** of the plane with triangles (see Figure 9.38). Find any other values of R and D that have this property. Identify the corresponding tesselation.

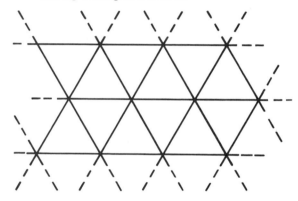

Figure 9.38

7. Verify Euler's Formula for the solids pictured in Figure 9.39.

8. Calculate $V - E + F$ for the solids in Figure 9.40. These solids, with polygonal faces, have one "hole" through them. Note that no single face has a hole; rather, the hole is surrounded by several polygons.

9. Let H be the number of holes in a solid. (For the solids in Figure 9.39, $H = 0$; for those in Figure 9.40, $H = 1$; and so on.) Determine a formula for solids that relates V, E, F, and H. (See Exercises 7 and 8.)

10. Show that each of the graphs representing the five regular solids is a Hamilton graph.

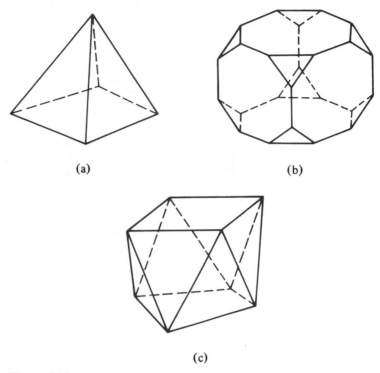

(a)

(b)

(c)

Figure 9.39

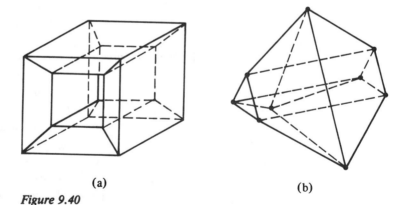

(a)

(b)

Figure 9.40

WRITING EXERCISES

1. In examining the relationship among V, E, and F, Table 9.4 compares $E - V$ to F. Describe in detail a similar comparison between $E - F$ and V. What relationship can be discovered by looking at this data? Comment on how and why this is similar or dissimilar to Euler's Formula.

2. Explain the etymology of the names for the five regular solids. (What are the root mean-

ings of the parts of these words? What language(s) do they come from?)

3. In the proof of Euler's Formula, there are two types of steps that, in combination, can describe the construction of any polygonal graph. Suppose, instead, that we describe the construction of a polygonal graph by first locating all vertices, then drawing in all the edges. Discuss the pros and cons of this as a method of describing the construction, then as a tool for proving Euler's Formula.

4. In the discussion about regular solids, the text asserts that the formula $R \cdot F = 2E$ follows by duality. Write a paragraph justifying this assertion.

5. Write a short paper in which you:

   (a) Determine a formula that relates the degree of a vertex to the number of faces at that vertex.

   (b) Identify and define precisely the other quantities that need to be counted for your formula.

   (c) Give at least two examples to demonstrate your formula.

   (d) Try to prove that your formula is always correct. (See Exercise 5.)

9.7   Looking Back

In reflecting on what you have studied in this chapter, it is important to observe that some aspects of our exploration of graph theory typify common general approaches and procedures in the exploration of mathematical ideas. Here we summarize a few parts of our discussion that are significant in this regard.

The method by which graphs with edge paths were identified in Section 9.3 is a common one in mathematics. Recall that we were looking for characteristic properties of graphs so that we could find an edge path. What we did was to assume that we had a solution (an edge path) and then investigate the "hypothetical" graph with that path. In a sense, we assumed the "answer" and worked backwards to examine the "question."

In Sections 9.3 and 9.4 we examined the problem of finding edge paths and vertex paths, respectively. The two problems are closely analogous, each dealing in exactly the same way with one of two fundamental components of a graph. But the results were anything but analogous. Edge paths were completely characterized in terms of connectedness and the degrees of vertices, but vertex paths defied characterization in terms of any combination of properties we could find. We were left with only partial results about vertex paths, and hence with a problem that remains unsolved (as it has been for some two centuries). Such disparity of solution or theory for apparently similar problems is not unusual in mathematics.

On the other hand, the problem of crossing curves, which at first seemed quite different from the path problems, turned out to be virtually identical to that of edge paths. This, too, is not uncommon. In fact, a fundamental mathematical process is the attempt to abstract the essential structure from a given situation. By stripping away misleading or irrelevant details, mathematics often can identify the true nature of a problem and then (hopefully) use known facts from previously solved problems which are essentially the same. Even when solutions are not known, the identification of abstract similarities in two or more problems paves the way for the crossover of techniques from one problem to another, leading to a possible solution of all the problems involved. In many such instances, as in the one we encountered, the similarities between the problems are so pervasive that each is the **dual** of the other; that is, each statement can be formed from the other simply by exchanging one or two key terms (and their related phrases) in one statement for corresponding key terms in the other.[9] Typically, the basic connection between the duals is relatively simple, even though many particular details may correspond as a consequence. In several areas of mathematics, the transfer of ideas via duality is a major source of new ideas and theorems.

Finally, the historical development of graph theory, from puzzles and curious diversions to a variety of applications, is again typical of much of mathematics. At various times in history, new concepts (such as negative numbers or complex numbers) or even an entire branch of mathematics (such as non-Euclidean geometry) were considered merely as curiosities, of value at best to clarify the understanding of "real" mathematics. Only later did they take on profound practical significance as valuable tools for describing and analyzing real situations. So it is with graph theory, which is now beginning to flourish in the mathematical garden of powerful, useful ideas.

Exercises 9.7

WRITING EXERCISES

1. Explain how the use of equations to solve a word problem in elementary algebra is an example of assuming an "answer."

2. Write a short paper defending the assertion that the analogy between the edge-path problem and the vertex-path problem is illusory. (Are there only two components to a graph? Are they analogous? What does Section 9.5 suggest?)

---

[9] For instance, "Any two distinct points determine a unique line" and "Any two distinct lines determine a unique point" are dual statements, using *point* and *line* as the corresponding terms. In Euclidean geometry, only one of these statements is true; in projective geometry, however, they are both true.

LINK: 9.8

Digraphs and Project Management

Digraphs can be used to manage major projects, such as those arising in the construction industry.

We mentioned at the outset that graph theory has been applied in a number of areas. Some of these areas are internal to mathematics. For example, as we saw in Exercise 9 of Section 9.6, solids have a formula relating the number of vertices, edges, faces, and holes. There are many complicated figures for which it is difficult to determine whether or not something that appears to be a hole should be counted as such. The known formula becomes a way of deciding this and thereby classifying the solid. A similar technique helps topologists to classify the dimension and shape of various figures. Our discussion of edge paths and vertex paths can be applied in a variety of situations. Sometimes such problems can be figured out (or avoided) by common sense, without appealing to graph theory; however, by abstracting the problem to its essential features, a solution that may be hard to see in its "real" setting often can be made much clearer in its graphic representation.

One area of application involves the construction of planar graphs with specific conditions as to which vertices are to be joined by edges. A classic puzzle illustrates this type of problem: Three eccentric recluses live in three houses. Each has a grudge against the other two and refuses even to walk on the same ground, but each one needs to do business at each of three local shops. Is it possible to find paths from each house to each of the three shops in such a way that no recluse crosses another's path? In graph-theoretic terms, this is equivalent to constructing a planar graph with an edge connecting each of vertices A, B, and C to each of vertices X, Y, and Z. (We leave the solution of this problem to you.) Such problems find practical application, for instance, in the design of printed circuits. The general theoretical form of this problem has been completely solved, but practical difficulties often arise in specific examples. In other words, the theory may guarantee that a certain planar graph exists, but the actual drawing of such a graph can be very tedious.

Many applications of graph theory require the added notion of restricting the *direction* of allowable paths by assigning a direction to some or all edges. (This is like making some streets one-way.) A **directed edge** is denoted by an arrowhead; a graph with all edges directed is a **directed graph**, or simply a **digraph**. A **directed path**, as you would expect, is a path in a digraph which follows the prescribed directions of the edges.

Example 9.14   Figure 9.41 shows two examples of digraphs. In Graph (a), $ABED$ is a directed path from A to D. There is no directed path from D to A. In Graph (b), there are two odd vertices (P and S), but there is no directed edge path, since P and S each have two exiting edges and only one entering edge.   □

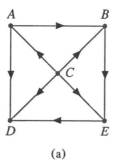 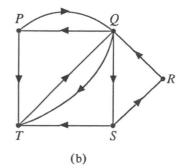

(a)   (b)

Figure 9.41   *Two digraphs.*

As Example 9.14 suggests, the statements about paths in Section 9.3 invite revision for directed paths, but pursuing these revisions in detail would distract us from our purpose in introducing digraphs. The great utility of the digraph comes from its applicability in so many areas. The vertices of a digraph may represent individuals or groups and the directed edges could signify their financial, social, or political influence on one another. The corresponding digraph would allow one to analyze the interrelations among the individuals or groups. Or the vertices may represent locations in a computer's memory, and the directed edges may represent the ways in which information can be transferred from one location to the other. Such a digraph can be extremely useful in data processing, where efficient manipulation of, and access to, stored information is essential. Digraphs can represent the interlocking food chains in some ecosystem; they can model the results of athletic contests; they can represent the organizational structure of a business; there are applications in virtually any field that involves relationships of any kind.

Graphs, directed or not, can be coupled with arithmetic to meet other needs. An obvious application of this kind uses the assignment of numbers to represent lengths of edges. The numbers might represent distance (as on a road map) or they might measure time, seniority, and so on. Alternatively, numbers can be assigned to vertices (which, for historic reasons, is known as **coloring** a graph). For example, each vertex of a digraph might represent a portion of a computer program, with a number that specifies the time required to execute that portion; directed edges between some vertices might correspond to the fact that certain things must be done before others, as when the results of one computation are necessary for another. Analysis of such a graph can predict, within limits, how long the program will take to run in certain circumstances. This is critical, for example, in designing computer programs for a missile defense system.

In another important application of colored digraphs, each vertex represents a task to be accomplished in some project. Numbers represent the time required for each task, and the directed edges represent the required order of precedence among the tasks, where appropriate. The digraph is then a planning tool to determine the most efficient means of allocating resources and personnel. This is the heart of the management technique known as PERT (Program Evaluation and Review Technique). We shall examine this application by means of an extended, somewhat contrived example. Although the story may seem a bit facetious, the process it portrays can be of serious value in planning and managing large, complex projects.

Example 9.15   Lester Leezure is a wealthy eccentric who lives on a remote private island in the Pacific Ocean. He contacts a contractor in California to get an estimate for turning his unfinished basement into a recreation room. Lester wants a paneled, carpeted entertainment space, including a wet bar. The contractor provides an itemized estimate for the cost of materials and for the cost of labor based on the time needed for each task, which is listed in Table 9.6.

Task	Days
A. Frame Walls	3
B. Panel Walls	4
C. Hang Acoustic Ceiling	3
D. Install Wiring	2
E. Install Electrical Fixtures	1
F. Do Plumbing	3
G. Install Wet Bar	2
H. Lay Carpet	2
Total	20

Table 9.6 *Tasks for renovating Lester's basement.*

Lester is not bothered by the cost of the materials, but he expresses some concern about the projected time span for completing the project. The contractor explains that the labor estimate is based on person-days of work. For reasons of safety and availability, he always uses two-person crews, and thus he would plan to have the job finished in ten days. Lester still has a problem, though; he would like to minimize the upset to his home and habits and get the work done as quickly as possible. "If one person can do the job in twenty days, or two can do it in ten days, why not put more workers on the job and get it done more quickly? How about employing ten workers and getting it done in two days?" he asks.

The contractor points out that increasing the number of workers does not automatically reduce the time proportionately — too many people just get in each other's way. Through years of experience, he has found that putting more than two men to work on any task becomes counterproductive.

"Even so," Lester retorts, "the most time-consuming task is the paneling, which should take two workers only two days; the rest of the work could be done by other workers at the same time."

The contractor (by now contemplating physical violence) points out that it would not be possible to do *all* of the tasks simultaneously with different workers. For example, the paneling can only be done after the framing, and the carpeting should not be laid while other work is still being done. To put it in black and white for Lester, the contractor produces another list, which is shown in Table 9.7.

"Well, how quickly can you do the job?" asks Lester.

"I've never worked that way before," replies the contractor in exasperation, "so I don't know. If you want to pay for them, I'll put on a larger crew and we'll find out how long it takes. But since you live so far from civilization, once a crew is assigned to your job, they'll have to stay with it for the whole time; I can't afford to ferry them back and forth."

Before starting task	we must complete task
B	A, D, F
D	A
E	A, B, C, D
G	F
H	A, B, C, D, E, F, G

Table 9.7   *Task ordering for the basement project.*

Impatient but uncertain, Lester decides to compromise. He agrees to pay for three two-person crews — a total of six workers — in the hope that the job can be done in a little more than 3 days. The contractor, who is honest and competent, puts his people to work in the most efficient ways he can find, and none of the individual tasks takes longer than its estimated time. Nevertheless, the job takes six days. Needless to say, Lester is upset about having to pay for 36 person-days of labor for a job estimated at 20 person-days, and he complains about the amount of time some of the workers sat around doing nothing. The contractor is upset because the extra labor time seems to cast doubt on his ability and his integrity. If they had only consulted a graph theorist, they would have avoided all this wasted money, bad temper, and distress!

You see, a digraph and PERT can be used to come up with an answer to Lester's question. Using the contractor's two charts, we produce a single digraph, shown in Figure 9.42. The vertices represent the tasks, and the numerical "coloring" records the time for each; the directed edges convey the required precedence relations listed in Table 9.7.

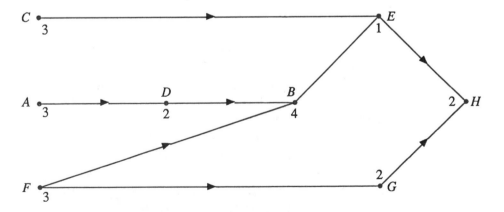

Figure 9.42   *The digraph for Lester's
basement renovations.*

Determining the time to do the construction job translates into adding up the numbers in the graph. The various paths in the graph correspond to *possible* different workers (or pairs of workers, in half the time) accomplishing different sequences of tasks. Several distinct, intertwining sequences of tasks must be carried out in order to accomplish the job — specifically, *CEH*, *ADBEH*, *FBEH*, and *FGH*. The total lengths (in days) of these paths are 6, 12, 10, and 7, respectively. The longest path, *ADBEH*, is called a **critical path**; its length sets a lower limit for the time in which the total job can be finished.

This graph shows Lester that the shortest work-time he can hope for is six days (with a pair of workers following the tasks on the critical path and others working to do the remaining tasks). In fact, then, the contractor had done the most efficient scheduling job possible for three pairs of workers. He began with each pair of workers simultaneously doing tasks *A*, *C*, and *F*. However, when they finished their first tasks (after a day and a half), while one pair proceeded to *D* and another pair to *G*, the third pair had to sit around idle. Overall, an optimal schedule for a three-pair crew is *ADBEH* for one pair, *FG* for another, and *C* for the third; the total job takes six days with a lot of wasted time for most of the workers. Moreover, because there are only three separate paths at any stage of the graph, adding a fourth pair of workers would have made no time difference whatsoever!

This little bit of graph theory could have saved Lester a lot of money by convincing him that he would lose no time using a two-pair crew: One pair would follow the critical path while the other pair would do task *F* first, then *C* and *G*, in either order. Again the total job would take the minimum time of six days (there is no way to escape the maximal length of a critical path), but Lester would have cut his labor cost to 24 person-days. The critical-path analysis shows that to minimize time, Lester could do no better than six days, with some workers necessarily idle some of the time. To minimize cost, the workers would have to be fully employed at all times. This can only be done by using two workers (or one) for the entire job, doing the tasks in any order that does not violate the precedence requirements; as the contractor said, this would take ten (or twenty) days.

One final observation about this case is instructive: If the second pair made the mistake of doing *C* before *F*, they would force a delay for the other pair before the latter could do task *B*. Even with the assistance of a digraph, there is more to efficient planning than just identifying the critical path.   □

Generalizing from this example, it is not really surprising that putting more workers on a job does not necessarily translate in a simple, proportional way into getting the job done faster. The example also makes clear that there are times when putting more workers on a job has absolutely no effect on the total length of time to do the job. This particular example does not show some counterintuitive features, such as the fact that assigning available workers successively to tasks in the longest critical path not yet begun is sometimes a very inefficient way of scheduling, or that efficiencies in performing individual tasks (such as reducing the time to do the paneling, lay the carpet, etc.) can sometimes actually increase the total time to do the job. Nevertheless, it does give some flavor of the use of digraphs to represent information in a scheduling problem, of the limits set by critical paths, and of the fact that, even with a digraph, determining a schedule that optimizes time or cost can still be a challenge. And ours was a trivial example. Consider the problem of scheduling the more than 20,000 tasks in the Apollo space mission!

Topics for Papers – Chapter 9

1. Investigate the potential use of dual graphs for pursuing the problem of characterizing those graphs that contain vertex paths. How does the problem carry over to its dual? How do the concepts defined in Section 9.4 transfer to their dual forms? Create appropriate terminology and give precise definitions. Look for insights into solving the dual problem; if you identify any, how do they translate back to the original problem via the dual process? Give your assessment of the value, real or potential, of the use of dual graphs to solve the original problem.

2. Generalize Euler's Formula to three-dimensional "polyhedral graphs." Note that there are *four* quantities to count: vertices, edges, faces, and three-dimensional regions separated by faces. Make any necessary definitions to clarify your discussion; consider any appropriate restrictions (e.g., what would correspond to the restriction "polygonal"?); generate data from several examples; formulate a conjecture; if possible, generalize the proof of Euler's Formula to prove your conjecture. There may be different answers, depending on how you establish restrictions and define quantities; don't try to pursue *all* possibilities, but carefully explain the generalization you see.

For Further Reading

1. Behzad, Mehdi, Gary Chartrand, and Linda Lesniak-Foster. *Graphs & Digraphs.* Boston, MA: Prindle, Weber & Schmidt, 1979.

2. Biggs, N. L., E. K. Lloyd, and Robin J. Wilson. *Graph Theory 1736–1936*. Oxford: Clarendon Press, 1976.

3. Chartrand, Gary. *Introductory Graph Theory*. New York: Dover Publications, Inc., 1977.

4. Devlin, Keith. *Mathematics: The New Golden Age*. London: Penguin Books, 1988, Chapter 7.

5. Flores, Ivan. *Data Structures and Management*. Englewood Cliffs, NJ: Prentice Hall, Inc., 1970, Chapter 2.

6. Harary, Frank. *Graph Theory*. Reading, MA: Addison-Wesley Publishing Company, Inc., 1969.

7. Maldevitch, Joseph, and Walter Meyer. *Graphs, Models, and Finite Mathematics*. Englewood Cliffs, NJ: Prentice-Hall, Inc., 1974.

8. Steen, Lynn Arthur, ed. *Mathematics Today, Twelve Informal Essays*. New York: Springer-Verlag, 1978.

9. Tutte, W. T. *Graph Theory*. Menlo Park, CA: Addison-Wesley Publishing Company, Inc., 1984.

APPENDIX

BASIC LOGIC

A.1   Statements and Their Negations

We present here a brief summary of elementary mathematical logic. In order to find a simple starting point, we leave to the philosophers and scientists many relevant, but difficult, questions, such as "What is truth?" and we examine, instead, just the *form* of the reasoning process. Consequently, our study of logic focuses on how the known truth or falsity of some statements can be used to guarantee the truth or falsity of others. Thus, we begin the study of formal logic by regarding the words **true** and **false** just as labels applied (somehow) to some sentences. These two words are called **truth values**; it is a basic principle of our logical system that there are only two of them. (There are other systems of logic that use more than two truth values, but they are not as simple or as widely used as the system presented here.)

All reasoning is based on "statements" of one sort or another, so we consider first what makes a statement meaningful in the context of logic. A **statement** is a sentence that has a truth value; that is, a sentence that can be labeled either *true* or *false*. This truth value may come from various sources. Often we rely on our common-sense view of reality to tell us whether a statement is true or false; in some cases the truth value is just assigned (as in formal axiom systems). Sometimes the truth value of a compound or complex sentence can be derived from knowing the truth values of its simpler parts. In any case, a sentence *must* have a truth value in order for it to be called a statement.

Example A.1 "There is printing on this page" is a true statement. "No one passes mathematics" is a false statement. □

Example A.2   "Today is Tuesday" is a statement. It is either true or false, depending on when you are reading this.   □

Example A.3   The sentences "What's your name?" and "Close the book!" are not statements because there is no reasonable way to assign either of them a truth value.   □

Standard logic is based on two fundamental assumptions about truth values:

The Law of the Excluded Middle: There are only two truth values, *true* and *false*.

The Law of Contradiction: No statement may be both true and false at the same time (in the same context).

These laws may seem too obvious to mention, but they are crucial to the proof process. Specific examples of their use appear later in this Appendix and throughout the book.

Example A.4   Every sentence in the second paragraph of this section is a statement; each one is either true or false. (The fact that you might consider them all true stems, no doubt, from your complete faith in the accuracy of the authors. Such faith is flattering, but not always wise.)   □

Example A.5   The sentence "This statement is false" is *not* a statement in standard logic because it cannot have either truth value. If we consider it true, then we must accept what it says, which makes it also false, violating the Law of Contradiction. If we consider it false, then the Law of the Excluded Middle implies that the only alternative to what it says is that it is true, again violating the Law of Contradiction.   □

Two types of statements deserve special consideration because of their importance in the reasoning process; they are the two basic kinds of "quantified" statements:

A **universal statement** asserts that all things of a certain kind satisfy some condition.

An **existential statement** asserts the existence of at least one thing that satisfies some condition.

Example A.6   "All mice are animals" and "Every building has a flat roof" are universal statements.   □

Example A.7   "There is a fly in my soup," "Wizards exist," and "There really are unicorns" are existential statements.   □

Note

The word *some* is used in mathematics as a shorter form of "there is at least one." It does not necessarily refer to more than one thing, nor does it rule out the possibility that *all* things might satisfy the condition being discussed. Both "Some trout are fish" and "All trout are fish" are true statements.

Example A.8   "Today is Thursday" and "The moon is made of green cheese" are neither universal nor existential statements. They are called **particular statements**, because they refer to properties of specifically designated things. □

Universal, existential, and particular statements can be either true or false.

Example A.9   "$x + 3 = 7$" is *not* a statement because not enough is known about x to decide whether the sentence is true or false. However,

$$\text{"For all numbers } x, x + 3 = 7\text{"}$$

is a false universal statement;

$$\text{"There exists a number } x \text{ such that } x + 3 = 7\text{"}$$

is a true existential statement;

$$\text{"There exists a number } x \text{ such that } x + 3 = x\text{"}$$

is a false existential statement;

$$\text{"}4 + 3 = 7\text{"}$$

is a true particular statement.   □

Every statement has a *logical opposite*. Having said this, let us quickly caution you that "opposite" must be interpreted very carefully. For instance, "This car is going north" is *not* the *logical* opposite of "This car is going south." To reinforce that caution, we shall label this idea with a less ambiguous word defined in terms of truth values, considering the values *true* and *false* as logical opposites.

DEFINITION   The **negation** of a statement is a statement whose truth value is always opposite to that of the original statement.

Note

The word "always," as used in the definition of negation, means "in every circumstance." Thus, *whenever* a statement is true, its negation *must* be false; and *whenever* the statement is false, its negation *must* be true.

Example A.10   "The wall is completely white" and "The wall is not completely white" are negations of each other. "The wall is completely white" and "The wall is completely green" are *not* negations of each other because, although they cannot both be true at the same time, they can both be false at the same time. (Such statements are called **contraries**.)   □

NOTATION   In algebra, the task of writing numerical expressions is often simplified by using letters to represent numbers. Similarly, we simplify (and shorten) the writing of logical expressions by using letters, such as s, p, and q, to represent statements. If s represents a statement, then its negation is denoted by $\sim s$.

Example A.11   If s represents "Jack owns a car," then $\sim s$ represents "Jack does not own a car." Suppose s is true; that is, suppose that Jack actually owns a car. Then the statement $\sim s$ is false. On the other hand, suppose s is false. This means that Jack does not own a car, which means that $\sim s$ is true.   □

The truth-value relationships among several statements may be represented conveniently by a diagram called a **truth table**. A very simple truth table can be used to show the relationship between a statement s and its negation $\sim s$. For instance, suppose s is the statement "Jack owns a car," as in Example A.11. Then $\sim s$ is "Jack does not own a car." We use a column of the table for each of the two statements (Table A.1(a)). Because s may be either true or false, we enter both possibilities (symbolized by T and F) in the first column (Table A.1(b)). The entries in the second column indicate that the truth value of $\sim s$ must always be the opposite of the truth value of s (Table A.1(c)).

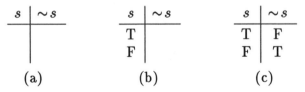

s	$\sim s$

(a)

s	$\sim s$
T	
F	

(b)

s	$\sim s$
T	F
F	T

(c)

Table A.1   Negation.

As another example, the relationships among a statement s, its negation $\sim s$, and the negation of that, $\sim(\sim s)$, are shown by Table A.2.

s	$\sim s$	$\sim(\sim s)$
T	F	T
F	T	F

Table A.2   Double negation.

Notice that the first and third columns have identical truth-value entries. This means that s and $\sim(\sim s)$ are the same logically, even though they might be grammatically different. If s is "Jack has a car," then $\sim(\sim s)$ can be phrased "It is not the case that Jack doesn't have a car."

DEFINITION Two statements are **logically equivalent** (or simply **equivalent**) if they always have the same truth values.

Thus, Table A.2 shows that, for every statement s,

$$s \text{ and } \sim(\sim s) \text{ are logically equivalent.}$$

Because they have identical truth values, equivalent statements are interchangeable at any time in any logical argument. They are just different ways to say the same thing. The choice of which form to use is merely a matter of convenience or taste.

Example A.12   The negation of the universal statement "All animals have fur" is

"Not all animals have fur."

It is logically equivalent to the existential statement

"Some animals do not have fur." □

Example A.13   The negation of "All mice are animals" may be phrased as "There is a mouse that is not an animal," or "Some mice are not animals," or "Not all mice are animals." The statement "No mice are animals" is *not* a negation of the first statement; it and the first statement are contraries. □

Examples A.12 and A.13 illustrate why we refer to *the* negation of a statement, despite previous examples of several apparently different negations of the same statement. The forms of the negation may be *grammatically* different, but they are *logically* equivalent; that is, *from the viewpoint of logic, every statement has exactly one negation.* These two examples also illustrate a basic relationship between universal and existential statements:

The negation of a universal statement is (equivalent to) an existential statement.

The negation of an existential statement is (equivalent to) a universal statement.

Example A.14   "There are red fire engines" and "No fire engines are red" are negations of each other. "All birds fly" and "Some birds do not fly" are negations of each other.   □

Example A.15   "It is false that the wall is not green" is equivalent to "The wall is green." (Why?)   □

EXERCISES A.1

In Exercises 1–10, which of these sentences are statements? Why? Label each statement *true* or *false*.

1. The sun is shining.
2. This book is written in French.
3. Nobody loves me.
4. Where are you going?
5. Join the Pepsi generation.
6. Meet me in St. Louis.
7. $3 + 5 = 8$
8. $3 + x = 8$
9. $3 + 5 = 7$
10. For some number x, $3 + x = 8$

In Exercises 11–18, classify each statement as *universal*, *existential*, or *particular*.

11. Some flowers are yellow.
12. All dogs have fleas.
13. There is a unicorn on the lawn.
14. The sun is shining.
15. Some elephants fly.
16. These trees are green.
17. Not all students are sophomores.
18. No textbooks are useful.

In Exercises 19–26, express (an equivalent form of) the negation of each statement.

19. This is a mathematics book.
20. Today is Sunday.
21. All roses are red.
22. Some violets are blue.
23. $8 + 3 = 11$
24. No rabbits chase mice.
25. Some cars do not get good gas mileage.
26. There are at least three trees in the field.

27–34. Express (an equivalent form of) the negation of each statement in Exercises 11–18.

A.2   Conjunctions and Disjunctions

We next consider truth values for combinations of statements. We focus on pairs of statements because more complex combinations can be analyzed two at a time. The truth value of a combination can be deduced from the truth value of its component parts, according to several simple rules that are generally in agreement with our common sense. These rules are actually the *definitions* of the connective words used. In this section we consider the formal logical usage of *and* and *or*.

DEFINITION   The **conjunction** of two statements p and q is a statement of the form "*p* and *q*"; it is true if p and q are both true, and is false otherwise. "*p* and *q*" is often written "$p \wedge q$."

DEFINITION   The **disjunction** of two statements p and q is a statement of the form "*p* or *q*"; it is false if p and q are both false, and is true otherwise. "*p* or *q*" is often written "$p \vee q$."

Note

This usage is sometimes called the *inclusive or* because it allows both statements to be true. If we were to define "*p* or *q*" to be false when both statements are true, it would be called the *exclusive or*. This latter usage is discussed in Exercise 20.

The truth values for these two compound statement forms are summarized in Table A.3. Notice that we must consider four different cases, corresponding to the four different truth-value possibilities for the *pair* of statements p, q. Those four pairs of possible truth values are listed in the first two columns of the table.

p	q	p and q	p or q
T	T	T	T
T	F	F	T
F	T	F	T
F	F	F	F

Table A.3   *Conjunction and disjunction.*

Example A.16   Consider the statements

p: Today is Friday.

q: The sun is shining.

Then

"*p* and *q*" is "Today is Friday and the sun is shining."

"*p* or *q*" is "Today is Friday or the sun is shining."

To see how the truth tables behave in this situation, observe from Table

A.3 that "*p* and *q*" is true only on sunny Fridays, but "*p* or *q*" is true every Friday, regardless of the weather, and is also true on every sunny day.   □

At this point it is natural to ask how negation affects these compound-statement forms. In particular, how are the negations of the conjunction and disjunction of two statements related to the negations of the separate statements? Notice that everyday English usage is based on a precise logical structure, but we seldom pay specific attention to the structure itself. For instance, if you tell someone who wants you to take a message that you do not have paper or pen, you are saying that you do not have paper *and* you do not have a pen. Often this appears as a prepositional phrase — such as "without paper or pen," meaning "without paper and without pen" — in which the preposition itself conveys the negative sense. This shift from *or* to *and* when the compound statement is negated typifies an important general principle governing the negation of conjunctions and disjunctions.

Consider the four combination forms involving negation:

$$\sim(p \text{ and } q) \qquad \sim(p \text{ or } q) \qquad (\sim p) \text{ and } (\sim q) \qquad (\sim p) \text{ or } (\sim q)$$

As usual in mathematical symbolism, we first perform the operations within the parentheses. Thus, in "$\sim(p \text{ and } q)$" we first consider "*p* and *q*" and then negate it; on the other hand, in "$(\sim p) \text{ and } (\sim q)$" we first form the negations $\sim p$, $\sim q$ — then we put them together by conjunction.

Now examine Table A.4, which gives us the truth values for these four combination forms. The tables show that, as negation is distributed over the separate statements in these combinations, conjunction is changed to disjunction and vice versa.

p	q	p and q	$\sim(p$ and $q)$	p or q	$\sim(p$ or $q)$
T	T	T	F	T	F
T	F	F	T	T	F
F	T	F	T	T	F
F	F	F	T	F	T

p	q	$\sim p$	$\sim q$	$(\sim p)$ or $(\sim q)$	$(\sim p)$ and $(\sim q)$
T	T	F	F	F	F
T	F	F	T	T	F
F	T	T	F	T	F
F	F	T	T	T	T

Table A.4 *Negation with conjunction and disjunction.*

By examining all the truth-value possibilities, Table A.4 actually *proves*:

(A.1)	"$\sim(p$ and $q)$" is logically equivalent to "$(\sim p)$ or $(\sim q)$";
(A.2)	"$\sim(p$ or $q)$" is logically equivalent to "$(\sim p)$ and $(\sim q)$."

These two statements are called **De Morgan's Laws**, after the 19th-century British mathematician Augustus De Morgan.

Example A.17   A simple, but slightly cumbersome, example of De Morgan's Law (A.1) might be phrased:

"It is false that (both) today is Friday and the sun is shining"
is equivalent to
"Either today is not Friday or the sun is not shining."

An example of De Morgan's Law (A.2) is:

"It is false that either today is Friday or the sun is shining"
is equivalent to
"Today is not Friday and the sun is not shining."   □

One more question about the relationships among *and*, *or*, and $\sim$ is worth pursuing at this point, if only because it is an obvious one to ask: For a single statement s, what can be said about "s and $(\sim s)$" and about "s or $(\sim s)$"? Table A.5 provides the answer.

s	$\sim s$	s and $(\sim s)$	s or $(\sim s)$
T	F	F	T
F	T	F	T

Table A.5   A statement combined with its own negation.

Notice that, regardless of the truth value of s,

$$s \text{ or } (\sim s)$$

is always true, and

$$s \text{ and } (\sim s)$$

is always false. These are examples of two important types of logical statements:

DEFINITION   A statement that is always true, regardless of the truth values of its component parts, is called a **tautology**, and a statement that is always false is called a **contradiction**.

Using the form "s or $(\sim s)$," we can write many obvious tautologies, such as "Either today is Friday or it's not Friday" and "Either I'll go out or I won't." Similarly, we can construct obvious contradictions, such as "Today is Monday and it's not Monday" and "I'm reading this page and I'm not reading it." Some far more useful instances of tautologies and contradictions occur in formal logic. They form the basis for establishing the validity of logical arguments. Examination of that part of the theory is beyond the scope of this appendix; however, important examples of the use of contradictions in proofs occur in Chapters 3 and 6 and in Appendix B.

EXERCISES A.2

In Exercises 1–10, consider the statements

p: This book is interesting.
q: I am falling asleep.

Write each of the following as a grammatically correct sentence.

1. p and q

2. p or q

3. $(\sim p)$ and q

4. p and $(\sim q)$

5. $(\sim p)$ or q

6. p or $(\sim q)$

7. $(\sim p)$ and $(\sim q)$

8. $(\sim p)$ or $(\sim q)$

9. $\sim (p$ and $q)$

10. $\sim (p$ or $q)$

In Exercises 11–19, let p, q, and r represent statements. Use truth tables to prove:

11. "p and q" is logically equivalent to "q and p."

12. De Morgan's Law (A.2).

13. "$\sim \Big(p$ and $(\sim q)\Big)$" is logically equivalent to "$(\sim p)$ or q."

14. "$\sim \Big(p$ or $(\sim q)\Big)$" is logically equivalent to "$(\sim p)$ and q."

15. "p or $\sim (p$ and $q)$" is a tautology.

16. "p and $\sim (p$ or $q)$" is a contradiction.

(*Warning*: In the next three exercises, your truth table will require eight rows. Why?)

17. "$(p$ and $q)$ and r" is logically equivalent to "p and $(q$ and $r)$."

18. "$(p$ and $q)$ or r" is *not* logically equivalent to "p and $(q$ or $r)$."

19. "p or $(q$ and $r)$" is logically equivalent to "$(p$ or $q)$ and $(p$ or $r)$."

20. Although in mathematics *or* is usually interpreted in the *inclusive* sense unless specifically indicated otherwise, this is not always the case in everyday usage. For each of the following statements, determine from the context whether *or* is being used in the *inclusive* sense or the *exclusive* sense.

(a) The first prize for the contest is a trip to Hawaii or the cash equivalent.

(b) A student will qualify for financial aid if he/she can demonstrate financial need or high scholastic ability.

(c) Decide whether each statement is true or false.

(d) That tree is either an oak or a maple.

(e) To receive a passing grade you either must have a class average of B or better before the final, or you must earn an A on the final exam. (Interpret both *or*'s.)

(f) To be exempt from the tax, you must be a senior citizen or have an annual income less than $10,000.

(g) He intends to join the Army or the Navy next week.

In Exercises 21–24, let s represent any statement, let t represent a tautology, and let c represent a contradiction. Use truth tables to prove:

21. "s and t" is logically equivalent to s.

22. "s or c" is logically equivalent to s.

23. "s and c" is logically equivalent to c.

24. "s or t" is logically equivalent to t.

25. Exercise 21 says that any statement formed by conjunction with a tautology can be simplified by eliminating the tautology. Give the analogous interpretations of Exercises 22, 23, and 24.

A.3 Conditionals and Deduction

One of the most important ways to combine two statements is by the condition-consequence linkage, sometimes called the "if-then" form. As with the other compound logical statements, the formal truth-value structure coincides with normal usage, provided we think of such statements as contracts or agreements between two parties. For instance, suppose you and I have an agreement:

> If you get a perfect score on the final, then I will give you an A for the course.

Let us analyze what it means to say that this statement is true (the agreement has been kept) or false (the agreement has been broken). First of all, observe that this is a compound statement made up of two simple statements:

> p: You get a perfect score on the final.

> q: I give you an A for the course.

Each of these statements might be either true or false, so there are four cases to consider:

(1) p and q might both be true.
(You get a perfect score on the final and I give you an A.)

(2) p might be true and q false.
(You get a perfect score on the final, but I don't give you an A.)

(3) p might be false and q true.
(You don't get a perfect score on the final, but I still give you an A.)

(4) p and q might both be false.
(You don't get a perfect score on the final and I don't give you an A.)

Clearly, the agreement is upheld in Case (1) and is broken in Case (2). But what about the other two cases? Since the condition of getting a perfect score on the final is not fulfilled in those cases, I can give you an A or not, as I wish, without violating our agreement. Thus, the agreement is unbroken (true) *in all cases except* (2). In general:

DEFINITION   A **conditional statement**, or just a **conditional**, is a statement that can be put in the form "if p, then q," where p and q are themselves statements; it is false if p is true and q is false, and it is true otherwise. The statement p is called the **hypothesis** of the conditional, and q is called the **conclusion**.

NOTATION   The conditional "if p, then q" is symbolized by "$p \rightarrow q$," which can be read "p implies q."

Example A.18   "If it is spring, then the grass is green" is a conditional statement. Its hypothesis is "It is spring"; its conclusion is "The grass is green."   □

The statement form "if p, then q" is logically equivalent to "q if p." The word "if" *always* labels the hypothesis.

Example A.19   "I shall be happy if I get an A" is a conditional statement. Its hypothesis is "I get an A"; its conclusion is "I shall be happy."   □

Example A.20   "All dogs are animals" is a universal statement. It is easily rephrased as a conditional:

"If something is a dog, then it is an animal."   □

Example A.21   "If today is Saturday, then the moon is made of green cheese" is a conditional statement. Note that the hypothesis and conclusion need not have any real causal connection. The *form* of the statement determines whether or not it is a conditional.   □

The truth-value possibilities for the conditional are summarized in Table A.6. The four rows correspond exactly to the four cases in the example at the beginning of this section. In every conditional statement, a condition (hypothesis) is asserted to guarantee a consequence (conclusion); this assertion is false *only* if the condition is satisfied but the consequence does not follow.

p	q	$p \to q$
T	T	T
T	F	F
F	T	T
F	F	T

Table A.6   The conditional.

If the condition is not satisfied, then the conditional statement itself is considered true "by default." In other words, *a conditional statement is true whenever its hypothesis is false.* In such cases we say the conditional is **vacuously true**. (Recall that in our logical system we *must* assign either *true* or *false* to each case.) Thus, to test whether a conditional $p \to q$ is true or false, *it suffices to assume that the hypothesis p is true.* If it is possible for q to be false at the same time, then the conditional $p \to q$ is false; otherwise, $p \to q$ is true.

The claim that a conditional statement is *false*, then, says that there are instances when the hypothesis is true and the conclusion is false. *It does not say that this is the only possible circumstance that can exist.* For example, to say that the statement "If a number ends in 3, then it is divisible by 3" is false means that *there exists* a number that satisfies the hypothesis (ends in 3) but not the conclusion (is not divisible by 3). Thus, to prove the statement false, all we have to do is find one such number.

DEFINITION   An example that proves a statement false is called a **counterexample**.

Example A.22   In the example of the preceding paragraph, 13 is a counterexample for the given statement. The fact that there are other numbers, such as 63, that satisfy both the hypothesis and the conclusion does not matter; if the hypothesis does not *guarantee* the conclusion, the conditional statement is false.   □

Example A.23   [*Example A.20 again*] "All dogs are animals" can be regarded as a conditional disguised as a universal statement. We consider this to be a general truth in the sense that, if Phydeau is a dog, we are confident that it is also an animal. On the other hand, if Phydeau turns out to be a cat or a boat, the fact that Phydeau may or may not also be an animal does not alter the truth of the statement, "All dogs are animals."   □

Example A.24   [*Example A.21 again*] "If today is Saturday, then the moon is made of green cheese" is a conditional statement that is (vacuously) true six days out of every week because from Sunday through Friday both the hypothesis and the conclusion are false. Only on Saturdays is the hypothesis true, and then the falsity of the conclusion makes the entire conditional statement false.   □

Unlike conjunctions and disjunctions, the order of the two simple statements in a conditional makes a difference. "If a figure is a square, then it is a rectangle" says something quite different from "If a figure is a rectangle, then it is a square." Conditionals involving the negations of the hypothesis and the conclusion can complicate matters further. Since these forms and the relationships among them are often useful, it is helpful to have special terms to distinguish one form from another.

As you read through the following definitions, it might help you to think of a simple example of a conditional. For instance:

$$p: \text{Nancy wins.} \qquad q: \text{Tom loses.}$$

Then the original conditional $p \to q$ is

If Nancy wins, then Tom loses.

DEFINITION   The **converse** of a conditional statement $p \to q$ is the statement

$$q \to p$$

formed by interchanging the hypothesis and the conclusion. (If Tom loses, then Nancy wins.)

The **inverse** of $p \to q$ is the statement

$$(\sim p) \to (\sim q)$$

formed by negating both the hypothesis and the conclusion. (If Nancy doesn't win, then Tom doesn't lose.)

The **contrapositive** of $p \to q$ is the statement

$$(\sim q) \to (\sim p)$$

formed by interchanging the hypothesis and the conclusion and also negating them. (If Tom doesn't lose, then Nancy doesn't win.)

We can summarize these definitions as follows:

a conditional	if p, then q	$p \rightarrow q$
its converse	if q, then p	$q \rightarrow p$
its inverse	if not p, then not q	$(\sim p) \rightarrow (\sim q)$
its contrapositive	if not q, then not p	$(\sim q) \rightarrow (\sim p)$

Example A.25 *Statement*: "If it is spring, then the grass is green."

Converse: "If the grass is green, then it is spring."

Inverse: "If it is not spring, then the grass is not green."

Contrapositive: "If the grass is not green, then it is not spring." □

Universal statements can be put into conditional form. Thus, we can write the converse, inverse, and contrapositive of a universal statement. These related forms can, in turn, be translated back into a universal form, if desired.

Example A.26 *Universal statement*:
"All dogs are animals."

Conditional form:
"If something is a dog, then it is an animal."

Converse:
"If something is an animal, then it is a dog."
"All animals are dogs."

Inverse:
"If something is not a dog, then it is not an animal."
"Anything that is not a dog is not an animal."

Contrapositive:
"If something is not an animal, then it is not a dog."
"Anything that is not an animal is not a dog." □

The foregoing examples suggest interesting logical connections among a conditional and its converse, inverse, and contrapositive. Since these statements are themselves conditionals, their truth values are easily determined, as shown in Table A.7.

This table shows that

A conditional statement and its contrapositive
are logically equivalent

p	q	$p \to q$	$q \to p$	$\sim p$	$\sim q$	$(\sim p) \to (\sim q)$	$(\sim q) \to (\sim p)$
T	T	T	T	F	F	T	T
T	F	F	T	F	T	T	F
F	T	T	F	T	F	F	T
F	F	T	T	T	T	T	T

Table A.7 *Conditional, converse, inverse, and contrapositive.*

(because the truth values in those two columns are exactly the same). Thus, "If Nancy wins, then Tom loses" is equivalent to "If Tom doesn't lose, then Nancy doesn't win."

> The converse and the inverse are also logically equivalent statements

(for the same reason). "If Tom loses, then Nancy wins" is equivalent to "If Nancy doesn't win, then Tom doesn't lose." These interrelationships give us convenient alternative forms of conditional statements because logically equivalent statements are interchangeable.

A true conditional may or may not have a true converse. For example, "If something is a rabbit, then it eats lettuce" is true (normally), but its converse, "If something eats lettuce, then it is a rabbit," is not. Thus, a special situation exists when a conditional $p \to q$ and its converse $q \to p$ are simultaneously true. We might describe this by saying that the truth of either of the conditions p or q guarantees the truth of the other. This is sometimes abbreviated by saying, "p is true if and only if q is true."

DEFINITION   A **biconditional statement**, or simply a **biconditional**, is a statement that can be put in the form "$(p \to q)$ and $(q \to p)$" (that is, "p if and only if q"), where p and q are themselves statements. It is true precisely when p and q have the same truth values.

NOTATION   The biconditional "p if and only if q" is denoted by "$p \leftrightarrow q$" or "p iff q."

Notice from Table A.7 that $p \leftrightarrow q$ is true precisely when p and q have the same truth values. Thus,

> Saying that $p \leftrightarrow q$ is true is the same as asserting the logical equivalence of p and q.

Example A.27   No odd whole number is divisible by 2, and every whole number that is not divisible by 2 is odd. Thus, "A whole number is odd" and "A whole number is not divisible by 2" are equivalent statements. That is, a whole number is odd *if and only if* it is not divisible by 2. ☐

Example A.28 The statements "If the sun is shining, then it is daytime" and "Either it is daytime or the sun is not shining" are logically equivalent (even though they seem to be quite different). We can analyze them with a truth table, but first they should be written in terms of simpler statements. Let p and q be

$$p: \text{The sun is shining.} \qquad q: \text{It is daytime.}$$

Then "If the sun is shining, then it is daytime" is the conditional $p \to q$, and "Either it is daytime or the sun is not shining" is the disjunction "q or $(\sim p)$." Now the truth-table analysis is easy; it is shown in Table A.8, which establishes a general form of equivalent statements. ☐

p	q	$\sim p$	$p \to q$	q or $(\sim p)$
T	T	F	T	T
T	F	F	F	F
F	T	T	T	T
F	F	T	T	T

Table A.8 Equivalent statements.

Because the last two columns of Table A.8 agree in all cases, the statements "$p \to q$" and "q or $(\sim p)$" are logically equivalent. In other words, "$(p \to q)$ iff $(q$ or $(\sim p))$" is a tautology. This gives us a simple way to form the negation of $p \to q$ in terms of p, q, and their negations. We can just form the negation of q or $(\sim p)$, which, by De Morgan's laws, is equivalent to p and $(\sim q)$.

Notice that this approach agrees with what we should expect from our earlier discussion about the truth values for $p \to q$. That is, recall that $p \to q$ is false if and only if p is true and q is false. Thus, the truth of "p and $(\sim q)$" guarantees the falsity of $p \to q$, making "p and $(\sim q)$" a likely candidate for the negation. To prove that it is, we construct the appropriate truth table (Table A.9).

p	q	$\sim q$	$p \to q$	$\sim (p \to q)$	p and $(\sim q)$
T	T	F	T	F	F
T	F	T	F	T	T
F	T	F	T	F	F
F	F	T	T	F	F

Table A.9 The negation of a conditional.

In somewhat oversimplified terms, the connection between formal logic and deductive reasoning (including mathematical proofs) comes by way of conditional statements that are tautologies. The hypothesis of an argument is often the conjunction of several (or perhaps many) statements (referred to in the plural as "hypotheses"), and these hypotheses may themselves be complex statements derived from prior arguments. An argument is called **valid** if its conclusion is true whenever all of its hypotheses are true; in other words, if "(conjunction of hypotheses) → (conclusion)" is a tautology.

Thus, in order for the final conclusion of a valid deductive argument to be guaranteed true, the hypotheses of that argument *must* be true. This principle is useful in two ways. The more obvious one is a straightforward application: To prove a statement true, we construct an argument starting with hypotheses that are known to be true and ending with the statement in question. This type of argument is called a **direct proof** of the statement. There are many examples of direct proofs throughout the book, from simple number-theoretic statements in Chapter 2 to the fairly complex proof of Lagrange's Theorem in Chapter 7.

The second use of the principle stated at the beginning of the previous paragraph is based on the definition of the negation of a statement and the Law of the Excluded Middle. Because a true hypothesis and a valid argument together must yield a true conclusion,

> a valid argument that yields a false conclusion must proceed from a false hypothesis.

Therefore, we may also prove the truth of a statement s by forming its negation, $\sim s$, and using $\sim s$ as the hypothesis of a valid argument whose conclusion is known to be false. This implies that $\sim s$ must itself be false, and hence s must be true. This procedure is known as **proof by contradiction**, or **indirect proof**. A simple example of this type of argument is the proof that $\sqrt{2}$ is not a fraction, which appears in Appendix Section B.3. Other examples of indirect proof appear in several chapters. Perhaps the most important instances occur in the discussion of axiomatic independence and the non-Euclidean geometries (in Chapter 3) and in the proofs of Cantor's diagonalization process and Cantor's Theorem (in Chapter 6).

EXERCISES A.3

In Exercises 1–10, state the hypothesis, the conclusion, and the converse of each statement.

1. If there is snow on the ground, then this is winter.

2. If birds fly, then fish swim.

3. Rabbits must have good eyesight if they eat carrots.

4. If all men are mortal, then experience is the best teacher.

5. All turkeys are birds. (*Convert to conditional form first.*)

6. All roads lead to Rome. (*Convert to conditional form first.*)

7. No airplanes fly like birds. (*Convert to conditional form first.*)

8. If there are sufficient funds and the carpenters do not strike, the entire building will be renovated by September.

9. A conditional is true if its hypothesis is false.

10. I will be able to sleep through this lecture if nobody asks a question.

11–20. State the inverse and the contrapositive of each statement in Exercises 1–10.

21. Construct a truth table to show that "$\sim(p \rightarrow q)$" is equivalent to "p and $(\sim q)$."

22. Write the negation of "If birds have wings, then they can fly" in the form "p and $(\sim q)$."

23. Do the same as in Exercise 22 for (the conditional form of) "All elephants have trunks."

24. Use De Morgan's Laws and the fact that any statement s is equivalent to its double negation $\sim(\sim s)$ to write "$p \rightarrow q$" in the form of a disjunction. Check your answer by making a truth table.

25. Using Exercise 24, write the original conditionals of Exercises 22 and 23 as disjunctions.

26. Construct truth tables for the converse, the inverse, and the contrapositive of a conditional.

Topics for Papers – Appendix A

We have seen that in mathematical logic there are only two truth values, *true* and *false*, and that every legitimate statement must have one of these truth values. However, in everyday discourse we sometimes run across statements that are not decidably true or false. (A simple example: "It will snow in Omaha, Nebraska, on December 25, 2050." Must we wait until the middle of the 21st century before we can regard this as a statement?) Suppose that we altered the Law of the Excluded Middle so that our logic system had three truth values — *true*, *false*, and *maybe*. (The name of the third value is irrelevant.) Write a paper tracing the effects of such a change on the rules of logic presented in this Appendix. When you find a rule that breaks down, try to alter it so that it comes as close as possible to capturing the spirit of the original while allowing for the third truth value. (For example, a new Law of Contradiction might be: "No statement may have more than one truth value at the same time (in the same context).") Each time you adjust a rule, use the new form from there on in your discussion.

For Further Reading

1. Copi, Irving M. *An Introduction to Logic*, 7th ed. New York: Macmillan Publishing Co., 1986.

2. Exner, Robert M., and Myron F. Rosskopf. *Logic in Elementary Mathematics*. New York: McGraw-Hill Book Co., 1959.

3. Hodges, Wilfred. *Logic*. London: Penguin Books, 1980.

4. Lucas, John F. *Introduction to Abstract Mathematics*, 2nd ed. New York: Ardsley House, Publishers, Inc., 1990, Chapters 1 and 2.

5. Morash, Ronald P. *Bridge to Abstract Mathematics*, 2nd ed. New York: McGraw-Hill Book Co., 1990, Chapters 2–6.

6. Smith, Douglas, Maurice Eggen, and Richard St. Andre. *A Transition to Advanced Mathematics*, 3rd ed. Monterey, CA: Brooks-Cole Publishing Co., 1990, Chapter 1.

A BRIEF
HISTORY OF
MATHEMATICS

B.1 Introduction

The investigation of any field of human achievement may be placed in perspective by relating individual accomplishments to the overall development of that field. For this reason it is appropriate in an introductory mathematics course to present a brief outline of the history of mathematics. No attempt is made here to detail mathematical theories; we merely offer a historical skeleton of names, places, discoveries, and inventions to provide a context for the material found elsewhere in the book. Our process of selection is unavoidably subjective, but choices and opinions advocated by recognized mathematical historians are followed wherever possible.

We confine our attention predominantly to the mathematics of Western Civilization, because mathematical development in the Far East was largely independent of the West until modern times and because the genealogy of mathematics as we know it today can be traced almost entirely through Europe and the Near East. This emphasis is not intended to minimize the mathematical achievement of the Orient, but rather, to indicate the relatively small influence it had on present-day Western mathematics.

The chronological sequence is divided into eight major parts, determined mainly by the characteristics of the developments during each period. In more recent times this corresponds roughly to a division by centuries, but the lengths of the intervals prior to the Renaissance vary greatly. Of course, the partitioning is by no means absolute. Ideas and patterns of thought tend to overlap each other, intermingling with the passage of time; consequently, this separation of history into pieces is only an approximation to aid in the analysis of specific events.

As you read through this retrospective overview, you might be struck by the relatively few women whose names appear among the mathematical luminaries of previous centuries. Before drawing any inappropriate conclusions from this observation, consider the following facts. The "place" of women in most of the societies and cultures of Western civilization has, for the most part, denied women access to significant formal education, particularly in the sciences. Moreover, even when a woman succeeded in learning enough about mathematics to make an original contribution to the field, often her results were published anonymously or not at all (for fear of social criticism), or were subsumed under the name of a male mathematician (who had access to the learned societies, the university faculties, and other traditional academic outlets for research). Only in very recent years have historians begun to uncover the full extent of these obscured contributions of woman mathematicians.[1] Although most of the formal barriers to women in the sciences had been dissolved by the middle of the 20th century, some effects of this "uneven playing field" in mathematics and science persist even today. Thus, the perception of mathematics as a male domain has been a remarkably resilient self-fulfilling prophecy that is only now succumbing to the twin realities of careful historical research and outstanding current achievement by 20th-century women in mathematics.

B.2 From the Beginning to 600 B.C.

Somewhere in prehistoric times, perhaps during the Middle Stone Age, two general concepts began to emerge from the countless diverse phenomena of the physical world. They were *quantity* and *form*, the ancestors of two great families of thought, algebra and geometry, whose true kinship would remain obscure until the seventeenth century A.D. Although accounts of this period are based largely on conjecture, it is believed that ideas about quantity developed from attempts to compare collections of objects by counting, and they slowly evolved into a variety of primitive number systems. The earliest of these systems were quite simple, often based on the idea of two or three things, with collections containing more than five or six things classified by "much" or "heap" or some other equally imprecise expression, until the need for exchange and barter gave rise to more extensive number systems. The emergence of form began as primitive art with plaited rushes, woven patterns in cloth, and simple ornamentation on pottery and buildings. The mathematical aspects of form did not become apparent for a while; what we now call geometric figures began simply as decorative designs.

---

[1]See, for example, the descriptions of the work of Ada Lovelace and Grace Chisholm Young in [8] in the list *For Further Reading* at the end of this Appendix.

At the beginning of the historic period, about 5000 B.C., mathematics was well into its second stage of development. The quantitative needs of early societies had become so widespread and constant that it was desirable to develop general methods for calculation and to record these rules and results for future use. Fortunately, the Babylonians wrote on almost indestructible clay tablets and the Egyptian papyrus stayed well preserved in the dry climate of northern Africa, so there are enough relics to provide a fairly detailed picture of this work in the early civilizations of the Near East.

The earliest evidence of organized mathematical knowledge seems to indicate the existence of an Egyptian calendar in 4241 B.C., and possibly a Babylonian one before that. By 3000 B.C., Sumerian merchants had a workable arithmetic, and texts from the Third Dynasty, during the reign of King Hammurabi, show that by 1950 B.C. the Babylonians had an algebra capable of handling linear and quadratic equations in two unknowns and some equations of higher degree. Their geometry consisted of "recipes" for finding simple areas and volumes.

Egyptian progress was not far behind that of Babylonia. One of the earliest sources still in existence is the Ahmes Papyrus, written in 1650 B.C.[2] It is a practical handbook containing methods of solving types of linear equations, material on fractions with numerator 1 (a unique feature of Egyptian

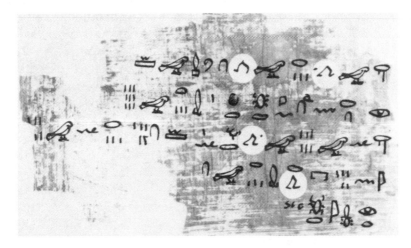

Pictured is a calculation on the Ahmes Papyrus. In modern algebraic symbolism:

$$x + \frac{2}{3}x - \frac{1}{3}\left(x + \frac{2}{3}x\right) = 10$$

---

[2]This is also known as the Rhind Papyrus, named after A. Henry Rhind, the 19th-century archaeologist who brought the manuscript to England.

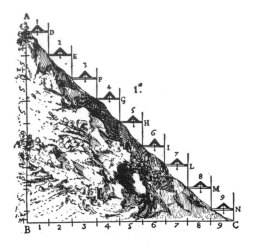

This method of leveling was used by the Egyptians and later by the Greeks and Romans. From Pompodoro's La Geometria Prattica, *Rome, 1624.*

mathematics), measuring techniques, and problems in elementary series. The scribe Ahmes stated that he was copying an earlier work that had been written about 1800 B.C., so it may well be a compilation of the mathematical knowledge of that time. Thus, the Babylonians and the Egyptians may be credited with producing the first great period of mathematics.

Our knowledge of early Chinese and Indian mathematics is comparatively poor. These peoples wrote on bark or bamboo, so all their manuscripts were highly susceptible to decay. These natural inconveniences sometimes were compounded by man's perversity. For example, in 213 B.C. the Chinese Emperor Shi Huang-ti of the Ch'in Dynasty ordered all existing books burned and had protesting scholars buried alive in order that he might be considered the creator of a new era of learning. Nevertheless, transcriptions of several ancient treatises were preserved, among them the *Chou-pei*, a dialogue concerning astronomy and mathematics. It was written shortly before 1100 B.C., and some of the material in it is attributed to a much earlier period. The *Chou-pei* contains material on the geometry of measurements, the computational principle of the Pythagorean Theorem, some elementary trigonometry and a discussion of instruments for astronomical measurements.

We know even less about Indian mathematics of this period. All that can be said with certainty is that there is evidence of a workable number system used for astronomical and other calculations, and of a practical interest in elementary geometry.

The outstanding feature of all pre-Hellenic mathematics is the absence of deductive reasoning. In general, attempts were not made to justify statements; the rules were given because they had worked in the past and they were expected to work in the future. Trial-and-error methods were the sources of knowledge, and successful results were noted and passed on to succeeding generations as formulas. Not until the blossoming of Greek civilization did mathematics come of age.

B.3   600 B.C. to A.D. 400

Early in the first millennium before Christ, sweeping changes occurred in the lands around the Mediterranean Sea. The Age of Iron brought with it increased travel and trade; new towns sprang up along the coasts of Asia Minor and Greece, and the supremacy of the landlord gave way to the rising stature of wealthy merchants. The exchange of goods was accompanied by the exchange of ideas, and wealth begot leisure, so that by the sixth century B.C. there was a class of men whose affluence allowed the luxury of intellectual speculation. With the first "Why?" mathematics entered a new phase of development, a science studied for its own sake.

Credit for that first speculative impulse is given to Thales of Miletus (c. 640 to c. 546 B.C.), a merchant whose travels to Babylon and Egypt acquainted him with Near Eastern mathematics. Until this time geometry had been confined to its literal meaning, measure of the earth, and its propositions were simply rules for accomplishing this task. Thales, however, chose six statements, including

A circle is bisected by any of its diameters.

When two lines intersect, the vertical angles are equal.

The sides of similar triangles are proportional.

and he demonstrated that they followed logically from previous ones. The statements themselves were well known, but his approach to them was a radical departure from that of traditional mathematics.

One of the most interesting early Greek scholars was Pythagoras, whose life was as mysterious as his work was brilliant. Both the date and the place of his birth are uncertain, but it is generally held that he lived from about 570 to 500 B.C. Much of his life is obscured by myths, for he was a legendary figure in Greece. His followers were banded together in a secret society that worshipped the idea of *Number* and hoarded knowledge as if it were gold. It is alleged that Pythagoras founded his own school at Crotona, a town of Magna Graecia on the southeast coast of the Italian peninsula. There he taught a philosophy based on the immutable elements of nature, embodied in and represented by whole numbers and their ratios. Despite their tendency to indulge in number mysticism, the Pythagoreans contributed greatly to number theory, the theory of music, astronomy, and geometry. They insisted upon deductive proof; the famous theorem about right triangles (shown in Figure B.1 and discussed in Section 3.2) still bears their name. The theory of perfect numbers, discussed in Chapter 2, was begun by the Greeks of this era.

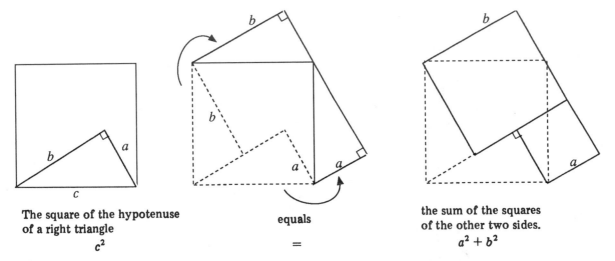

The square of the hypotenuse of a right triangle
$$c^2$$

equals
$$=$$

the sum of the squares of the other two sides.
$$a^2 + b^2$$

Figure B.1   *The Pythagorean Theorem.*

Strangely enough, one of the most significant ideas to come from the Pythagorean School was the verification of a concept that destroyed the Pythagorean philosophy. The Pythagoreans discovered the existence of incommensurable line segments; that is, they proved the existence of irrational quantities, lengths that cannot be expressed as ratios of whole numbers. Specifically, one of the Pythagoreans proved that the diagonal of a square one unit on a side cannot be expressed as a ratio of whole numbers (in current terminology, as a fraction). The proof, outlined here and illustrated in Figure B.2, is interesting for its logical form as well as for its content. It is a clear, brief example of *proof by contradiction,* a type of argument described at the end of Appendix A.

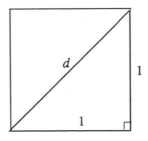

Figure B.2 $d^2 = 1^2 + 1^2 = 2$

Suppose that a square one unit on a side has a diagonal d that can be expressed as a fraction. Then d can be written in lowest terms, say as $\frac{a}{b}$. By the Pythagorean Theorem,

$$\left(\frac{a}{b}\right)^2 = 1^2 + 1^2$$

so

$$\frac{a^2}{b^2} = 2$$

implying

$$a^2 = 2b^2$$

This means that a^2 is even, so a must be even. Write a as $2s$, where s is some whole number. Then, substituting $2s$ for a in the last equation, we have

$$(2s)^2 = 2b^2$$

or

$$4s^2 = 2b^2$$

so

$$2s^2 = b^2$$

which implies that b^2 is even. But this means b must be even. The fact that a and b must both be even contradicts the hypothesis that $\frac{a}{b}$ is in lowest terms. The only logical escape from this absurdity is to recognize the original hypothesis to be false. That is, the diagonal d *cannot* be expressed as the ratio of whole numbers!

The unfortunate Pythagorean who uncovered this numerical heresy was drowned in a shipwreck. His colleagues tried to suppress the offensive idea he had found, but eventually word leaked out and people began shopping for a new philosophy of nature.

Another disturbing figure in the turbulent world of Greek thought was Zeno of Elea, a philosopher of the 5th century B.C., who speculated on the nature of motion of change of any kind. His contribution to mathematics consists of four paradoxes, which he posed for the thinkers of his day. They involved two diametrically opposite viewpoints of infinity and motion, with two paradoxes for each viewpoint. We give here one example of each type:

Achilles — Achilles is running to catch a plodding tortoise. He must first reach the place where the tortoise started. However, by the time he does this, the tortoise has already left, so it is ahead of him. This process must recur again and again.

The Arrow — Each moment is an indivisible point in time. A moving arrow, at any given moment, is either moving or at rest. The arrow cannot move within a moment, or else the moment would be divisible. But time is made up of moments. If the arrow cannot move in any one moment, then it cannot move at any time, and so it is always at rest.[3]

Any attempt to resolve these paradoxes involves a consideration of limits. Zeno's contemporaries were unsuccessful in their attempts to unravel his verbal tangles, but a seed had been planted that would eventually blossom into calculus. It would, however, require a growing season of 2000 years.

The works of Plato and Aristotle exemplify the type of thought that was the Greeks' greatest contribution to mathematics. Their development of logical principles and axiomatic methods of demonstration put mathematics on a foundation that was considered unshakable until our own century. Under their guidance mathematics partook of the glory of the Golden Age as a philosophical science second to none. In this period arose the "problems of antiquity," perhaps the most famous mathematical problems of all time. They were geometric-construction problems, which allowed only an unmarked straightedge and a collapsing compass as tools for their solution. (See Section 3.2.) With these tools one was asked to:

- Divide a given angle into three equal parts (*Trisection of the Angle*);
- Find the side of a cube whose volume is twice that of a given cube (*Duplication of the Cube*);
- Find a square whose area equals that of a given circle (*Quadrature of the Circle*).

All three remained unsolved until modern times, when the constructions were shown to be impossible. Nevertheless, the very fact of their existence bred a tantalizing kind of annoyance which led scholars to many new discoveries.

One of these scholars was Eudoxus (c. 408 to c. 355 B.C.), a pupil of Plato, a physician, a legislator, and a mathematician. He is credited with developing a theory of proportions that overcame the difficulties of dealing with incommensurable quantities. He introduced the "method of exhaustion," a way of dealing with areas and volumes that is remarkably similar to some of the basic concepts of integral calculus.

As Alexander the Great set out to conquer the world in 334 B.C., Western Civilization began to change. Greek culture and thought mingled with that of the Orient, and the mathematical center of the Western world shifted to Alexandria. That city's period of ascendancy began with Euclid (c. 300 B.C.), "the most successful textbook writer the world has ever known" and "the only man to whom there ever came or can ever come again the glory of

---

[3] A discussion of all four paradoxes may be found on page 24 of [1].

having successfully incorporated in his own writings all the essential parts of the accumulated mathematical knowledge of his time."[4]

Euclid's works were so comprehensive that they superseded all previous writings, and for this reason very few pre-Euclidean Greek manuscripts were preserved. Most of the information regarding work prior to the third century B.C. has had to be reconstructed from second-hand source material. Euclid's greatest work, the *Elements,* is a 13-book treatise that covers plane geometry, Eudoxus' theory of proportions, number theory, a geometric treatment of square roots, and solid geometry. Some of this material is undoubtedly Euclid's own work and he freely acknowledges his indebtedness for other parts, but the exact division is difficult to ascertain. However, even if he had contributed nothing original, the *Elements* would be significant, for it was a remarkable attempt to organize all of mathematics into a system logically deduced from a single axiomatic foundation.

This second great period of mathematics reached its peak with the flourishing of the School of Alexandria. Shortly after Euclid came Archimedes (287–212 B.C.), an astronomer and pioneer in physics and applied mathematics as well as a speculative mathematician. He studied at Alexandria for a short time and then returned to Syracuse, where he had been born. Archimedes masterminded an ingenious scientific defense of that city against the hordes of Marcellus; he was killed by an impetuous Roman soldier when the city was finally taken. His works include major contributions to number theory and algebra, and an enormous amount of geometrical work. Among his greatest achievements was his use of the "method of exhaustion" to solve geometric problems now handled by calculus.

A contemporary of Archimedes was Apollonius of Perga (c. 260 to c. 210 B.C.), "the great geometer." Seven of his masterful eight books on conic sections have survived intact. In these he systematically developed many basic properties of the ellipse, the parabola, and the hyperbola, investigating such topics as congruence and similarity criteria for these curves, tangent figures, and inscribed and circumscribed polygons. The world was not to see another pure geometer of his stature until Jacob Steiner in the nineteenth century.

With the passing of Apollonius, the tide of Greek mathematics crested and ebbed slowly into oblivion with the rest of Greek Civilization. Only two great waves appeared above the ripples of minor writers of this period. The astronomer Ptolemy (c. A.D. 85 to c. 165) wrote a comprehensive treatise on astronomy known as the *Almagest,* in which the frontiers of computational mathematics were extended to plane trigonometry and the beginnings of spherical trigonometry. About a century later, Diophantus of

---

[4][15], Vol. 1, pp. 103-104.

Alexandria (c. 275) wrote the *Arithmetica*, an unusual blend of Greek and Oriental mathematics, of which six books have survived. It is a milestone in the development of number theory, containing a treatment of indeterminate equations and problems requiring fractional solutions. It is the first treatise in which a type of algebraic symbolism is used systematically. In this regard Diophantus was several centuries ahead of his contemporaries.

One of the last great mathematical figures of this era was Hypatia (c. 370 to 415), daughter of the Alexandrian mathematician Theon. Hypatia wrote commentaries on the works of both Apollonius and Diophantus; she is the first woman known to have written on mathematical subjects. A celebrity in Alexandria at the beginning of the fifth century, she became entangled in the power struggle between Orestes, the Prefect, and Cyril, the Archbishop. Her torture and murder in the cathedral of Alexandria by followers of the Archbishop brutally signaled the end of a period of prolific scholarship in the Western world and the beginning of what some historians have called the Dark Ages.

The Oriental mathematics of this period in both India and China remained mostly computational and devoid of proof. With the exception of the book-burning incident in 213 B.C., the Chinese made slow but steady progress in the field of manipulative arithmetic, as did the Indians.

B.4   A.D. 400 to 1400

Despite its productivity in many other areas, the Roman Empire was mathematically barren, and its conquest of the Mediterranean world did little for speculative science. Whatever activity there was resulted from surviving Greek and Oriental influences. With the fall of Rome and the dissolution of Latin political domination in 476, Western Civilization entered a period of intellectual stagnation. The only person worthy of note at this time is Anicius Boethius (c. 475–524), a Roman citizen, statesman, philosopher, and mathematician, whose works include an arithmetic, a geometry, a book on astronomy, and a treatise on music (which was considered to be part of mathematics then). Although his texts lacked originality and were not especially rich in content, they were considered authoritative in schools and universities for many centuries. Indeed, the progress of ideas in Western Europe reached its nadir in the sixth century.

Meanwhile, Hindu mathematics was beginning to bear fruit. The influences of Babylonian and Greek science blended with native Indian ideas and there emerged significant results in algebra and arithmetic. Hindu mathe-

Arithmetic and Boethius; Pythagorus and Music. From an illuminated manuscript, Bavaria, 1240.

maticians of the fifth century worked with *plane* and *solid numbers* (squares, cubes, triangle numbers, etc.), obtaining both theoretical and computational results, including an approximation of π as $\frac{62,832}{20,000}$, or 3.1416. In the next century this work was extended to a treatment of algebraic equations, following the methods of Diophantus.

By far the most important achievement of Hindu mathematics was the development of the numeration system we use today, a place system based on ten and including a symbol for zero. Decimal systems and place systems had been developed previously by other peoples, but this was the first combination of the two concepts. Evidence of its use dates from A.D. 595; but the zero symbol in Indian arithmetic cannot be traced back beyond the

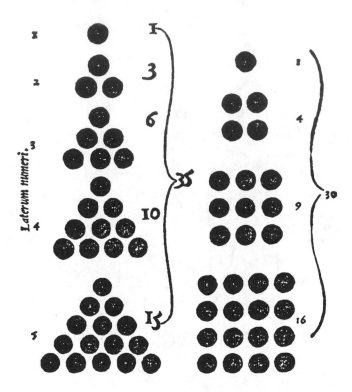

Pyramidal Numbers. From Ringelbergius, Opera, *1531.*

ninth century with any certainty, although the Babylonians had a somewhat similar notational device long before the Christian era.

Starting with Mohammed's Hegira in 622, the Moslems became the dominating influence in Western mathematics. The soldiers of Islam swept across North Africa, into western Asia, and through parts of Europe. Both Greek and Hindu mathematics were absorbed and merged by Arab scholars, who translated most of the important manuscripts into Arabic. They contributed little that was original, but they refined existing material and preserved the continuity of thought. The Mohammedan caliphs of the eighth and ninth centuries were great patrons of the exact sciences, especially astronomy and mathematics, so scientific learning spread throughout the Arab world.

Among the Moslem scholars of this period, one deserves special mention. He is Mohammed ibn Musa al-Khowarizmi (c. 825), who wrote two signif-

Arab Mathematician. *Although they contributed little that was original, Arab scholars refined and preserved existing material.*

icant books, one on arithmetic and the other on algebra. Only the second still exists in the original Arabic, but both were translated into Latin by 12th-century European scholars. The title of the first in translation was *Algorithmi de numero Indorum* (literally, "Al-Khowarizmi on Indian Numbers"). This book was one of the means by which Europe was introduced to the Hindu number system, and its title is the source of the word *algorithm*.

Seated Monk Writing. *Scholastic philosophers
speculated on the nature of infinity.
From the* Mer des Hystoires, *Paris 1488–89.*

The second book was entitled *al-jabr w'al muqabalah* (literally, "Restoration and Opposition"); it was devoted entirely to the study of linear and quadratic equations. By Latinization, the key word of the title became *algebra*, and, because of the widespread popularity of this text in Europe, the term soon became synonymous with the science of equations.

It was not until the latter part of the eleventh century that the Greek mathematical classics started to filter into Europe, and with its slow emergence from feudalism, the West began to stir from intellectual somnolence. During the 12th century the great mathematical works that had been translated into Arabic a few centuries earlier were translated from Arabic into Latin, especially in Spain after the defeat of the Moors in 1085, when many Jewish scholars were employed for this purpose. As trade between East and West expanded, powerful commercial centers were established on the coasts of Italy. European merchants began to visit the Orient to seek and put to practical use whatever scientific information they could find. The first of these to do significant mathematical work was Leonardo of Pisa (c. 1170 to c. 1250), more commonly known as Fibonacci. His main work, the *Liber Abaci*, was instrumental in acquainting Western Europe with the Hindu-

St. John's College, Oxford. The first universities were chartered in the 13th century.

Arabic numeration system. He also wrote on algebra and geometry, and investigated the infinite sequence that still bears his name.[5]

Not all mathematics of this period was done solely for its practical value. Scholastic philosophers speculated on the infinite, and their ideas influenced mathematicians of the 17th and 19th centuries. Clerical scholars also worked in the fields of geometry and algebra. Outstanding among these men was the Bishop of Lisieux, Nicole Oresme (c. 1323–1382), who was also an excellent economist. His mathematical work includes the first known use of fractional exponents and the location of points by numerical coordinates.

As cities sprang up throughout Europe, the church schools began to assume a new role. They became universities, empowered to grant degrees recognized by both Church and State. The first universities were at Paris, Oxford, Cambridge, Padua, and Naples, all chartered in the 13th century. This innovation held out great promise for the progress of learning during the 14th century. That promise was only partially fulfilled, for although other universities were established, the Hundred Years' War (1337–1453) and the Black Death (1347–1351) staggered the emerging European culture and postponed the revival of mathematical creativity.

B.5   The Fifteenth and Sixteenth Centuries

The renaissance of European art and learning began in earnest during the 15th century. When Constantinople fell in 1453, many Greek scholars moved to the cities and new universities of the West. Their migration coincided with the invention of movable-type printing, an invaluable aid in the dissemination of information. The ever-increasing demands of trade, navigation, astronomy, and surveying spurred mathematical study, but at the same time confined it somewhat. The dominating theme of 15th- and 16th-century mathematics was computation, and great strides were made in achieving accuracy and efficiency.

The leading mathematician of the 15th century was Johannes Müller (1436–1476), also known as Regiomontanus. Besides translating many classical Greek works and studying the stars, he wrote *De triangulis omnimodis*, the first treatise to be devoted solely to trigonometry. This book differs very little from the trigonometry of today except in notation; it marked the beginning of trigonometry as a study independent of astronomy.

---

[5]$0, 1, 1, 2, 3, 5, 8, 13, \ldots$. Each number (after the second) is the sum of the two preceding terms.

The first printed mathematics books were a commercial arithmetic that appeared in 1478, and a Latin edition of Euclid's *Elements*, which appeared in 1482. Mathematics derived its first real benefits from printing with the appearance in 1494 of *Summa de Arithmetica* by Luca Pacioli, a Franciscan monk. The book was written in Italian (rather than Latin) and was a complete summary of all arithmetic, algebra, and trigonometry known at that time, ending with the remark that the general cubic equation was insoluble with existing mathematical techniques. As if in direct response to this challenge, mathematicians at the University of Bologna attacked the problem and disposed of it within the first quarter of the next century.

The foremost scientist of this age was Leonardo da Vinci (1452–1519), a man of universal brilliance, whose fields of activity included painting, sculpture, biology, architecture, mechanics, and optics. His mathematical work was centered about geometry and its applications to art and the physical sciences. The application of mathematics to art was by no means confined to da Vinci's work. Several famous artists of the 16th century were excellent geometers. Notable among them was Albrecht Dürer. He wrote the first printed work dealing with higher plane curves, and his investigation of perspective and proportion is reflected both in his paintings and in the artistic work of his contemporaries.

The father of English mathematics was Robert Recorde (c. 1510–1558), a medical doctor, educator, and public servant. He published four books on mathematics, written in English dialogue with clarity, precision, and originality: *The Ground of Artes*, an arithmetic that went through 29 editions; *The Castle of Knowledge*, the first English exposition of the Copernican theory in astronomy; *The Pathwaie to Knowledge*, an abridged version of Euclid's *Elements*; and *The Whetstone of Witte*, an algebra, in which the equality sign "=" appears for the first time.

In the second half of the 16th century, a French lawyer by the name of François Viète (1540–1603) began to devote his leisure time to mathematics, and he advanced algebraic technique considerably. He was the first to use alphabetical coefficients in solving equations, and his consistent symbolism, including the signs + and −, helped him to pioneer the development of general methods for solving equations. The last name of importance in the 16th century is that of Simon Stevin of the Netherlands, developer of the theory of decimal fractions in 1585. Thus ended 200 years of preparation for a scientific explosion that signaled the dawn of the third great mathematical era.

B.6 The Seventeenth Century

Two previously destructive trends began to bear fruit in the 17th century. The first and more obvious of these was the frequent occurrence of political and religious wars, both large and small. Ever since the Crusades, Europe had been in turmoil; hardly a year had passed without open conflict somewhere. As has been the case throughout history, war and its ever-insistent need for newer and better weapons pressed the best minds of the age to compete in devising machines for battle. Once the weapons were made, however, the truly great minds that had devised them turned to the investigation of machines in general; Leonardo da Vinci's scientific work is a prime example of this. Slowly, Europe embarked upon the first stages of mechanization, and in an effort to develop the necessary engineering skill, its scientists and scholars turned to the study of motion and change.

The second trend was just as powerful, but it began more subtly. The religious reformation championed by Martin Luther on the Continent and by Henry VIII and his daughter Elizabeth I in England unleashed an intellectual revolt against authority and tradition that had been brewing for many years. The general skepticism rejuvenated philosophy, and the explorers of ideas pushed ahead in all directions. This composite age of science and rationalism produced some of the greatest men of history — astronomers Galileo and Kepler; philosophers Hobbes, Spinoza, and Locke; authors Dryden, Milton, and Shakespeare. Mathematics experienced a period of growth unparalleled until modern times. The number of people who contributed significantly to the advance of science becomes so large from this time on that henceforth we shall be compelled to confine our attention primarily to creative mathematicians of the first rank.

The Tudors had just passed the British crown to the Stuarts and William Shakespeare was in his prime when John Napier reached the peak of his scientific career. Born in 1550, he was a Scottish contemporary of Galileo and Kepler, and a true product of his time. He spent much of his life alternating between attacks on the Catholic Church and the contemplation and design of military devices, envisioning such futuristic fantasies as submarines and self-propelled cannon. His claim to immortality was established just a few years before his death with the publication of *Mirifici Logarithmorum Canonis Descriptio*, in which he set forth the theory of logarithms. This work was followed by another on logarithms, one on computing rods, and one on algebra. Napier's system of logarithms was a bit unwieldy, but several years after his death in 1617 it was perfected by a friend and colleague, Henry Briggs, who had originally suggested the use of 10 as the base.

Not to be outdone by its neighbors across the Channel, France produced four brilliant mathematicians within half a century. The first of these was the philosopher-scientist René Descartes (1596–1650). His many years of study and reflection convinced him that all scientific investigations are related and the key to that relation is mathematics. In his famous *Discourse on Method*, published in 1637, Descartes stated:

> The long chains of simple and easy reasonings by means of which geometers are accustomed to reach the conclusions of their most difficult demonstrations, had led me to imagine that all things, to the knowledge of which man is competent, are mutually connected in the same way, and that there is nothing so far removed from us as to be beyond our reach, or so hidden that we cannot discover it, provided only we abstain from accepting the false for the true, and always preserve in our thoughts the order necessary for the deduction of one truth from another.[6]

He explained that his method was a fusion of logic, the "Analysis [geometry] of the Ancients" and the "Algebra of the Moderns." He set forth four basic rules of procedure, saying,

> In this way I believed that I could borrow all that was best in both Geometrical Analysis and in Algebra, and correct all the defects of one by the help of the other.

To say that Descartes succeeded in unifying all of science would be an exaggeration, but he did achieve a marriage between quantity and form in an appendix to the *Discourse on Method*, simply titled *La Geometrie*. This was the first publication of analytic geometry. In it Descartes applied the methods of algebra to geometry, expressing and classifying various curves and other geometric figures by means of algebraic equations relative to a coordinate system. He used this algebraic approach to settle a number of geometric questions, including some classical problems that hitherto had been insoluble. (Some modern consequences of Descartes' work appear in Chapter 8.)

A countryman and acquaintance of Descartes was Pierre de Fermat (1601–1665), called by some the greatest pure mathematician of the 17th century. He was certainly one of the foremost scientific amateurs in history. Fermat was a quiet, unobtrusive lawyer and civil servant who indulged in mathematics sheerly for the fun of it, publishing little, but exhibiting his creativity in exchanges of correspondence with Descartes, Pascal, Mersenne, and others. He invented analytic geometry independently of Descartes, conceived the tangential approach to differential calculus before either Newton

---

[6]This and subsequent quotes from *Discourse on Method* (trans. John Veitch; London and Washington: L. Walter Dunne, 1901).

or Leibniz was born, and was one of the creators of the mathematical theory of probability. (See Chapter 4.)

Despite these monumental achievements, Fermat is best known for his work in number theory on the properties of primes. Ironically, his name has been permanently associated with a statement he neither originated nor proved. In his copy of a translation of Diophantus, one of Fermat's many marginal jottings was next to a problem asking for values x, y, and a to satisfy the equation

$$x^2 + y^2 = a^2$$

In this note he states that for any power greater than 2 there are no integer solutions, claiming,

> I have discovered a truly marvellous demonstration [of this] which
> this margin is too narrow to contain.[7]

Unfortunately, he apparently never wrote it down anywhere else, and "Fermat's Last Theorem" still ranks among the most puzzling unsolved problems of mathematics.

The third man in this group was Gerard Desargues (1593–1662), soldier, engineer, and geometer. During his lifetime much of his work was eclipsed by the general interest in Descartes' writings, but two centuries later his treatise on conics was republished and was immediately hailed as a classic in pure geometry. Desargues introduced the geometric treatment of points at infinity and worked extensively with perspective, thus becoming the founder of modern projective geometry.

Completing this quartet of famous Frenchmen is Blaise Pascal (1623–1662). From the age of twelve he regarded geometry as recreation, and by sixteen he had proved one of the most beautiful and far-reaching theorems in all of geometry, applying it then to consolidate and extend much of the previous work in this field. The theorem states:

> If a hexagon is inscribed in a conic section, then the points of
> intersection of the three pairs of opposite sides are collinear. [An
> example is shown in Figure B.3.]

Pascal invented a computing machine when he was nineteen, and in his twenties was recognized as a competent physicist. He was cocreator with Fermat of probability theory, a study to which both were led in a joint attempt to answer questions from members of the gambling nobility. How-

---

[7] *Œuvres de Fermat*, ed. Tannery, P., and C. Henry. (Paris, 1891–1912), Vol. 3, p. 241.

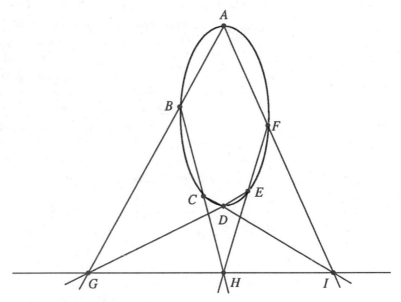

Figure B.3 *Pascal's Theorem for an ellipse. Hexagon ABCDEF is inscribed in the ellipse. Points G, H, and I are collinear.*

ever, at the age of twenty-five Pascal became a nearly fanatical proponent of Jansenism (a religious movement devoted to personal holiness and mysticism). As a consequence, he considered mathematics as a trifle to be toyed with only occasionally. Most of his later life was devoted to a study of philosophy and religion, from which emerged two literary classics, *Pensees* and *Provincial Letters*.

During the first part of the seventeenth century, the final building block for the foundation of calculus was fashioned in Italy. Bonaventura Cavalieri, a professor at the University of Bologna, proposed the "principle of indivisibles" in 1629. This principle sets forth a criterion for comparing areas and volumes of certain geometric figures, based on the assertion that a planar region can be considered as an infinite set of parallel line segments and a solid figure can be considered as an infinite set of parallel planar regions. His work circulated throughout the European scientific community, and in the latter part of the century two men combined it with Cartesian geometry to erect a towering mathematical structure with a very shaky ground floor.

One of these was Isaac Newton (1642–1727), whose achievements as a physicist are well known. When he was a boy, Newton showed little formal mathematical ability, and his early scholastic record was less than outstanding, but after a brief absence from school he returned to his studies with renewed enthusiasm. He entered Trinity College at Cambridge as a student at the age of nineteen, and in eight years became a professor of mathematics. In the decade between 1666 and 1676, Newton developed his "theory of fluxions" in three treatises. Although they were not published until many years later, these papers provided a foundation for a subsequent work, *Philosophiae Naturalis Principia Mathematica*, an axiomatic development of physics, in which Newton first set forth a complete mathematical formulation of his famous laws of motion. The *Principia* quickly assumed the role of an indispensible prerequisite for future scientific and technical progress; it ranks among those few mathematical works that have radically affected the history of civilization.

Among Newton's contemporaries and successors, however, acceptance and acclaim were far from unanimous. The man of genius had known what he wanted, and he allowed intuition to suppress a few logical scruples, accepting the theory primarily because it worked. But some of his peers were justifiably skeptical. None of these critics was as caustically witty and incisive as George Berkeley, Anglican Bishop of Cloyne. In *The Analyst*, published in 1734, he contended that scientists who criticized faith in mysteries of religion had precisely the same difficulties in their own field:

> And what are these fluxions? The velocities of evanescent increments. And what are these same evanescent increments? They are neither finite quantities, or quantities infinitely small, nor yet nothing. May we not call them the ghosts of departed quantities? Certainly...he who can digest a second or third fluxion...need not, methinks, be squeamish about any point in Divinity.

Newton's archrival was Gottfried Wilhelm von Leibniz (1646–1716), Germany's universal genius. His extraordinary talents benefited law, diplomacy, religion, philosophy, physical science, and mathematics. He independently developed differential and integral calculus shortly after his British counterpart had done the same. This last fact was hotly disputed for many years, with charges of plagiarism from both camps of admirers flying back and forth across the English Channel and engendering a bitter patriotic partisanship among the scientific community that was to last for many decades. Far less publicized, yet quite important, is Leibniz's work in combinatorial analysis, the study of specialized counting formulas that are basic to probability and statistics. Moreover, in searching for a "universal characteristic" unifying all of mathematical thought, he became one of the founders of symbolic logic, a study that would not be investigated intensively until two centuries later.

Baron Gottfried Willhelm von Leibniz. He independently developed calculus shortly after Newton did.

B.7 The Eighteenth Century

Calculus dominated mathematical development in the 18th century. Exploration of this powerful new theory proceeded in two directions — application to other parts of mathematics and to physics, and examination of its theoretical foundation. Research during this era was carried on mainly by Royal Academies subsidized by the "enlightened despots" of the age, while the universities played only a minor part in the production of ideas. The most prominent academies were at Berlin, London, Paris, and St. Petersburg. France held the lion's share of mathematical talent, but Switzerland also contributed significantly to the field.

To avoid religious persecution, a family named Bernoulli fled from Belgium in 1583 and ultimately settled in Switzerland. Their descendants provide a strong argument for the inheritance of intellectual ability. Within three generations they had produced eight eminent mathematicians, of whom four achieved international renown! The first and best known of these were two brothers, Jacob (1654–1705) and Johann (1667–1748). Lured away from careers in theology and medicine by the fascination of Leibniz's pioneering work in calculus, Jacob and Johann Bernoulli entered into a fierce mathematical rivalry with each other and with Leibniz himself that produced much of the material now contained in courses on elementary calculus, as well as results in the theory of ordinary differential equations. Much of the Bernoulli brothers' work focused on the properties of various special curves, such as the catenary and the logarithmic spiral. Besides his contributions to calculus, Jacob also did outstanding work in geometry and wrote the first book devoted to the theory of probability. Two of Johann's sons also achieved fame — Nicholaus (1695–1726) for his work in geometry, and Daniel (1700–1782) for his extensive publications in the fields of astronomy, mathematical physics, and hydrodynamics.

The most productive mathematician of the 18th century was Leonhard Euler, born in Basel, Switzerland, in 1701. He was instructed in mathematics by Johann Bernoulli, and he also studied theology, medicine, Oriental languages, astronomy, and physics. In 1727 he accepted a post at the St. Petersburg Academy, became head of the Berlin Academy in 1747, then returned to St. Petersburg twenty years later, where he remained until his death in 1783. He was an indefatigable worker whose labors were not appreciably diminished by partial blindness at the age of twenty-eight, nor by complete blindness at fifty-nine. He wrote almost 900 important books and papers, including works on arithmetic, algebra, analysis, mechanics, music, and astronomy. Euler is responsible for much of our modern notation in algebra and calculus, and he put trigonometry into its present form. One of his papers is considered to mark the beginning of graph theory, a branch of

Leonhard Euler. *He established calculus as a purely analytic theory independent of geometry.*

mathematics that is finding widespread applications in this century.[8] But perhaps Euler's most consequential achievements resulted from his efforts to establish calculus as a purely analytic theory independent of geometry.

In France, much of the mathematics in the early part of the century focused on the rival viewpoints of Newton and Leibniz about the mathematical basis of "natural philosophy" (physics). Pierre de Maupertuis (1698-1758) was an ardent, outspoken supporter of Newtonian physics. He found an early ally in the writer–philosopher Voltaire (1694–1778), and together they labored to move French thought beyond the philosophy of Descartes. Their work was profoundly influenced by Emilie du Châtelet[9] (1706–1749), a lover first of Maupertuis and then, for the last 16 years of her life, Voltaire's lover, coauthor, and intellectual companion. Châtelet was the first prominent female mathematical scholar since Hypatia (some 1300 years earlier). She and Voltaire wrote the *Éléments de la Philosophie de Newton* in 1738. Two years later she published *Institutions de physique*, a detailed exposition of physics based on the principles of Leibniz. In 1745, she turned back to Newtonian physics; the remaining four years of her intellectual life were spent writing *Principes mathématiques de la philosophie naturelle*, a two-volume French translation of and commentary on Newton's *Principia* that made Newton's work accessible to French readers. One historical commentator[10] summarizes Châtelet's mathematical achievements by saying:

---

[8]The paper was a mathematical treatment of the "Königsburg Bridge Problem," presented in 1736. See Section 9.1 for a little more about Euler and the Königsburg Bridge Problem, and see Chapter 9 in general for an introduction to graph theory.

[9]Her full name was Gabrielle-Émilie Le Tonnelier de Breteuil, Marquise du Châtelet-Lomont.

[10]Garry J. Tee in [8], page 24.

...her work of translation, commentary, and synthesis was a valuable contribution to the triumph of Newtonian science in the eighteenth century.

Meanwhile, just across the border in northern Italy, another woman was making her mark in European mathematical history. The first child of a mathematics professor, Maria Gaetana Agnesi (1718–1799) was born into a post-Renaissance culture somewhat more tolerant of educated women than was 18th-century France. The development of her prodigious intellectual skills was encouraged and supported by her parents, and by 1748 (at only thirty years of age) she had completed and published *Instituzioni Analytiche* ("Foundations of Analysis"), a 1020-page, two-volume exposition of algebra, analytic geometry, calculus, and differential equations. Since this book brought order and clarity to a large body of hitherto fragmented mathematical writings, it was immediately recognized as an important contribution to the mathematical literature and quickly translated into French and English. *Instituzioni Analytiche* gained for Agnesi immediate recognition as a first-class mathematician. In 1750, Pope Benedict XIV named her honorary professor at the University of Bologna, but she never taught there. Shortly after 1750, Maria Agnesi turned away from mathematics and dedicated the rest of her life to religious and humanitarian activities, spending almost a half century caring for the poor, the sick, and the dying.

The latter part of the 18th century was a period of political turmoil. Great Britain was having a bit of difficulty disciplining her rowdy American colonies, and France's nobility had lost their knack of controlling the peasantry. The upheaval in France was so drastic that it is incredible that scientific investigation not only endured but even thrived throughout the entire period. In fact, French mathematics retained its position of superiority. The Continental mathematicians had a strong advantage over the English in that the differential and integral calculus of Leibniz was much easier to understand and apply than Newton's unwieldy theory of fluxions, and their achievements indicate that they put it to good use.

"The lofty pyramid of the mathematical sciences," according to Napoleon Bonaparte, was Joseph Louis Lagrange (1736–1813), a brilliant, but modest, mathematician who was honored extravagantly by Napoleon and two foreign monarchs, and whose career was advanced immeasurably by the selfless support of Euler. Lagrange improved and organized much of Euler's calculus and worked extensively in the theory of equations, number theory, and mechanics. He was also responsible for inaugurating intensive mathematics curricula in France's two newly established schools, the École Normale and the École Polytechnique.

Pierre Simon Laplace (1749–1827) was the 18th century's applied mathematician *par excellence*. Born of a peasant couple, he used his exceptional mathematical talent to rise through the social strata until he was made a count by Napoleon. His most famous works are *Théorie analytique des probabilités* and the five-volume *Mécanique céleste*, which reviewed, unified, and extended all previous work in the fields of probability and celestial mechanics, respectively. Despite an annoying tendency to borrow ideas without acknowledgement, Laplace is generally recognized as an outstanding creative scientist. The size of his works belies their succinctness; in the words of one of his translators[11]:

> I never come across one of Laplace's "Thus it plainly appears" without feeling sure that I have hours of hard work before me to fill up the chasm and find out and show how it plainly appears.

Brief mention must be accorded to Adrien-Marie Legendre and Gaspard Monge. Both of these men reached the peaks of their careers at the end of

Adrien-Marie Legendre. He worked in analysis and applied mathematics, as well as in geometry.

---

[11] Nathaniel Bowditch (1773–1838), American astronomer.

the 18th century, but their works differ radically from each other. Although he was responsible for a text that provided a thorough reorganization of Euclidean geometry, a major portion of Legendre's work was in analysis and applied mathematics, the dominant fields of that time. Monge, on the other hand, was strictly a geometer, and geometric ideas pervade all his writings. As one of the first modern mathematical specialists, he is a herald of the next age.

B.8 The Nineteenth Century

The modern era began with Carl Friedrich Gauss (1777–1855), "prince of mathematicians." Just as Archimedes' influence pervaded the science of the Hellenistic age and Newton's achievements overshadowed the post-Elizabethan period, so did the genius of Gauss dominate 19th-century mathematics. A child prodigy, he could do arithmetic at the age of 3 and was familiar with infinite series by the time he was 10. The modern theory of numbers traces its ancestry back to his monumental *Disquisitiones Arithmeticae*, published in 1801. As a result of his writings on celestial mechanics, Gauss became generally recognized as the leading mathematician of Europe. His work was clear and concise, and was characterized by rigorous proof. Landmarks in almost every branch of mathematics bear his name, but Gauss expressed a strong personal inclination toward one particular field, saying,

> Mathematics is the queen of the sciences, and the theory of numbers is the queen of mathematics.

With the dawn of the new century, mathematical creativity began to increase exponentially, and by 1900 it had produced about five times as much original mathematics as had been accomplished in all previous ages. With this phenomenal wealth of material, mathematics had become a subject so vast that its totality defied the comprehension of a single mind. Except for a few individuals at the highest level of brilliance, mathematicians were forced to confine their efforts to one major branch of the subject, such as algebra, geometry, or analysis (the study of concepts generalized from calculus). At the same time, many other changes were occurring in the mathematical landscape. The dominance of royally supported scientific academies declined rapidly, and research became an important function of the universities. Within mathematics itself, practitioners became increasingly critical of previously accepted results and techniques. Their demands for a new level of rigor in proofs and their distrust of intuition led to the study of symbolic logic and axiomatics.

From this point on, a strictly chronological succession of biographies is no longer sufficient to outline mathematical progress. It must give way to a series of narratives, each following some topical subdivision. There is an unavoidable overlapping of names, times, and ideas, but every effort has been made to preserve the perspective and continuity of the overall picture.

As physical science and engineering reaped the rewards of calculus, the mathematicians of France continued to play an important role. Among them was Sophie Germain (1776–1831), a Parisian contemporary of Gauss who overcame her society's prejudice against female intellectuals[12] to become one of France's finest mathematical scholars. Germain did considerable research in number theory, much of it detailed in correspondence with Gauss and Legendre, including a partial proof of Fermat's Last Theorem. However, her most famous work — for which she won first prize in the 1816 competition held by the Institut de France — was on the mathematical theory of elastic surfaces, a calculus-based solution to an important physical problem. Her work provided the basis and direction necessary for other researchers to perfect the theory of elasticity, which was applied so successfully in constructing the Eiffel Tower later in the century.[13]

In Scotland, too, a woman overcame the obstacles of her society to become one of the century's foremost mathematical scientists. Denied access to formal mathematical education and faced with the active disapproval of relatives and acquaintances, Mary Fairfax Somerville (1780–1872) nevertheless pursued for almost 80 years an interest in mathematics that began by a chance encounter with an algebra puzzle in a women's fashion magazine. During that time, aided and encouraged by some of England's foremost scientific scholars, she taught herself enough mathematics and science to do important experimental work on the physical effects of the sun's rays and to write four major books on science. The first of these books, *The Mechanism of the Heavens* (1831), an annotated and expanded translation of Laplace's *Mécanique céleste*, gave the English scientific community access to French mathematical methods, which were more advanced than the English methods of the day. Her second and perhaps most influential book, *On the Connexion of the Physical Sciences* (1834), viewed physical science as a

---

[12] Molière's satire, *La Femme Savante*, provides a painfully clear illustration of the dominant French attitude in Sophie Germain's time.

[13] Nevertheless, the name of Sophie Germain is *not* inscribed on the base of the Eiffel Tower among the 72 scholars whose work made possible its construction!

single entity and, in fact, was instrumental in shaping the scope of the term "physical science."[14]

Meanwhile, theoretical mathematics had begun to benefit from the spirit of revolution that was sweeping the Western world. In France the overthrow of the monarchy and the subsequent Napoleonic period provided an ideal climate for the cultivation of new ideas. Into that climate was born Évariste Galois, a brilliant, temperamental boy, who spent a large part of his short life being thrown out of schools and into jails. Despite these frequent educational

Evariste Galois. The night before he was killed in a duel, he dashed off a letter in which he outlined much of the development of group theory.

and political embroilments, he devoted much of his time to algebra, which at that time was little more than generalized arithmetic, but his writings went unnoticed. In 1832, shortly before his twenty-first birthday, Galois became involved in a duel that cost him his life. The night before the "affair of honor," he dashed off a hurried letter to a friend, sketching some of his recent mathematical ideas. He wrote:

---

[14] John Couch Adams, who predicted in 1845 the discovery of the planet Neptune (a year later) by calculating its orbit, mass, and size from otherwise unexplained irregularities in the movements of Uranus, claimed to have gotten his idea of looking for Neptune from reading a passage in this book.

> I have made some new discoveries in analysis.... Later there
> will be, I hope, some people who will find it to their advantage
> to decipher all this mess.

"This mess" was the theory of groups, the foundation of modern algebra
and geometry. (Group theory is explored in Chapter 7.)

The liberation of algebra from its dependence on arithmetic took a giant stride forward with the work of William Rowan Hamilton (1805–1865),
an Irish astronomer and mathematician. A child prodigy, who at the age
of twelve had a working knowledge of twelve languages,[15] Hamilton spent
much of his early scientific career applying mathematics to physical theories, especially mechanics and optics. In 1835 he turned his attention to
algebra, and eight years later discovered "quaternions." Roughly speaking,
the system of quaternions is a generalization of the complex-number system.
Quaternion multiplication was the first worthwhile example of a noncommutative operation. General classes of algebras were soon forthcoming, and the
subject was well on its way to abstraction. England was the center of this
19th-century algebra with its geometric applications, which flourished under
the active guidance of Arthur Cayley (1821–1895), the originator of matrix
theory, and James Joseph Sylvester (1814–1897). Sylvester was also a prime
mover in the early development of American mathematics, as a result of a
lengthy visit to Johns Hopkins University.

The work of the French mathematician Augustin Louis Cauchy (1789–
1857) and his contemporaries made analysts more aware of the need for
strictly logical demonstration. Cauchy gave a workable definition of the
limit concept, and proceeded to establish a firm foundation for calculus. He
also developed the theory of functions of a complex variable at about the
same time that Gauss published his complex arithmetic. Bernhard Riemann
(1826–1866) of Germany also pioneered in the theory of complex variables;
much of his work had decidedly geometric overtones. The most important single contribution of Gauss, Abel, Cauchy, Riemann, and the other
mathematicians of the early 19th century was their meticulous insistence on
rigorous proof. Their work paved the way for Karl Weierstrass (1815–1897),
a mathematician famous for his deliberate, painstaking reasoning. He clar-
ified the notions of *function* and *derivative*, and eliminated all remaining
obscurity from calculus.

A protégé of Weierstrass was Sofia Kovalevskaia (1850–1891), a brilliant
Russian woman, who moved to Berlin to study with him at the prestigious
University of Göttingen. Denied her access to formal classes because she
was female, she studied under Weierstrass' private tutelage for three years
and became the first woman to receive a doctorate in mathematics from

---

[15]English, Latin, Greek, Hebrew, French, Italian, Arabic, Sanskrit, Syriac, Persian,
Hindustani, and Malay

Göttingen, in 1874. Kovalevskaia's effective use of Weierstrass' function-theoretic techniques in solving problems of mathematical physics brought her recognition as one of the world's leading mathematicians. Tragically, she died of pneumonia at the age of 41, at the peak of her professional career.

Insistence on rigor and logical simplicity was typified by Leopold Kronecker (1823–1891), who asserted,

> All results of the profoundest mathematical investigation must ultimately be expressible in the simple form of properties of the integers.

He was a number theorist, but is best known for his prolonged ideological feuds with Weierstrass, Dedekind, and Cantor, whose theories were based on the concept of infinite progressions. Kronecker, on the other hand, would not admit the mathematical existence of anything not actually constructible in a finite number of steps. (See Section 6.9 for a discussion of these ideas.)

Diametrically opposed to this view were Richard Dedekind (1831–1916) and Georg Cantor (1845–1918). Dedekind rigorously developed the concept of irrational number, thus enabling the real-number system to become the basis for all of analysis. Cantor, in his *Mengenlehre* (Set Theory), based the concept of number on that of set, and proceeded to develop different types of infinity, called transfinite numbers, that behave somewhat like the whole numbers of elementary arithmetic. This, in Kronecker's opinion, was a dangerous travesty of mathematics; he attacked both theory and person with such vehemence that Cantor suffered a series of breakdowns and ultimately died in a mental hospital. Set theory, however, has remained as a prominent, though controversial, part of mathematical thought. (Cantor's work is explored in Chapter 6.)

The revolution in geometry was foreshadowed as early as 1733 by the work of Girolamo Saccheri. Ever since Euclid set forth his *Elements*, the fifth postulate (often called the "Parallel Postulate") had been questioned by those who thought it might be provable from the other four. Knowing that all previous attempts to establish this had failed, Saccheri proposed a radically different approach to the problem. He replaced the questionable postulate with its negation, hoping to arrive at two contradictory statements within the new system. If he had done this, it would have meant that the original fifth postulate must be a necessary consequence of the other four; but the new system yielded no contradictions. Disappointed, Saccheri turned back — just when one more step would have yielded the discovery of the century — and his work was promptly forgotten.

Early in the 19th century, three men in three different countries used Saccheri's approach, and they had the insight to realize the meaning of their "failures" as well as the courage to publish their findings. Nicolai Lobachevsky in 1829 (in Russia), Janos Bolyai in 1832 (in Hungary), and Bernhard Riemann in 1854 (in Germany) all published consistent non-Euclidean systems of geometry, each independently of the other two. Gauss had found some of these ideas several decades before, but had refrained from publishing them for fear of criticism. These ideas conflicted with the prevailing Kantian philosophy, which contended that space conception is Euclidean *a priori*, so they remained obscure for several decades more. But the logical floodgates had been opened. No longer were axioms regarded as statements that were intuitively evident; they were viewed as the initial assumptions whose choice was logically arbitrary, subject to no preconditions. This was the start of formal axiomatics. (Chapter 3 discusses these ideas in depth.)

Now that geometry was no longer confined to visual images, it expanded at a fantastic rate. Grassmann's *Ausdehnungslehre* (Theory of Extension) gave the world a fully developed geometry of n dimensions for metric spaces.[16] This work established him as one of the founders of vector analysis (along with Hamilton). Jacob Steiner (1796–1863), a pure geometer who detested algebra and analysis, developed much of projective geometry. On the other hand, Felix Klein (1849–1925) unified all geometries by means of modern algebra, declaring that every geometry is the study of properties of a set which remain invariant with respect to a group of functions.

The trend toward unification of mathematics was personified by Henri Poincaré (1854–1912). His almost superhuman memory and powers of logical apprehension enabled him to produce valuable contributions to arithmetic, algebra, geometry, analysis, astronomy, and mathematical physics. He also wrote popularizations of mathematics and was actively interested in the psychology of creativity. Poincaré profoundly influenced the development of topology, a relatively new branch of mathematics, sometimes referred to as "rubber sheet geometry."

The formalistic treatment of algebra in England and the abstract axiomatic approach to geometry on the Continent triggered a sudden interest in logic and the foundations of mathematics, an interest that was redoubled after the appearance of Cantor's controversial set theory. The first significant mathematical studies of logic were *The Mathematical Analysis of Logic* (1847) and *The Laws of Thought* (1854), both by George Boole (1815–1864). In these works, Boole exhibited a completely symbolic approach to logic and laid the foundation for future extensions of the field. In 1884, Gottlob Frege (1848–1925) published *Die Grundlagen der Arithmetik*, which offered a derivation of arithmetical concepts from formal logic and thus greatly stimulated efforts to unify logic and mathematics. (See Appendix A for a brief treatment of elementary logic.)

---

[16]*Metric space* is a generalization of the usual concept of physical space. It is a set of points on which some idea of distance is defined.

B.9 The Twentieth Century

In 1900, members of the International Congress of Mathematicians, assembled in Paris, listened as one of their foremost colleagues lectured on mathematics in the newborn century. David Hilbert (1862–1943) had just completed his now-famous *Grundlagen der Geometrie* (Foundations of Geometry), a complete renovation of Euclid's *Elements* using modern axiomatic methods. Hilbert outlined twenty-three unsolved problems, a challenge for the new century. His insight was so accurate that every one of these problems has led to important new results, and the solution of even part of one of Hilbert's problems carries with it international recognition for the solver. Most of the problems have now been solved; a few remain important open questions in contemporary mathematics.

Even Hilbert, however, could not foresee how mathematics would mushroom in the 20th century. The phenomenal growth rate that began in the 1800s has continued, with mathematical knowledge doubling every twenty years or so. Lest the implications of the last statement slip by unnoticed, consider that it means this: Almost as much original mathematics has been produced *after* astronauts first walked on the moon as there had been in all previous history! In fact, it is estimated that 95% of the mathematics known today has been produced in this century! More than 300 periodicals published in various parts of the world devote a major share of their attention to mathematics. The abstracting journal *Mathematical Reviews* publishes each year more than 8000 synopses of recent articles containing new results. The present century is justifiably called the "golden age of mathematics."

Quantity alone, however, is not the key to the unique position the 20th century occupies in mathematical history. It is essential to understand that beneath this astounding proliferation of knowledge, there is a fundamental trend toward unity, a unity even more profound and productive than that envisioned by Descartes and Leibniz. The basis for this unity is abstraction. Although the non-Euclidean geometries of the nineteenth century paved the way for an abstract axiomatic treatment of mathematics as a whole, many of the fundamental interrelationships among the major branches of the subject did not begin to appear until the 1940s. The recent recognition of these unifying theories and of the vast unexplored fields they unlocked has led some prominent mathematicians to regard the end of World War II as the beginning of a new era in mathematics.[17]

As we approach the present, accurate evaluation of the relative significance of new mathematical results becomes almost impossible, so we have

---

[17]See, for example, "Recent Developments in Mathematics," by Jean Dieudonné in the *American Mathematical Monthly*, **71** (3): 1964, p. 240.

made little effort to compare individual contributions with one another, leaving that task to a later generation. Hence, although all mathematicians mentioned here have achieved prominence through acclaim from the modern scientific world, no claim is made for the completeness of the listings, nor is the topical coverage intended to represent a comprehensive survey of 20th-century mathematics. The primary purpose of this section is to indicate briefly the scope and power of contemporary mathematical activity.

As a result of Boole's work and the recognition of formal axiomatics following the birth of the non-Euclidean geometries, interest in the logical foundations of mathematics began to spread rapidly. The most notable successor to Boole's efforts in mathematical logic is the *Principia Mathematica*, a monumental two-volume work that appeared during the years 1910–1913, in which the philosopher-mathematicians Bertrand Russell (1872–1970) and Alfred North Whitehead (1861–1947) attempted to fulfill Leibniz' dream by expressing all of mathematics in a universal logical symbolism. Hilbert, too, dreamed of unifying mathematics, and he labored for many years to find a single provably consistent set of axioms upon which all mathematics could be based. Thus, mathematics entered a fourth phase. Starting only with an animal skin, it had first fashioned some serviceable garments and then acquired an elaborate wardrobe. Now, elegantly attired, mathematics began looking into its own pockets for unseen holes.

The holes started to appear as Russell found inconsistencies in Cantor's theory of sets, a theory upon which all of mathematics could be based. This made Hilbert's goal all the more desirable, and hence the mathematical community was profoundly shocked when Kurt Gödel (1906–1978) proved in 1931 that this goal was unattainable. Gödel also laid the groundwork for one of the century's most spectacular mathematical discoveries.

In 1964, Paul J. Cohen, basing his work on Gödel's, proved that both the Continuum Hypothesis and the Axiom of Choice are independent of the currently accepted axioms of set theory.[18] Gödel and Cohen showed that these two statements are unprovable from the other axioms, and that the inclusion of their negations in set theory can lead to entirely new theories of mathematics as a whole!

Once they recovered from their initial surprise, mathematicians accepted the fact that mathematics was not provably consistent from within, and continued to explore their own branches of the subject. In fact, the development of most parts of mathematics was virtually unaffected by the sudden tremor

---

[18]The Continuum Hypothesis is discussed in Section 6.8. Roughly speaking, the Axiom of Choice states that, given any collection of pairwise disjoint sets, we can choose exactly one element from each set in the collection. Although not recognized explicitly as an axiom until 1904, this assumption is basic to much of topology, modern analysis, and abstract algebra.

that had rocked its foundation. Algebra became far more general than it had ever been before, and similar tendencies toward abstraction in geometry led to extensive advances in both fields.

Analysis is undergoing an extraordinary metamorphosis during this century. When Henri Lebesgue (1875–1941) revolutionized the theory of integration in 1902, he opened the way for a much more abstract and unified treatment of analysis, as exemplified by the theories of abstract spaces developed by E. H. Moore in 1906 and Maurice Fréchet in 1928. This new generality has linked analysis with both algebra and topology.

According to at least one expert commentator on contemporary mathematics,

> the main fact about our time which will be emphasized by future historians of mathematics is the extraordinary upheaval which has taken place in and around what was earlier called algebraic topology.[19]

Algebraic topology, a mixture of the ideas of "rubber sheet geometry" and group theory, began to be a major field of investigation with the work of Henri Poincaré at the end of the last century. During the first half of this century it became the spawning ground for some of the most powerful tools in all of mathematics. Besides extending the frontiers of mathematical knowledge in traditional areas of research, the methods of algebraic topology became the basis for an even newer field, homological algebra, that has forged powerful bonds of unity among previously separate theories in algebra, analysis, and geometry. Homological algebra and a close relative called category theory (which treats entire classes of abstract structures as individual objects) did not begin to appear in print until the 1940s and 1950s, but their techniques have already invaded much of mathematics. Other areas of topology, such as the study of knots and manifolds (generalized surfaces), have also become very active, particularly in the latter part of the century.

The use of mathematics in the physical and social sciences has been so widespread and penetrating in this century that it would be futile to attempt any summary of modern advances in applied mathematics. Nevertheless, there are several men whose works must be mentioned because of their profound effect on the world we live in. Albert Einstein (1879-1955) at the age of twenty-six revolutionized physical science with his theory of relativity, a radically different analysis of change based in part on non-Euclidean geometry. Einstein's work made him the best known scientist of his day. John von Neumann (1903-1957) was a Hungarian-born American who directed the development of some of the first electronic computers at Princeton's Institute for Advanced Study, helped to develop quantum theory in physics,

---

[19]Dieudonné, "Recent Developments...," p. 243.

and is considered to be the founder of game theory, a mathematical analysis of strategies that is widely applied in fields such as economics and psychology. Von Neumann also made important contributions to the development of the atomic and hydrogen bombs and to long-range weather forecasting. Finally, we come to the father of automation, Norbert Wiener (1894-1964), whose work on information processes resulted in a field named by the title of his book, *Cybernetics, or Control and Communication in the Man and the Machine.*

The unique feature of 20th-century applied mathematics has been the development of electronic computers. Their ability to perform routine calculations at speeds of many millions of operations per second has radically altered problem-solving methods in mathematics, the physical sciences, the social sciences, business, and many areas of daily life. The social sciences are currently undergoing a fundamental shift of emphasis from qualitative to quantitative considerations, a transition made extremely difficult by the many variables that are usually present in social situations. The proper handling of these variables would be virtually impossible by traditional techniques, but computers can simulate and analyze complex situations involving hundreds, or even thousands, of items, including people, keeping track of random effects and other pertinent data. Thus, electronic computers are fast becoming indispensable research and instructional tools in economics, sociology, psychology, education, and many other fields. (See Chapter 5 for a discussion of computers.)

The "computer age" is affecting pure mathematics as well. It has created the need for a new branch of mathematical logic concerned with problems of machine design and coding (discussed in Section 2.8), and it has aided and encouraged the development of numerical analysis. Rapid advances in machine logic, program efficiency, and hardware technology during the second half of the 20th century have provided computers with astounding capabilities. These increasingly powerful machines have been used to make major breakthroughs in settling some of the longstanding open questions of pure mathematics. Here are two of the famous problems that fell to the power of the computer in this century.

- **The Four-Color Problem** — In 1852, a young British mathematician by the name of Francis Guthrie posed a seemingly straightforward question. Simply put, he asked whether four colors were enough to permit every possible (planar) map to be colored so that no two "countries" with a common border would be the same color. It is easy to show that three colors are not enough, and a proof that five colors always suffice had emerged before the end of the 19th century. However, the critical case of four colors resisted the efforts of the world's best mathematicians for more than a century and thereby spawned some major advances in graph theory. (See Chapter 9 for a discussion of graph theory.) It was finally resolved in 1976 by Kenneth Appel and Wolfgang Haken, who used powerful, intricate computer techniques to show that four colors were indeed enough. Their computer-dependent argument generated quite a bit of controversy about what is acceptable as a mathematical proof, and many mathematicians are still somewhat uncomfortable with procedures whose accuracy cannot be checked "by hand," as it were.[20]

- **Classification of Simple Groups** — The problem here is a little harder to state, except by analogy. A type of algebraic structure called a *group*, which entered the mathematical scene in the 19th century, has become very important in many areas of pure and applied mathematics. (See Chapter 7 for a detailed description of groups.) Among all the possible finite group structures, there are some — called *simple groups* — that can be used as "building blocks" for all the rest, like the prime numbers in number theory. As in the case of prime numbers, the natural question of finding some comprehensive classification that described all simple groups turned out to be a very hard problem. Its solution was finally completed in 1980, after some 40 years of computer-assisted work by more than a hundred mathematicians around the world. The published proof stretches over nearly 15,000 pages of journal articles![21]

While these and other previously intractable mathematical problems have fallen to the power of the computer in this century, a few famous questions have so far resisted even the most sophisticated combined efforts of mathematician and machine. Among them are two in number theory that are discussed elsewhere in this book — the characterization of perfect numbers (See Chapter 2) and Fermat's Last Theorem (See Section B.6). In each of these cases (as in other problems like them), computers have been able to test the relevant conjectures out to astronomically large numbers. However, since such testing can never exhaust the infinite set of integers, it appears that disposition of these questions will have to await an ingenious logical breakthrough in the theory.

Besides settling old questions, the computer age has begun to generate new areas of mathematical investigation. Perhaps the most striking of these, in conceptual originality and in sheer visual impact, is the theory of fractals. The term *fractal* refers to the notion that, rather than being 1-, 2-, 3-, or

---

[20] See Chapter 7 of [5] for a detailed discussion of the Four-Color Problem.

[21] See Chapter 5 of [5] for more details about this problem and its solution.

4-dimensional (as traditional geometry would lead us to expect)[22], the geometric objects of this theory are considered to be of fractional dimension, such as $\frac{4}{3}$ or 1.2618. Fractal geometry was created almost singlehandedly in the decade 1967–1977 by the French mathematician Benoit Mandelbrot, a researcher working for IBM in the United States. Perhaps the most striking feature of this theory is that it reveals an underlying mathematical order in much seemingly chaotic behavior that occurs in nature. Since 1980, the theory and its applications have been extended by increasing numbers of mathematicans, scientists, and even artists; but we are still much too early in the development of this beautiful concept to assess accurately its power or its limitations. Suffice it to say that fractal geometry typifies the modern symbiotic relationship between human logical theory and electronic computational power.[23]

The growth of mathematics in the United States during this century has indeed been phenomenal. American mathematical scholarship is usually considered to have truly begun with the extended visit of British mathematician James Joseph Sylvester at Johns Hopkins University late in the nineteenth century; his presence brought a new level of respectability to mathematical research in this country. Sylvester was also instrumental in providing women equal access to higher education in mathematics. Through his efforts, Johns Hopkins grudgingly admitted Christine Ladd-Franklin (1847–1930) to graduate study in 1878, but it did not award her a doctorate for the work until 1926, 44 years after she had finished her dissertation! Throughout her career as a lecturer at Johns Hopkins and then at Columbia, Ladd-Franklin, in turn, devoted much of her time, energy, and money to giving women equal access to graduate education.

Similar efforts in late 19th-century England were also to affect the American mathematical scene. The algebraist Arthur Cayley at Cambridge actively encouraged promising young mathematics students from nearby Girton College, England's first college for women.[24] Among the Girton mathematics students helped by Cayley were Grace Chisholm Young (1868–1944), who went on to work with Felix Klein and David Hilbert and get her doctorate at Göttingen, and Charlotte Angas Scott (1858-1931), who came to America in 1885 as the (only) mathematician among the eight found-

---

[22]See Chapter 8 for a discussion of the usual notion of dimension.

[23]A beautiful visual display of this geometry appears in *The Beauty of Fractals*, by H. O. Peitgen and P. H. Richter (Springer-Verlag, 1986). A description of the theory and its application can be found in *The Fractal Geometry of Nature*, by Benoit B. Mandelbrot (W. H. Freeman & Co., 1982).

[24]Girton College opened in 1869 with five students. In 1873, it relocated to within 3 miles of Cambridge, where they were allowed to attend lectures by some (but not all) of the professors. See page 194 of [8].

ing faculty members of Bryn Mawr College in Pennsylvania. Bryn Mawr was a college dedicated to providing undergraduate and graduate education for women. Scott's long, productive career there had a profound impact on American mathematical education. She spent 40 years building Bryn Mawr's mathematics program into one with a well-deserved reputation of excellence at all levels, including the granting of doctorates; she personally supervised the dissertations of seven doctoral students. She was instrumental in attracting to the Bryn Mawr faculty Emmy Noether (1882-1935), one of the truly great algebraists of modern times. Noether, who earned her doctorate at Erlangen, had been invited by David Hilbert and Felix Klein to Göttingen in 1915, where she stayed until the rise of the Nazis in 1933. She accepted a guest professorship at Bryn Mawr that year, and lectured both there and at Princeton's Institute for Advanced Study until her sudden, premature death two years later.

American mathematics became truly self-supporting with the emergence in 1913 of George David Birkhoff (1884-1944) as an internationally renowned mathematician. Birkhoff was the first American-educated mathematician to achieve such stature, and he, in turn, directed and guided many of the most productive American mathematicians of this century through his position as a professor, first at Princeton and then at Harvard from 1912 until his death. As the prestige of the United States has grown, more and more Asian and European mathematicians have migrated to the major American research centers, such as Princeton's Institute for Advanced Study (whose original faculty included Einstein and von Neumann). Today, the combined membership of the three major American mathematical societies[25] is approximately 30,000, as compared with a mere handful in 1913. Moreover, the employment focus of this mathematical talent has shifted and broadened over the years. In the early part of this century, almost all mathematicians worked in academic settings. Now, however, a variety of industrial, commercial, and government needs, often occasioned by the ever-widening use of computers, has drawn more and more mathematicians away from the universities and into applied-mathematical endeavors. In this way, mathematics is riding the historical pendulum, swinging back from the abstractions of the past hundred years toward applications of these theoretical results to "real" problems in other areas. Perhaps this remarriage of mathematics and reality will carry with it an appropriate social attitude change toward the subject, and mathematics once again will be viewed not as an area of irrelevant esoterica, but as a vital subject whose pursuit is essential to the progress of the human race.

---

[25]The American Mathematical Society, the Mathematical Association of America, and the Society for Industrial and Applied Mathematics.

Topics for Papers – Appendix B

1. Choose a mathematical event or mathematician mentioned in this Appendix and write a paper that describes what was going on in the world at that time in history. Do not confine yourself to the region of the person or event you choose, but consider all parts of the world. Also consider all major historical features of the time — political changes, social trends, scientific breakthroughs, religious movements, literature, art, etc. Then discuss how (if at all) the mathematics of your person or event influenced and/or was influenced by the historical situation you described. It is legitimate to present your own opinions about this, but they should be accompanied by supporting arguments based on the historical evidence.

2. Do most women think about mathematics differently than most men do? If so, how? (Your *opinion* is solicited here; there are no definitive psychological research findings on this question yet.) If your answer to the first question was *no*, then this paper topic is not for you. If you answered *yes*, read on. Assuming your opinion is correct, in what ways might the mathematics that you studied in this course have developed differently if it had been originated primarily by women? Provide specific illustrations from the course; then try to generalize your comments to encompass the development of mathematics as a whole.

For Further Reading

1. Bell, E. T. *Men of Mathematics*. New York: Simon & Schuster, Inc., 1937.

2. Boyer, Carl B. *History of Mathermatics*, 2nd ed. New York: John Wiley and Sons, Inc., 1989.

3. Burton, David M. *The History of Mathematics*. Boston: Allyn and Bacon, Inc., 1985.

4. Dantzig, Tobias. *Number, The Language of Science*, 4th ed. New York: Macmillan Publishing Co., Inc., 1954.

5. Devlin, Keith. *Mathematics: The New Golden Age*. London: Penguin Books, 1988.

6. Dunham, William. *Journey Through Genius: The Great Theorems of Mathematics*. New York: John Wiley and Sons, Inc., 1990.

7. Eves, Howard. *An Introduction to the History of Mathematics*, 4th ed. New York: Holt, Rinehart and Winston, 1976.

8. Grinstein, Louise S., and Paul J. Campbell, eds. *Women of Mathematics*. New York: Greenwood Press, 1987.

9. Keyser, C. J. *Mathematics as a Culture Clue*. New York: Scripta Mathematica, Yeshiva University, 1947.

10. Kline, Morris. *Mathematical Thought from Ancient to Modern Times*. New York: Oxford University Press, 1972.

11. —— . *Mathematics in Western Culture*. New York: Oxford University Press, 1964.

12. Newman, James R., ed. *The World of Mathematics*, Vols. 1–4. New York: Simon & Schuster, Inc., 1956.

13. Osen, Lynn M. *Women in Mathematics*. Cambridge, MA: The MIT Press, 1974.

14. Perl, Teri. *Math Equals*. Reading, MA: Addison-Wesley Publishing Co., 1978.

15. Rossiter, Margaret. *Women Scientists in America*. Baltimore: Johns Hopkins University Press, 1982.

16. Smith, David Eugene. *History of Mathematics*, Vols. 1 and 2. Boston: Ginn and Company, 1923.

ANSWERS

TO ODD-NUMBERED EXERCISES

Chapter 1

Section 1.2, page 8

1. *Angle*: A figure formed by two line segments with a common endpoint.   *Angle measure* (degrees): A multiple of $\frac{1}{90}$th of a right angle.   *Decagon*: A plane figure bounded by ten line segments.   *Sum*: The result of addition.   **7.** $8 \times 180° = 1440°$ **11.** If n is an integer, then 6 is a factor of $n(n+1)(n+2)$. **13.** Prove that the product has both 2 and 3 as factors.   **15.** Either n or $n+1$ is a multiple of 2, so the product is, as well; either n or $n+1$ or $n+2$ is a multiple of 3, so the product is, as well. Thus, the product is a multiple of 6.   **19.** A spiral starts 5 cm from a point P and ends 30 cm from that point after 750 cycles around P. What is its length?

Section 1.3, page 13

1. If two copies of Figure 1.1 are fit together along their "saw-tooth" edges, the result is a rectangle 999,999 by 1,000,000 units; the area is 999,999,000,000 square units, so each half has area 499,999,500,000 square units.   **3.** The sum of all positive integers from p to q, inclusive, is $\frac{q(q+1)}{2} - \frac{(p-1)p}{2}$.   **5.** How many squares are determined by the lines of an n-by-n chessboard? What is the sum of $1^2 + 2^2 + 3^2 + \ldots + n^2$?   **7.** 1296   **9.** 1225   **11.** $2^{64} - 1$ moves; about 58.5 billion years.

Chapter 2

Section 2.1, page 24

1. 200   **3.** 469   **5.** 289   **7.** 10,000   **9.** $994^2 = 988,036$ **11.** $2161^2 - 994^2 = 3,681,885$

13. 21   **15.** 600   **17.** 19   **19.** 598   **21.** 52   **23.** 1403   **25.** They record the number of points in a triangular array.   **27.** $\frac{n(n+1)}{2}$ **29.** 180

Section 2.2, page 27

1. True   **3.** False   **5.** False   **7.** True   **9.** True   **11.** True   **13.** True   **15.** False
17. 1   **19.** b and c are even; so $2 \mid b$, $2 \mid c$; so $2 \mid (b+c)$; that is, $b+c$ is even.   **21.** One of the two must be even, so the product is even.

Section 2.3, page 38

1. 7×13 **3.** 13×23 **5.** 347   **7.** $2^2 \times 11 \times 19$ **9.** $3 \times 13 \times 107$ **11.** $7 \times 11 \times 13^2 \times 23$
13. 4   **15.** 4   **17.** 2   **19.** 12   **21.** 8   **23.** 24   **25.** 4; 5; $k+1$ **27.** For each divisor a of n, there is a "partner" divisor b, determined by $a \times b = n$. Each divisor can be paired with its partner; so $D(n)$ is even unless n is the square of c, in which case c is its own partner, and $D(n)$ is odd.   **29.** p^9 or $p^4 \times q$ **31.** p^{11} or $p^5 \times q$ or $p^3 \times q^2$ or $p^2 \times q \times r$ **33.** $180 = 2^2 \times 3^2 \times 5$
35. $7 \times 11 \times 13$; a number whose digits are $pqrpqr$ is $1001 \times pqr$, so division by 7, 11, and 13 necessarily yields the quotient pqr.   **37.** (a) $3 \times 7 \times 13 \times 37$ (b) Pick any two-digit number; form a six-digit number by repeating the two-digit number twice; divide this number by 13; divide the quotient by 21; divide this quotient by 37; the final quotient will equal the original number.

Section 2.4, page 42

1. $1, 7, 13, 7 \times 13$ **3.** $1, 2, 2^2, 2^3, 2^4, 2^5, 3, 2 \times 3, 2^2 \times 3, 2^3 \times 3, 2^4 \times 3, 2^5 \times 3, 3^2, 2 \times 3^2, 2^2 \times 3^2, 2^3 \times 3^2,$
$2^4 \times 3^2, 2^5 \times 3^2$ **5.** $1, 2, 2^2, 2^3, 2^4, 31, 2 \times 31, 2^2 \times 31, 2^3 \times 31, 2^4 \times 31$ **7.** 112   **9.** 336   **11.** 348
13. 1680   **15.** 6048   **17.** 421,632   **19.** 63   **21.** 22,932   **23.** If $z = 1 + n + n^2 + \ldots + n^{k-1}$, then $n \times z = n + n^2 + n^3 + \ldots + n^k$. Subtracting the first from the second, $n \times z - n = n^k - 1$; dividing both sides by $n - 1$, we obtain the formula.

Section 2.5, page 44

3. 21   **5.** 37   **7.** 1   **9.** 844   **11.** 1875   **13.** 122,333   **15.** Deficient   **17.** Deficient
19. Deficient   **21.** Abundant   **23.** Deficient   **25.** Deficient   **27.** If n is prime, $P(n) = 1$; $1 < n$; so n is deficient.   **29.** Label the proper divisors of an abundant number n: $1, d_1, d_2,$ $d_3, \ldots, d_k$. Then $x, d_1 x, d_2 x, d_3 x, \ldots, d_k x$ are distinct proper divisors of nx, which implies that $P(nx) \geq x + d_1 x + d_2 x + d_3 x + \ldots + d_k x$, which equals $(1 + d_1 + d_2 + d_3 + \ldots + d_k)x > nx$. Thus, nx is abundant.   **31.** 945

Section 2.6, page 51

1. $2^5 \times 63 = 2016$; $P(2016) = 4536$ **3.** False; counterexample: $2 \times 5 \times 11$ **5.** True; $E_k = (2^k - 1)2^{k-1} = \frac{(2^k-1)2^k}{2}$; that is, the kth Euclidean number is the $(2^k - 1)^{\text{th}}$ triangle number.

Section 2.7, page 56

1. Composite   **3.** Prime   **5.** Composite   **7.** Prime   **9.** Not perfect   **11.** Perfect

13. Not perfect   **15.** Perfect   **17.** 2^{88} **19.** $M_{13} = 2^{13} - 1 = 8191$ **21.** 23 or 89
23. 524,287   **25.** 7 or 8191   **27.** 23, 89   **29.** 3, 5, 11, 31, 41

Chapter 3

Section 3.2, page 73

1. Construct the circle with center A, radius AB; construct the circle with center B, radius AB; label one point of intersection of the circles as C. Draw AC and BC; ABC is the desired triangle. (See Figure 1.)   **3.** Let A be the vertex of the given angle; let B be any point on one side of the angle. Construct C on the other side of the angle so that $AC = AB$; draw BC. Construct equilateral $\triangle BCD$; draw AD. $\triangle ABD \cong \triangle ACD$ by SSS, so $\angle BAD = \angle CAD$, and AD is the angle bisector. (See Figure 2.)

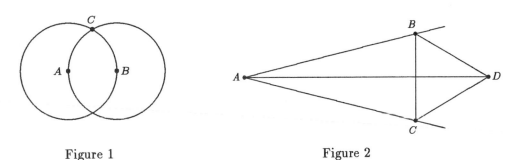

Figure 1   Figure 2

5. Let A be the given point; let B be any other point on the segment. Construct C on the segment (or its extension) on the opposite side of A from B such that $AC = BA$. Construct equilateral $\triangle BCD$. $\triangle DBC \cong \triangle DCB$ by SSS, so $\angle DBC = \angle DCB$. Draw DA. $\triangle DAB \cong \triangle DAC$ by SAS, so $\angle DAB = \angle DAC$. These angles are supplementary, so each is a right angle. (See Figure 3.)

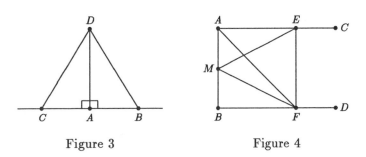

Figure 3   Figure 4

7. Given AB, construct $AC \perp AB$ at A and $BD \perp AB$ at B, with C and D on the same side of AB. Construct E on AC (or an extension of AC) such that $AE = AB$. Construct F on BD (or an extension of BD) such that $BF = AB$; draw EF. Construct M as midpoint of AB; draw ME and MF. Now, $\triangle AME \cong \triangle BMF$ by SAS, so $ME = MF$ and $\angle AEM = \angle BFM$. $\triangle MEF \cong \triangle MFE$ by SSS, so $\angle MEF = \angle MFE$. Then, by adding angles, $\angle AEF = \angle BFE = $ a right angle. Draw AF. $\triangle ABF \cong \triangle FEA$ by hypotenuse-leg, so $EF = AB$, implying $ABFE$ is a square. (See Figure 4.)

Section 3.3, page 77

1. Circular   **3.** Noncharacteristic; e.g., a right triangle   **5.** Noncharacteristic; e.g., any non-right triangle   **7.** Accurate   **9.** Noncharacteristic; e.g., a scalene triangle   **11.** Accurate   **13.** Noncharacteristic; e.g., an equilateral triangle

Section 3.4, page 84

1. A1: There is at least one brass ring (e.g., D).   A2: Every brass ring is attached to at least two wires (e.g., A is attached to 2 and 3, B to 1 and 5, etc.).   A3: Every wire is attached to at least two brass rings (e.g. 1 is attached to A and C, 2 to A and E, etc.).   A4: Given any two brass rings, there is exactly one wire attached to both of them (e.g., 1 is the only wire attached to both A and B, 7 is the only wire attached to both C and E, etc.).   **3.** Let the X's be the numbers 1–13, and the Y's the letters a–m; they are related by the following table:

1	2	3	4	5	6	7	8	9	10	11	12	13
a	b	c	d	e	f	g	h	i	j	k	l	m
b	c	d	e	f	g	h	i	j	k	l	m	a
d	e	f	g	h	i	j	k	l	m	a	b	c
j	k	l	m	a	b	c	d	e	f	g	h	i

5. *Cats*: 1, 3; *mice*: 2, 5; *x catches y* provided $x + y$ is even.   **7.** The model of Example 3.10 with *heart* as number, *spade* as set, and *trumps* as "is contained in."   **11.** a) Yes   b) No   c) No

Section 3.5, page 88

1. *Envelopes*: E, F; *letters*: a, b, c, d. E contains a, b, c; F contains a, b, d.   **3.** *Envelopes*: E, F; *letters*: none.   **5.** Yes; X's: p, q; Y's: A, B; *related to* (denoted $*$): $p * A$, $q * A$, $q * B$.   **7.** Yes; same as the answer to Exercise 5, but only $p * A$ and $q * A$.   **9.** Given (a) and (c), (b) necessarily follows, so (b) is dependent.   **11.** Axiom (e): There are at most 2 *crates*. Proof of inconsistency: By (b), there is a *crate*, A. By (d), there is some *box*, b, not in A. By (c), b is in two *crates*, B and C, different from A, so there are at least three *crates*, contradicting (e).   **13.** *Points*: 1, 2, 3, 4; *lines*: $\{1,2\}$, $\{3,4\}$, $\{1,2,3\}$; *on*: contains.   **15.** (e) is dependent. By (b), there are two *lines* with no *points* in common, and each, by (c), contains at least two *points*.   **17.** Dependent. By (b), there are *books*. By (d), there are 2 *students*, A and B, who *read books*. By (c), there is a *student*, C, distinct from A and B, who does not *read books*.   **19.** Vacuously true.

Section 3.6, page 98

1. Congruent triangles have all corresponding angles equal; thus, equal angle sum and equal excess.   **3.** $\mathrm{exc}(ABC) = \angle 1 + \angle 2 + \angle 4 + \angle 5 - 180° = \angle 3 + \angle 6 + \angle 4 + \angle 5 - 180° = \angle 3 + \angle 6 + \angle 4 + \angle 5 + 90° + 90° - 360° = \mathrm{exc}(BCHG)$.   **5.** Using the labeling and results of Exercise 4, triangles ABD and ACD cannot both have excess more than half of ABC.   **7–11.** The arguments and conclusions are entirely analogous to those for Exercises 1–5, respectively.   **13.** (a) A great circle (b) Any other circle (c) The arc between two points on any circle other than a great circle is not the shortest path between those two points.   **15.** (a) Greater than   (b) No   (c) Very close to π (d) No limit

Chapter 4

Section 4.2, page 121

1. $\{1, 3, 5, 7, 9\}$; U is the set of all digits   **3.** $\{0, 1, 2, 3, 4, 5, 6, 7, 8, 9, 10, 11, 12, 13, 14, 15, 16, 17, 18,$ $19, 20\}$; U is the set of all whole numbers   **5.** False   **7.** False   **9.** False   **11.** False
13. False   **15.** $A' = \{0, 6, 7, 8, 9\}$ **17.** $\{1, 2, 3, 4, 5, 6\}$ **19.** $\{1, 2, 3, 5, 6\}$ **21.** $\{3, 4, 5, 6\}$
23. $\{(1, 5), (2, 5), (3, 5), (1, 6), (2, 6), (3, 6)\}$ **25.** $\{0, 7, 8, 9\}$ **27.** $\{0, 1, 2, 4, 5, 6, 7, 8, 9\}$

29.
31.
33. By the definition of union, A is a subset of $A \cup$ any set, so, in particular, $A \subseteq A \cup \emptyset$. The reverse inclusion also follows from the definition of union. Anything in $A \cup \emptyset$ is either in A or in $\emptyset$; but $\emptyset$ contains no elements, so every

element of $A \cup \emptyset$ must be in A. Therefore, $A \cup \emptyset = A$, by the definition of equality.   **35.** By the definition of union, $U \subseteq A \cup U$. Conversely, any element of $A \cup U$ is either in A or in U. But every element of A is also in U, by the definition of universal set; so $A \cup U \subseteq U$, implying equality.

Section 4.3, page 127

1. $\{hhh, hht, hth, thh, htt, tht, tth, ttt\}$ **3.** $P(E) = \frac{1}{2},\ P(F) = \frac{1}{2},\ P(G) = \frac{1}{4}$
5. $E \cup F = \{hhh, hht, hth, tht, tth, ttt\}$; $P(E \cup F) = \frac{3}{4}$ **7.** $F \cup G = \{hhh, hth, tht, ttt, htt\}$;
$P(F \cup G) = \frac{5}{8}$ **9.** $E = \{2h, 4h, 6h, 2t, 4t, 6t\},\ F = \{5h, 5t\},\ G = \{1h, 2h, 3h, 4h, 5h, 6h\},$
$H = \{1t, 2t, 3t\}$ **11.** $E \cap G = \{2h, 4h, 6h\}$; $P(E \cap G) = \frac{1}{4}$ **13.** There are 36 of them.
15. $E = \{(1, 1)\},\ H = \{(1, 2), (1, 3), (2, 2), (3, 1), (2, 1)\}$
17. None; $(1, 1), (1, 6), (6, 1), (2, 5), (5, 2), (3, 4), (4, 3)\}$ **19.** $\{1, 1), (2, 2), (1, 3), (3, 1)\}$;
$\{(1, 1), (1, 2), (2, 1), (2, 2), (1, 3), (3, 1), (3, 3), (2, 4), (4, 2), (4, 4), (1, 5), (5, 1), (3, 5), (5, 3), (5, 5), (2, 6),$
$(6, 2), (4, 6), (6, 4), (6, 6)\}$ **21.** There are 64.   **23.** $P(E) = 0,\ P(F) = \frac{12}{64} = \frac{3}{16},\ P(G) = \frac{1}{2},$
$P(H) = \frac{1}{16}$ **25.** $0;\ \frac{12}{64} = \frac{3}{16}$ **27.** $\frac{3}{64};\ \frac{33}{64}$ **29.** It has 42 outcomes.   **31.** $\frac{19}{21}$

Section 4.4, page 135

1. 360   **3.** 720   **5.** 15   **7.** 120   **9.** 780   **11.** 11,881,376   **13.** 3125   **15.** 30,240
17. $3! = 6$ **19.** $6! = 720$ **21.** 2520   **23.** $\frac{1}{1,048,576}$

Section 4.6, page 145

1. 0   **3.** .55   **5.** 0   **7.** .30   **9.** 75%   **11.** 15%   **13.** 85%   **15.** They do not add up to

1.   **17.** $P(E \cup F)$ must be greater than or equal to both $P(E)$ and $P(F)$.   **19.** $E \subseteq F$ implies $P(F \mid E) = 1$, so they are not independent.   **21.** $\frac{3}{20}$ **23.** $\frac{4}{5}$ **25.** $\frac{11}{40}$ **27.** $\frac{3}{13}$ **29.** $\frac{3}{52}$ **31.** $\frac{1}{2}$ **33.** $\frac{1}{2}$ **35.** 40   **37.** 50   **39.** $\frac{2}{5}$

Section 4.7, page 153

1. $\frac{1}{4}$ **3.** $\frac{4}{17}$ **5.** $\frac{1}{17}$ **7.** $\frac{1}{2}$ **9.** $\frac{33}{100}$ **11.** $\frac{4}{25}$ **13.** $\frac{1}{4}$ **15.** $\frac{3}{4}$ **17.** $\frac{4}{33}$ **19.** $\frac{49}{100}$ **21.** $\frac{2401}{10,000}$ **23.** $\frac{1}{216}$ **25.** (a) *Hint*: Apply Property (7) twice.   (b) $P(E_1 \cap \ldots \cap E_n) = P(E_1) \cdot P(E_2 \mid E_1) \cdot \ldots \cdot P(E_n \mid E_1 \cap E_2 \cap \ldots \cap E_{n-1})$

Section 4.9, page 168

1. 4; 3; 1; 9; 2.7; 10.8; 3.3   **3.** 5.0; 4.6; each number (no useful mode); 10.5; 3.0; 15.0; 3.9   **5.** 693.3; 665; each number (no useful mode); 600; 160; 46,826.7; 216.4   **7.** (a) 4; $-3, -1, 3, 0, 1, 0, 5, -3, -2$ **11.** 70.1; 69; 20.2

Section 4.10, page 179

1. 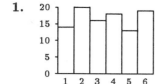   **3.** 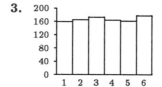   **5.**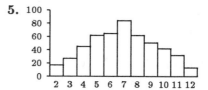

7.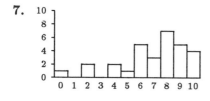

9. Since all six possible outcomes are equally likely, the Law of Large Numbers implies that the histograms for larger numbers of trials should become more like a horizontal line at a frequency level of $\frac{1}{6}$ the number of trials.

11. (b)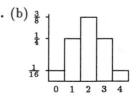

11. (a) $P(0h) = P(4h) = \frac{1}{16}$, $P(1h) = P(3h) = \frac{1}{4}$, $P(2h) = \frac{3}{8}$

Section 4.11, page 190

1. $(-1, 1)$ **3.** $(-3, 3)$ **5.** $(80, 120)$ **7.** $(6.85, 9.35)$ **9.** $(4.35, 11.85)$ **11.** $(-.08, .72)$ **13.** 95%   **15.** 99.7%   **17.** 95%   **19.** 197   **21.** 75   **23.** 10.4   **25.** 4.65   **27.** 1.90   **29.** .46   **31.** .50   **33.** (a) 1.5% (b) 197   **35.** (a) $\sigma \approx \sqrt{(.58)(.42)} \approx .49$, so $\bar{\sigma} \approx \frac{.49}{\sqrt{50}} \approx .07$. Thus, a 95% confidence interval about this sample mean of $\frac{29}{50} = 58\%$ is (44%, 72%). Since the break-even vote is well within this interval, the manager could not be 95% confident that the candidate would win. (b) The two phone polls together constitute a 150-voter sample, so this time $\bar{\sigma} \approx \frac{.49}{\sqrt{150}} \approx .4$. The corresponding 95% confidence interval is (50%, 64%). Although the 50% break-even vote is at one end of this interval, the likelihood that the actual vote for the candidate is *less* than 50% is only 2.5%, well inside the manager's desired confidence level.   **37.** (a) $\bar{\sigma} = \frac{.06}{3} = .02$. Thus, a 99.7% confidence interval for this sample is $(14.98, 15.10)$ — completely within the allowable variation. The company can be 99.7% confident.   (b) $\bar{\sigma} = \frac{.12}{3} = .04$. Thus, a 95% confidence interval for this sample is $(14.87, 15.03)$ and a 68% confidence interval is $(14.91, 14.99)$. In this case the company cannot be 95% confident, but it can be 68% confident.

Chapter 5

Section 5.2, page 212

11. 1, 10, 11, 100, 101, 110, 111, 1000, 1001, 1010, 1011, 1100, 1101, 1110, 1111, 10000   **13.** 11111
15. 111100   **17.** 18   **19.** 63   **21.** 357   **23.** 1984

Section 5.3, page 217

1. 7 + 4   **3.** 2 - 7   **5.** 5 * 8   **7.** -9 / 2   **9.** 3 ˆ 4 or 3 ** 4   **11.** 13 - 51
13. $2.4 - 3.5$ **15.** 17×-4 **17.** 17^{-4} **19.** $|6 \times -3|$ **21.** $|X - Y|$ **23.** $18 + 3 \times 2$;
24   **25.** $2^3 \times -2$; -16 **27.** $2^{(3\times-2)}$; 0.015625   **29.** $2 \times ((-3 + 5) \div 8)$; 0.5   **31.** $|2^3 \times -2|$;
16   **33.** Variable S has value "8"; nothing on CRT.   **35.** Nothing stored; error message on CRT.
37. Variable **F$** has value "FINAL EXAM"; nothing on CRT.   **39.** Nothing stored; "6" on CRT.
41. Nothing stored; error message on CRT.   **43.** Nothing stored; "1.21" on CRT.   **45.** Nothing
stored; error message on CRT.   **47.** Nothing stored; "1" on CRT.   **49.** Variable M has value "3";
variable N has value "2"; "6" on CRT.   **51.** Variable $B1$ has value "8"; variable $B2$ has value "$\frac{1}{3}$";
"2" on CRT.

Section 5.4, page 225

3. The program will prompt for input of a number as the radius of a circle, calculate and print the circumference of the circle.   **5.** Add line **25 INPUT N3** and add line **30 PRINT N1 + N2 + N3**
7. Change **100 INPUT A, B, C, D** and **DELETE** line **600**.   **9.** Add lines **45 A = PI * R ˆ 2** and **55 PRINT "AREA IS"; A**   **11.** In Example 5.4 change lines: **100 P = 1** and **300 P = P * S**

13.
```
10 INPUT "ENTER TWO NUMBERS" A,B
20 PRINT "THE SUM IS "; A+B
30 PRINT "THE PRODUCT IS"; A*B
40 END
```

15.
```
10 INPUT "ENTER LENGTH" L
20 INPUT "ENTER WIDTH" W
30 INPUT "ENTER HEIGHT" H
40 PRINT
50 PRINT "THE VOLUME IS "; L*W*H
60 END
```

17.
```
10 INPUT "ENTER THE DEGREE MEASURE OF TWO ANGLES" A1,A2
20 PRINT
30 PRINT "THE DEGREE MEASURE OF THE THIRD ANGLE IS "; 180-A1-A2
40 END
```

Section 5.5, page 231

1. 110 IF INT(S) = S THEN GOTO 200
120 PRINT "ENTRY MUST BE AN INTEGER"
130 GOTO 100

3. The program would PRINT the value entered. Add
140 IF S < 100 THEN GOTO 200
160 PRINT "ENTRY MUST BE LESS THAN 100"
170 GOTO 100

5. The program would PRINT "3880 IS NOT A SQUARE."   **7.** The program tests whether or not an entry is the square of an integer; if so, it PRINTs the square root.

9. 12 IF S = INT(S) THEN GOTO 20
14 PRINT S; " IS NOT AN INTEGER."
16 GOTO 10

11. 12 IF S >= 0 THEN GOTO 20
14 PRINT "THE NUMBER ENTERED MUST BE
POSITIVE."
16 GOTO 10

13. 10 is PRINTed.   **15.** 5050 is PRINTed.   **17.** See #11.

19. 5 INPUT "ENTER TWO INTEGERS"; A,B
10 FOR I = 1 TO A
20 FOR J = 1 TO B

21. The program PRINTs "12 HAS 2 FACTORS OF 2."
23. The program determines and PRINTs the number of times 2 occurs as a factor of the number ENTERed.

25. The program goes into an infinite loop.   **27.** 40 T = T/3 and 80 PRINT P: " HAS "; N; " FACTORS OF 3."   **29.** Change the program for Exercises 12–17: 30 FOR I = 1 TO 2*N-1 STEP 2

31. 10 F = 1
20 A = 0
30 FOR N = 1 TO 30
40 PRINT "NUMBER ";N;" IS ";F
50 B = F
60 F = A + B
70 A = B
80 NEXT N
90 END

Section 5.6, page 241

1. (3,10), (3,70), (17,10), (17,70)   **3.** Because the "I" at (17,10) and (17,70) cannot be underlined.

5. 10 CLS
20 FOR I = 1 TO 80
30 PRINTAT (1,I) "X"
40 PRINTAT (20,I) "X"
50 NEXT I
60 FOR J = 2 TO 19
70 PRINTAT (J,1) "X"
80 PRINTAT (J,80) "X"
90 NEXT J
100 END

7. There would be no erasure of the arrow. The CRT screen would show a short arrow extending to become an arrow the width of the screen, successively on each line.   **9.** Interchange lines 40 and 50; interchange lines 70 and 80.

11. 10 CLS
20 LET A$ = "--->"
30 LET B$ = "<---"
40 LET C$ = "   "
50 FOR I = 1 TO 20
60 IF I/2 = INT(I/2) THEN GOTO 120
70 FOR J = 1 TO 77
80 PRINTAT (I,J) A$
90 PRINTAT (I,J) C$
100 NEXT J
110 GOTO 160
120 FOR J = 77 TO 1 STEP −1
130 PRINTAT (I,J) B$
140 PRINTAT (I,J) C$
150 NEXT J
160 NEXT I
170 END

13. "X" is PRINTed AT $(10, 40)$.   **15.** "X" is PRINTed AT $(10, 40)$; overprinted by "A."   **17.** "X" is PRINTed AT $(10, 40)$; program ends.   **19.** X-RAY   **21.** LAX   **23.** L,0 or O,R,X   **25.** Impossible (See Exercise 17.)   **27.** When $H = 1$, $I = 39$. Tests in lines 310 and 340 fail. Lines 360–400 are executed. Visually, the "missile" hits the "arrow" with a "BOOM"; "YOU WIN!" appears. The program ends.   **29.** Add line: 310 PRINTAT (H, 40) " "   **31.** Add lines: 50 N = 1,   530 IF N = 10 THEN END, and 535 N = N + 1   **33.** Add, or replace, the following lines:

```
30 M = 0                        410 IF M = 3 THEN GOTO 550
50 S = 0                        415 PRINTAT (1,39) T$
70 F$ = "NO"                    420 FOR H = 2 TO 19
270 F$ = "YES"                  470 IF T$ = "BOOM" THEN GOTO 60
340 IF I<37 OR I>40 THEN GOTO 390   540 IF F$ = "YES" THEN GOTO 60
350 T$ = "BOOM"                 550 CLS
360 PRINTAT (1, I) T$           560 PRINT "END OF GAME"
370 S = S+1                     570 PRINT
380 GOTO 420                    580 PRINT "YOUR SCORE IS ";
390 T$ = "MISS"                 590 PRINT S
400 M = M+1                     600 END
```

Chapter 6

Section 6.1, page 253

1. True   **3.** False   **5.** True   **7.** True   **9.** False   **11.** True   **13.** False   **15.** False   **17.** True   **19.** False   **21.** False   **23.** False   **25.** True   **27.** True   **29.** False   **31.** False   **33.** True   **35.** False   **37.** False

Section 6.2, page 259

1. There are 24 in all.   **3.** Any 5-element set can be put in 1-1 correspondence with $\{0, 1, 2, 3, 4\}$; no other set can be.   **5.** Yes; yes; both represent $\mathbf{N}$.   **7.** Yes; match each element with itself.

9.
$$\{ \; 2, \quad 4, \quad 6, \quad \ldots, \quad 2n, \quad \ldots \; \}$$
$$\updownarrow \quad \updownarrow \quad \updownarrow \qquad \updownarrow$$
$$\{ \; 5, \quad 10, \quad 15, \quad \ldots, \quad 5n, \quad \ldots \; \}$$

Each of the following examples is only one of many possible correct answers: **11.** The days of the week   **13.** All members of the U. S. House of Representatives   **15.** $\{f, g, h, i, j\}$ **17.** $\{a, b, c\}$

19. $\{2, 4, 6, \ldots, 2n, \ldots\}$ **21.** $\{2, \frac{5}{2}, \frac{8}{3}, \frac{11}{4}, \ldots, 3 - \frac{1}{n}, \ldots\}$

23.
$$\{ \; 2, \quad 4, \quad 6, \quad \ldots, \quad 2n, \quad \ldots \; \}$$
$$\updownarrow \quad \updownarrow \quad \updownarrow \qquad \updownarrow$$
$$\{ \; 4, \quad 8, \quad 12, \quad \ldots, \quad 4n, \quad \ldots \; \}$$

25. $\{6, 12, 18, \ldots, 6n, \ldots\}$; $\{3, 6, 9\}$ **27.** $\{6, 7, 8, \ldots, n+5, \ldots\}$; $\{5, 6, 7, \ldots, 20\}$ **29.** $\{4, 16, 36, \ldots, (2n)^2, \ldots\}$; $\{9\}$

Section 6.3, page 264

Other correct answers to Exercises 1, 3, 5, and 13 are possible.   **1.** $\{5, 4, 3, 2, 1\}$, $\{2, 1, 3, 4, 5\}$, $\{1, 2, 4, 3, 5\}$ **3.** $\{2, 1, 3, 4, 5, 6, \ldots\}$, $\{2, 1, 4, 3, 6, 5, \ldots\}$, $\{3, 2, 1, 6, 5, 4, 9, 8, 7, \ldots\}$ **5.** Adjust the ordering determined by Figure 6.3 in ways similar to the answers to Exercise 3.   **7.** 50   **9.** -37
11. $2p$ **13.**

$$\{ \quad 1, \quad 2, \quad 3, \quad \ldots, \quad 17, \quad \ldots, \quad 50, \quad \ldots, \quad n, \quad \ldots \quad \}$$
$$\updownarrow \quad \updownarrow \quad \updownarrow \qquad \updownarrow \qquad \updownarrow \qquad \updownarrow$$
$$\{ \quad 30, \quad 60, \quad 90, \quad \ldots, \quad 510, \quad \ldots, \quad 1500, \quad \ldots, \quad 30n, \quad \ldots \quad \}$$

15. $\frac{4}{3}$ **17.** $-\frac{4}{3}$ **19.** 16   **21.** -19 **23.** Those same numbers (in lowest-terms form) are already matched with integers.

Section 6.4, page 270

1. 4; 1   **3.** 36; 3; yes   **5.** 6; 6; yes   **7.** 13; 0; no (if you consider repeated 0s)   **9.** 9989; 4; no   **11.** $0.4 = 0.4\overline{0}$ **13.** $0.6\overline{81}$ **15.** $0.\overline{153846}$ **17.** $0.\overline{428571}$ **19.** $3.\overline{142857}$ **21.** $0.0\overline{45}$
23. $\frac{2}{90}$ **25.** $\frac{2}{99}$ **27.** $\frac{14}{90}$ **29.** $\frac{61{,}554}{990}$ **31.** $\frac{40}{99}$ **33.** $\frac{68{,}446}{9990}$

Section 6.5, page 275

There are many correct answers for Exercises 1 and 3; typical examples are given here.   **1.** $\frac{1}{2}, \frac{1}{3}, \frac{1}{4}, \frac{1}{5}, \frac{1}{6}$
3. .51, .52, .515, .525, .525555...   **5.** (a) .121112...   (b) .775777..., .888848..., .555555...   The answers for Exercises 7–13 depend on the choice of P; the answers given here are for P chosen 3 inches beyond the 2-inch end of T.   **7.** $1\frac{1}{4}$ in.   **9.** $2\frac{5}{8}$ in.   **11.** $1\frac{5}{16}$ in.   **13.** $1\frac{13}{16}$ in.   **15.** 0 in.
17. No   **19.** Yes   **21.** Yes   **23.** No   **25.** 0.515151...   **27.** $0.73530303\ldots = 0.735\overline{03}$
29. $(0.777\ldots, 0.777\ldots)$ **31.** $(0.5, 0) = (\frac{1}{2}, 0)$ **33.** (a) $\frac{3}{4}, \frac{3}{2}, \frac{9}{4}, 1, 1.2, 0, 3$ (b) $0, \frac{1}{3}, \frac{2}{3}, 1, \frac{1}{6}, \frac{1}{9}, 0.081, \frac{5}{6}, 0.9$

Section 6.6, page 279

There are many correct answers for Exercises 1 and 5; typical examples are given here.
1. $\{a, b, c, d, e\}$, $\{p, q, r, s, t\}$, $\{a, e, i, o, u\}$ **3.** 26   **5.**

$$\{ \quad \spadesuit, \quad \heartsuit, \quad \diamondsuit, \quad \clubsuit \quad \}$$
$$\updownarrow \quad \updownarrow \quad \updownarrow \quad \updownarrow$$
$$\{ \quad a, \quad b, \quad c, \quad d, \quad e, \quad f, \quad g \quad \}$$

7. They are equal because they have equivalent reference sets.
9. No; yes   **11.** A reference set for 3, $\{a, b, c\}$, is equivalent to $\{1, 2, 3\}$, which is a proper subset of $\mathbf{N}$; so $\mathbf{N}$ represents the larger number, $\aleph_0$.   **13.** Any reference set for n is equivalent to $\{1, 2, \ldots, n\}$, which is a proper subset of $\mathbf{N}$; so $\mathbf{N}$ represents the larger number, $\aleph_0$.   **15.** For any natural number n, $n < \aleph_0$ and $\aleph_0 < c$; so $n < c$.

Section 6.7, page 283

There are many correct answers for Exercises 3–9 and 13; typical examples are given here.

1. $\{\emptyset, \{a\}, \{b\}, \{c\}, \{a, b\}, \{a, c\}, \{b, c\}, \{a, b, c\} \}$ **3.** $\{2, 4, 6\}$ **5.** $\{1, 2, 3, \ldots, 500\}$
7. $\{1, \frac{1}{2}, \frac{1}{3}\}$ **9.** $\{1, \frac{1}{2}, \frac{1}{3}, \ldots, \frac{1}{n}, \ldots\}$ **11.** $S = \{ \emptyset, \{a\}, \{b\}, A \}$. The set of all subsets of S
is $\{ \emptyset, \{\emptyset\}, \{\{a\}\}, \{\{b\}\}, \{A\}, \{\emptyset, \{a\}\}, \{\emptyset, \{b\}\}, \{\emptyset, A\}, \{\{a\}, \{b\}\}, \{\{a\}, A\}, \{\{b\}, A\},$
$\{\emptyset, \{a\}, A\}, \{\emptyset, \{b\}, A\}, \{\{a\}, \{b\}, A\}, \{\emptyset, \{a\}, \{b\}\}, A \}$. 4; 16
13. (a) $\{ \quad$ u, $\quad$ v, $\quad$ w, $\quad$ x, $\quad$ y, $\quad$ z $\}$ (b) $\{$v, w, z$\}$

$$\uparrow \qquad \uparrow \qquad \uparrow \qquad \uparrow \qquad \uparrow \qquad \uparrow$$

$$\{ \{u,v\}, \{w,x,y\}, \{z\}, \{x,z\}, \{v,y\}, \emptyset \}$$

Section 6.8, page 286

There are many correct answers for Exercises 1, 7, and part of 9; typical examples are given here.
1. (a) $[0,1]$ (b) $1, \frac{1}{2}, \frac{1}{3}$ (c) $\{1, \frac{1}{2}, \frac{1}{3}, \frac{1}{4}, \frac{1}{5}\}$ (d) $\{1, \frac{1}{2}, \frac{1}{3}, \ldots, \frac{1}{n}, \ldots\}$ **3.** $\aleph_6$ **5.** Subset; 1
7. $\{ \{x\} \mid x \in K \}$ **9.** Three $- \aleph_0, \aleph_1, \aleph_2$; $\mathbf{N}, \mathbf{R}, \{S \mid S \subseteq \mathbf{R}\}$

Chapter 7

Section 7.1, page 303

There are many correct answers for Exercises 1 and 3; typical examples are given here.   **1.** $3 \cdot 1 + 3 \cdot 2 =$
$3 \cdot 3$; $3 \cdot 2 + 3 \cdot 5 = 3 \cdot 7$; $3 \cdot 20 + 3 \cdot 5 = 3 \cdot 25$ **3.** $(1 + 2) + 3 = 1 + (2 + 3)$; $7 + 5 = 3 + 9$;
$25 + 10 = 20 + 15$ **5.** $2x = x + x$ **7.** $xy + xz = x(y + z)$

Section 7.2, page 308

1. b, d, a, c   **3.** 5   **5.** 5   **7.** 6   **9.** 4   **11.** 4   **13.** There are sixteen such tables in all;
six are given here:

	1	2
1	1	1
2	1	1

	1	2
1	1	2
2	1	1

	1	2
1	2	1
2	1	2

	1	2
1	2	2
2	2	2

	1	2
1	1	1
2	2	1

	1	2
1	1	2
2	2	1

15.

$\circ$	1	2	3	4	5
1	1	2	3	4	5
2	2	1	3	4	5
3	3	3	1	4	5
4	4	4	4	1	5
5	5	5	5	5	1

17. $1 + 3 = 4$; $3 + 7 = 10$

Section 7.3, page 312

Other correct answers are possible for Exercises 1 and 9.

1. $(12 \div 6) \div 2 = 1$; $12 \div (6 \div 2) = 4$ **3.** Commutative, associative **5.** Not commutative, associative **7.** Not commutative,
not associative (Check 2,4,6, for example.)

9.

	p	q	r
p	p	q	r
q	q	r	p
r	r	p	q

Section 7.4, page 316

1. 2 **3.** 6 **5.** 2 **7.** 2 **9.** 1 **11.** 0 **13.** 4
15. 1 **17.** Table I is the one that is not associative.
19. Possible **21.** Not possible **23.** Not possible (in this
case)

25. Determined entries:

*	1	2	3	4	5
1		1			2
2	1	2	3	4	5
3		3	2		
4		4	1		
5	2	5			

Section 7.5, page 322

7.

	r	s
r	s	r
s	r	s

9. Not possible **11.** Not possible **13.** Not possible **15.** Definition of
group; definition of operation; commutativity of $*$; associativity of $*$; definition
of inverse; definition of identity; definition of inverse; Property (7.3) (Why?)

Section 7.6, page 325

3. You should find 4 in all, of sizes 1, 2, 3, and 6 elements.

5. (a)

$\cdot$	1	i	-1	$-i$
1	1	i	-1	$-i$
i	i	-1	$-i$	1
-1	-1	$-i$	1	i
$-i$	$-i$	1	i	-1

5. (c) $\{1\}$, $\{1, -1\}$, $\{1, i, -1, -i\}$
7. 1
11. 1, 2, 3, 4, 6, 12; all factors of 12.

Section 7.7, page 332

1. a **3.** No; not closed — $b*b = c$ **5.** Three because 3 elements are their own inverses. **7.** No;
no identity **9.** No; not closed **11.** $\{0\}$, $\{0, 3, 6\}$, S **13.** (a) Y has the most; X has the fewest.
(b) 1, 2, 11, 22 in W; 1, 5, 25 in Z. (c) W and Y must; X and Z cannot.

Section 7.8, page 338

1. $\{2, 4, 6\}$ **3.** $\{2, 4, 6\}$ **5.** $\{1, 3, 5\}$ **7.** $\{2, 5, 8, 11\}$ **9.** $\{2, 5, 8, 11\}$ **11.** $\{1, 5, 9\}$
13. $\{3, 7, 11\}$ **17.** (a)

*	1	4
1	1	4
4	4	1

15. $\{a, c\}$, $\{b, d\}$

17. (b) Yes (c) $\{1, 4\}$, $\{2, 3\}$, $\{3, 6\}$, $\{4, 1\}$, $\{5, 2\}$
$\{6, 5\}$ (d) Not a group; the cosets of $\{1, 4\}$ are not
disjoint.

Section 7.9, page 343

3. Denote the eight rigid motions as:

   0 — rotation of 0°
   1 — rotation of 90°
   2 — rotation of 180°
   3 — rotation of 270°
   v — reflection on vertical median
   h — reflection on horizontal median
   d — reflection on left-right "down" diagonal
   u — reflection on left-right "up" diagonal

o	0	1	2	3	v	h	d	u
0	0	1	2	3	v	h	d	u
1	1	2	3	0	d	u	h	v
2	2	3	0	1	h	v	u	d
3	3	0	1	2	u	d	v	h
v	v	u	h	d	0	2	3	1
h	h	d	v	u	2	0	1	3
d	d	v	u	h	1	3	0	2
u	u	h	d	v	3	1	2	0

5. (a) There are only two possible rigid motions of this figure:

   0 — rotation of 0°
   f — flip (reflection) about the bisector of the right angle

o	0	f
0	0	f
f	f	0

(b) Yes   (d) It matches any of the subgroups {j, f}, {j, k}, {j, g}.

Chapter 8

Section 8.2, page 357

1. 4   **3.** 17   **5.** 0   **7.** 4   **9–15.**

$$-\tfrac{10}{3} \qquad\qquad 0 \quad 1 \qquad \tfrac{1}{2} \quad \sqrt{2} \qquad 3$$

17. 4   **19.** 7   **21.** 0   **23.** $2\sqrt{2}$ **25.** $|6 - x|$ **27.** $\{x \mid 2 \le x \le 3\}$; length 1

29. $\{x \mid -2 \le x \le 1\}$; length 3   **31.** $\{x \mid -5 \le x \le -\tfrac{1}{2}\}$; length $\tfrac{9}{2}$

33–37.

$$0 \quad 10 \qquad 40 \text{ min.}$$
$$37 \text{ min.} \qquad 1 \text{ hr.}$$

Section 8.3, page 366

1–7.

(−1, 3)   (2, 4)

(−5, 0)

(−1, −1)

9. 10   **11.** $\sqrt{2}$ **13.** 5

15–23.

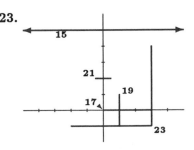

25. $\{(x, 0) \mid 0 \le x \le 3\} \cup \{(3, y) \mid 0 \le y \le 4\}$; $(3, 0)$
  $\{(0, y) \mid 0 \le y \le 4\} \cup \{(x, 4) \mid 0 \le x \le 3\}$; $(0, 4)$
27. $\{(x, 2) \mid 1 \le x \le 5\} \cup \{(5, y) \mid -1 \le y \le 2\}$; $(5, 2)$
  $\{(1, y) \mid -1 \le y \le 2\} \cup \{(x, 2) \mid 1 \le x \le 5\}$; $(1, -1)$

29. $\{(x,0) \mid -1 \le x \le 2\} \cup \{(-1,y) \mid -1 \le y \le 0\}$; $(-1,0)$
$\{(2,y) \mid -1 \le y \le 0\} \cup \{(x,-1) \mid -1 \le x \le 2\}$; $(2,-1)$

31. $\{(2,y) \mid \frac{-5}{2} \le y \le \frac{7}{3}\}$; no intersection points
$\{(x,\frac{-5}{2}) \mid 1 \le x \le 2\} \cup \{(2,y) \mid \frac{-5}{2} \le y \le \frac{7}{3}\} \cup \{(x,\frac{7}{3}) \mid 1 \le x \le 2\}$; $(2,\frac{-5}{2})$, $(2,\frac{7}{3})$

33. 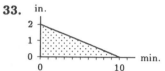 **35.**

37. $\{(3,y) \mid 0 \le y \le 1\} \cup \{(x,1) \mid \frac{1}{2} \le x \le 3\} \cup \{(\frac{1}{2},y) \mid 0 \le y \le 1\}$

Section 8.4, page 376

1. $\sqrt{66}$ $(= 8.124)$ **3.** 5   **5.** $\sqrt{3}$ **7.** 3   **9.** $\{(x,6,1) \mid x \in \mathbf{R}\}$ and $\{(x,6,1) \mid 4 \le x \le 7\}$
11. $\{(-1,2,z) \mid z \in \mathbf{R}\}$ and $\{(-1,2,z) \mid 0 \le z \le 2\}$
13. $\{(x,0,0) \mid 0 \le x \le 2\} \cup \{(2,y,0) \mid -5 \le y \le 0\} \cup \{(2,-5,z) \mid -1 \le z \le 0\}$; $(2,0,0)$, $(2,-5,0)$
15. $\{(x,3,5) \mid -4 \le x \le 7\} \cup \{(7,y,5) \mid -1 \le y \le 3\} \cup \{(7,-1,z) \mid -9 \le z \le 5\}$; $(7,3,5)$, $(7,-1,5)$
19. (a)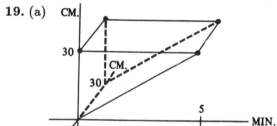

(b) $(0,0,0)$, $(30,0,0)$, $(0,30,0)$,
$(30,30,0)$, $(0,30,5)$, $(30,30,5)$

Many correct answers are possible for Exercises 21 and 23; these are typical examples.
21. $\{(1,3,z) \mid 0 \le z \le 1\} \cup \{(1,y,1) \mid 1 \le y \le 3\} \cup \{(1,1,z) \mid 0 \le z \le 1\}$
23. $\{(0,6,z) \mid 0 \le z \le 1\} \cup \{(0,y,1) \mid 1 \le y \le 6\} \cup \{(x,1,1) \mid 0 \le x \le 1\} \cup \{(1,1,z) \mid 0 \le z \le 1\}$

Section 8.5, page 383

1. 2   **3.** 4   **5.** $\{(x,6,4,2) \mid x \in \mathbf{R}\}$; $\{(x,6,4,2) \mid -1 \le x \le 5\}$ **7.** $\{-3,7,z,2) \mid z \in \mathbf{R}\}$;
$\{(-3,7,z,2) \mid -8 \le z \le -4\}$
9. $\{(x,0,0,0) \mid 0 \le x \le 1\} \cup \{(1,y,0,0) \mid 0 \le y \le 1\} \cup \{(1,1,z,0) \mid 0 \le z \le 1\} \cup \{(1,1,1,t) \mid 0 \le t \le 1\}$;
$(1,0,0,0)$, $(1,1,0,0)$, $(1,1,1,0)$
11. $\{(x,-3,4,-1) \mid 2 \le x \le 7\} \cup \{(2,y,4,-1) \mid -3 \le y \le 5\} \cup \{(2,5,4,t) \mid -1 \le t \le 0\}$;
$(2,-3,4,-1)$, $(2,5,4,-1)$ **13.** $\{(x,7,6,-1) \mid 2 \le x \le 5\} \cup \{(5,7,6,t) \mid -1 \le t \le 3\}$; $(5,7,6,-1)$
(*Note:* The answers to Exercises 15, 17, and 19 are given in space-time units of meters and hours.
They assume that $t = 0$ is the present time.)   **15.** From $(0,0,0,0)$ the path is 1 meter along the
positive x-axis, followed by 1 meter parallel to the y-axis in the positive direction, followed by 1 meter
(positive) parallel to the z-axis, followed by a 1-hour wait.   **17.** Start at $(7,-3,4)$ in the 3-space

of an hour ago. In that 3-space, go 5 meters in the positive x-direction and 8 meters in the positive y-direction. Then wait 1 hour, up to the present.   **19.** Starting at $(2,7,6)$ in the 3-space of an hour ago, 3 meters (positive) parallel to the x-axis to $(5,7,6)$, then a 4-hour wait, until 3 hours from now.   **23.** (a) $(0,0,0,0)$, $(2,0,0,0)$, $(0,2,0,0)$, $(0,0,2,0)$, $(2,2,0,0)$, $(2,0,2,0)$, $(0,2,2,0)$, $(2,2,2,0)$ (b) $\{(x,1,1,0) \mid 1 \le x \le 3\}$ intersects the cube wall at $(2,1,1,0)$.   (c) $\{(1,1,1,t) \mid 0 \le t \le 1\} \cup \{(x,1,1,1) \mid 1 \le x \le 3\} \cup \{(3,y,1,1) \mid 1 \le y \le 3\} \cup \{(3,3,z,1) \mid 1 \le z \le 3\} \cup \{(3,3,3,t) \mid 0 \le t \le 1\}$

Section 8.6, page 391

1. $t = 0$: $(0,0,0,0)$, $(2,0,0,0)$, $(0,2,0,0)$, $(0,0,2,0)$, $(2,2,0,0)$, $(2,0,2,0)$, $(0,2,2,0)$, $(2,2,2,0)$ $t = 1$: $(0,0,0,1)$, $(1,0,0,1)$, $(0,1,0,1)$, $(0,0,1,1)$, $(1,1,0,1)$, $(1,0,1,1)$, $(0,1,1,1)$, $(1,1,1,1)$ $t = \frac{3}{2}$: $(0,0,0,\frac{3}{2})$, $(\frac{1}{2},0,0,\frac{3}{2})$, $(0,\frac{1}{2},0,\frac{3}{2})$, $(0,0,\frac{1}{2},\frac{3}{2})$, $(\frac{1}{2},\frac{1}{2},0,\frac{3}{2})$, $(\frac{1}{2},0,\frac{1}{2},\frac{3}{2})$, $(0,\frac{1}{2},\frac{1}{2},\frac{3}{2})$, $(\frac{1}{2},\frac{1}{2},\frac{1}{2},\frac{3}{2})$ **3.** A $\frac{5}{4}$-meter cube with corners at $(0,0,0,\frac{3}{4})$, $(\frac{5}{4},0,0,\frac{3}{4})$, $(0,\frac{5}{4},0,\frac{3}{4})$, $(0,0,\frac{5}{4},\frac{3}{4})$, $(\frac{5}{4},\frac{5}{4},0,\frac{3}{4})$, $(\frac{5}{4},0,\frac{5}{4},\frac{3}{4})$, $(0,\frac{5}{4},\frac{5}{4},\frac{3}{4})$, $(\frac{5}{4},\frac{5}{4},\frac{5}{4},\frac{3}{4})$ **5.** A pyramid-type structure like Figure 8.31(a) with base corners $(0,0,0,0)$, $(2,0,0,0)$, $(0,2,0,0)$, $(2,2,0,0)$ and peak at $(0,0,0,2)$ **7.** As in Figure 8.31(b), with base corners at $(0,1,0,0)$, $(2,1,0,0)$, $(0,1,2,0)$, $(2,1,2,0)$ and "top" corners at $(0,1,0,1)$, $(1,1,0,1)$, $(0,1,1,1)$, $(1,1,1,1)$ **9.** A single square "sheet" with corners at $(0,2,0,0)$, $(2,2,0,0)$, $(0,2,2,0)$, and $(2,2,2,0)$ **11.** On   **13.** Outside   **15.** Outside   **17.** On   **19.** Outside   **21.** On   **23.** Outside   **25.** Outside   **27.** On   **29.** The sphere $\{(2,y,z,t) \mid y^2 + z^2 + t^2 = 21\}$; $(2,1,2,4)$ **31.** The sphere $\{(x,y,0,t) \mid x^2 + y^2 + t^2 = 25\}$; $(2,3,0,\sqrt{12})$

Section 8.7, page 402

Many correct answers are possible for Exercises 1 and 3; typical examples are given here.
1. $(3,4,6)$, $(-2,\sqrt{21},1)$ **3.** $(1,\sqrt{3},6)$, $(-2,\sqrt{12},2)$ **5.** $\{(x,y,z) \mid x^2 + y^2 = 9 \text{ and } 0 \le z \le 7\}$
7. $\{(x,y,z) \mid x^2 + y^2 = (4 - \frac{1}{5}z)^2 \text{ and } 0 \le z \le 20\}$ **9.** 4   **11.** 3   **13.** 0   **15.** 0   **17.** 10
19. 2   **21.** $\{(x,y,1) \mid x^2 + y^2 = 25\}$, a circle of radius 5 inches.
23. $\{(-4,y,z) \mid y^2 = 9 \text{ and } 0 \le z \le 12\}$, two parallel 12-inch line segments that are 6 inches apart.
25. $\{(x,y,z,3) \mid x^2 + y^2 + z^2 = 25\}$, a sphere of radius 5 inches.
27. $\{(x,0,z,t) \mid x^2 + z^2 = 25 \text{ and } 0 \le t \le 12\}$, a cylinder of radius 5 in. and height 12 in. (in xzt-space).
29. $\{(1,y,z,t) \mid y^2 + z^2 = 24 \text{ and } 0 \le t \le 12\}$, a cylinder of radius $\sqrt{24}$ in. and height 12 in.
31. $\{(5,0,0,t) \mid 0 \le t \le 12\}$, a 12-inch line segment.   **33.** $x^2 + y^2 = 25$ **35.** $x^2 + y^2 = 1$
Several different correct answers are possible for Exercises 37 and 39; we give two typical 2-dimensional cross sections for each exercise.   **37.** $\{(x,y,9) \mid x^2 + y^2 = 25\}$; $\{(-3,y,z) \mid y^2 = 16 \text{ and } 0 \le z \le 12\}$
39. $\{(x,\sqrt{7},z) \mid x^2 + 7 = (5 - \frac{1}{2}z)^2 \text{ and } 0 \le z \le 10\}$; $\{(x,y,2) \mid x^2 + y^2 = 16\}$ Many different correct answers are possible for Exercise 41; this is a typical example.   **41.** $\{(7,8,z) \mid 9 \le z \le 13\} \cup \{(x,8,13) \mid 0 \le x \le 7\} \cup \{(0,y,13) \mid 1 \le y \le 8\} \cup \{(0,1,z) \mid -1 \le z \le 13\} \cup \{(x,1,-1) \mid 0 \le x \le 7\} \cup \{(7,y,-1) \mid 1 \le y \le 8\} \cup \{(7,8,z) \mid -1 \le z \le 9\}$

Chapter 9

Section 9.2, page 422

1. All the drawings except (d) and (h) are graphs.   **3.** (b), (c), (g), and (i)   **5.** (a) Z_1WX
(b) $W_1ZYXW_2Z_1WX$ (c) W_2ZYXW_1Z (d) Six   (e) None   (f) Infinite   **7.** (a) True   (b) False
(c) True   (d) True   (e) False   (f) True   (g) True   (h) True   (i) False   (j) True   (k) True   (l) Cannot be determined

9. N – north bank, S – south bank, I – island: **11.** (a)   (b)

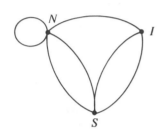

11. (c) Impossible; a vertex path guarantees a path from any vertex to any other; hence, the graph
must be connected.   (d) Impossible; if the graph is connected, then there must be edges joining at
least two (of the three possible) pairs of vertices. Thus, some vertex (say P) must be joined to each
of the others (say Q and R). Then there is a vertex path QPR.   (e) See the answer to Part (b).
(f) Impossible; by the reasoning of Part (c), the existence of a vertex path guarantees that the graph
is connected. With only three vertices, there must be an edge path, regardless of loops or multiple
edges.   (g) See the answer to Part (a).

Section 9.3, page 430

1. (a) Even: A, C, D, F; odd: B, E (b) Even: G, L; odd: H, J, K, M (c) All the vertices are even.
(d) All the vertices are even.   (e) Even: N, P; odd: K, L, M, Q, R, S (f) All the vertices are even.
(g) Even: F, G, H, J, K, L; odd: D, E **3.** (a) 2   (b) 4   (c) 0   (d) 0   (e) 6   (f) 0   (g) 2   **5.** Step
1 is modified so that a path is randomly constructed starting at any vertex, P. The path must
eventually end at P because all the vertices are even. The rest of the proof is the same.   **7.** Start
a path at any odd vertex, proceeding as described in Step 1. The path must end at some other odd
vertex. If there is a third odd vertex, start another path there; it will end at a fourth odd vertex.
Proceeding in this fashion, odd vertices can be paired, thus showing that there must be an even number
of them.   **9.** Adding or removing one edge in Figure 9.12 changes two odd vertices to even ones,
leaving only two odd vertices.   **11.** Three edges, joining distinct pairs of odd vertices   **13.** The
closed edge path must pass through every vertex because the graph is connected. If there are n vertices,
this path must contain at least n edges in order to pass through every vertex at least once and return
to its starting point.   **15.** •———————•   **17.** Two vertices, each with one loop.

Section 9.4, page 438

1. a   3. f   5. Yes. For example, if the vertices (intersections) are labeled alphabetically from left to right (top row: A, B, C; next row D through I; and so on) one possible path is *AB-CIONTUZYXRKQWVPJDEFLMSHGA*.   **7.** If a vertex path omits an H-edge at a vertex, it must use the (only) other edge at that vertex; thus, this vertex must be one end of the path. Since there can be only two ends to the path, there can be at most two H-edges omitted.   **9.** Suppose that A is a terminal vertex, connected by its one edge to B, which is in turn connected to other vertices in the graph. The deletion of B and its edges disconnects A from the rest of the graph, so B is critical.

Section 9.5, page 445

1.   **3.** (c), (d), (e), and (g)   **5.**

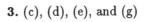

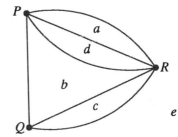

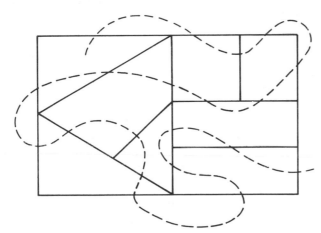

7. Any correspondence of faces and edges is correct.   **9.**

11. Impossible; the dual would have exactly one odd vertex. (See Exercise 9.3.7.)

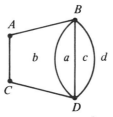

Section 9.6, page 452

1. $V = 7$, $E = 7$, $F = 3$; $V - E + F = 3$ **3.** $V - E + F = 4$ **5.** Not necessarily; a loop at a vertex increases the degree by two, but adds only one face; a terminal edge increases the degree by one, but adds no face.   **7.** (a) $V = 5$, $E = 8$, $F = 5$ (b) $V = 24$, $E = 36$, $F = 14$ (c) $V = 8$, $E = 16$, $F = 10$ **9.** $V - E + F + 2H = 2$

Appendix A

Section A.1, page 470

1. Statement   **3.** Statement   **5.** Not a statement   **7.** Statement   **9.** Statement   **11.** Existential   **13.** Existential   **15.** Existential   **17.** Existential   **19.** This is not a mathematics book.   **21.** Not all roses are red. (Some roses are not red.)   **23.** $8 + 3 \neq 11$ **25.** All cars get good gas mileage.   **27.** No flowers are yellow.   **29.** There is not a unicorn on the lawn. (No unicorns are on the lawn.)   **31.** No elephants fly.   **33.** All students are sophomores.

Section A.2, page 474

1. This book is interesting and I am falling asleep.   **3.** This book is not interesting and I am falling asleep.   **5.** (Either) this book is not interesting or I am falling asleep.   **7.** This book is not interesting and I am not falling asleep.   **9.** It is not true that this book is interesting and I am falling asleep.

11.

p	q	p and q	q and p
T	T	**T**	**T**
T	F	**F**	**F**
F	T	**F**	**F**
F	F	**F**	**F**

13.

p	q	$(\sim p)$ or q	$\sim(p$ and $(\sim q))$
T	T	**T**	**T**
T	F	**F**	**F**
F	T	**T**	**T**
F	F	**T**	**T**

15.

p	q	$\sim(p$ and $q)$	p or $\sim(p$ and $q)$
T	T	F	**T**
T	F	T	**T**
F	T	T	**T**
F	F	T	**T**

17.

p	q	r	p and q	$(p$ and $q)$ and r	q and r	p and $(q$ and $r)$
T	T	T	T	**T**	T	**T**
T	T	F	T	**F**	F	**F**
T	F	T	F	**F**	F	**F**
T	F	F	F	**F**	F	**F**
F	T	T	F	**F**	T	**F**
F	T	F	F	**F**	F	**F**
F	F	T	F	**F**	F	**F**
F	F	F	F	**F**	F	**F**

19.

p	q	r	p or (q and r)		(p or q) and (p or r)		
T	T	T	**T**	T	**T**	**T**	**T**
T	T	F	**T**	F	**T**	**T**	**T**
T	F	T	**T**	F	**T**	**T**	**T**
T	F	F	**T**	F	**T**	**T**	**T**
F	T	T	**T**	T	**T**	**T**	**T**
F	T	F	**F**	F	**T**	**F**	**F**
F	F	T	**F**	F	**F**	**F**	**T**
F	F	F	**F**	F	**F**	**F**	**F**

21.

s	t	s and t
T	T	**T**
F	T	**F**

23.

s	c	s and c
T	**F**	**F**
F	**F**	**F**

Section A.3, page 482

1. *hypothesis*: There is snow on the ground. *conclusion*: This is winter. *converse*: If this is winter, then there is snow on the ground. **3.** *hypothesis*: Rabbits eat carrots. *conclusion*: Rabbits have good eyesight. *converse*: Rabbits must eat carrots if they have good eyesight. **5.** *hypothesis*: Something is a turkey. *conclusion*: Something is a bird. *converse*: All birds are turkeys. **7.** *hypothesis*: Something is an airplane. *conclusion*: Something does not fly like a bird. *converse*: Anything that does not fly like a bird is an airplane. **9.** *hypothesis*: The hypothesis of a conditional is false. *conclusion*: A conditional is true. *converse*: If a conditional is true, then its hypothesis is false. **11.** *inverse*: If there is no snow on the ground, then this is not winter. *contrapositive*: If this is not winter, then there is no snow on the ground. **13.** *inverse*: Rabbits must not have good eyesight if they don't eat carrots. *contrapositive*: Rabbits must not eat carrots if they don't have good eyesight. **15.** *inverse*: Anything that is not a turkey is not a bird. *contrapositive*: Anything that is not a bird is not a turkey. **17.** *inverse*: Anything that is not an airplane flies like a bird. *contrapositive*: Anything that flies like a bird is not an airplane. **19.** *inverse*: A conditional is false if its hypothesis is true. *contrapositive*: If a conditional is false, then its hypothesis is true.

21.

p	q	~(p → q)		p and (~q)	
T	T	**F**	T	**F**	F
T	F	**T**	F	**T**	T
F	T	**F**	T	**F**	F
F	F	**F**	T	**F**	T

23. Something is an elephant and it does not have a trunk. **25.** (23) Either something is not an elephant or it has a trunk.

INDEX